POLYM

TRANSLOCATION

POLYMER TRANSLOCATION

Murugappan Muthukumar

CRC Press
Taylor & Francis Group
Boca Raton London New York

CRC Press is an imprint of the
Taylor & Francis Group, an **informa** business

CRC Press
Taylor & Francis Group
6000 Broken Sound Parkway NW, Suite 300
Boca Raton, FL 33487-2742

First issued in paperback 2019

© 2011 by Taylor and Francis Group, LLC
CRC Press is an imprint of Taylor & Francis Group, an Informa business

No claim to original U.S. Government works

ISBN-13: 978-1-4200-7516-8 (hbk)
ISBN-13: 978-0-367-38279-7 (pbk)

Visit the Taylor & Francis Web site at
http://www.taylorandfrancis.com

and the CRC Press Web site at
http://www.crcpress.com

DEDICATION

To Lalitha

CONTENTS

Preface   xiii

Acknowledgments   xv

Author   xvii

1   General Premise   1
     1.1   Biological Contexts   1
     1.2   Single-Molecule Experiments   4
     1.3   Nomenclature   7
     1.4   Entropic Barrier Idea   7
     1.5   Physics of Translocation   9
            1.5.1   Drift–Diffusion   9
            1.5.2   Capture   9
            1.5.3   Translocation   11
     1.6   Outlook   12

2   Size, Shape, and Structure of Macromolecules   13
     2.1   Measures of Polymer Conformations   13
     2.2   Universal Behavior   21
     2.3   Excluded Volume Interaction   23
     2.4   Coarse-Grained Models of Chain Connectivity   24
            2.4.1   Freely Jointed Chain   24
            2.4.2   Kuhn Chain Model   25
            2.4.3   Gaussian Chain   27
            2.4.4   Wormlike Chain   32
     2.5   Chain Swelling by Excluded Volume Effect   34
     2.6   Coil–Globule Transition   39
     2.7   Concentration Effects   40
     2.8   Summary   44

3   Electrolyte Solutions, Interfaces, and Geometric Objects   45
     3.1   Electrolyte Solutions   45
            3.1.1   Coulomb Interaction and Bjerrum Length   46
            3.1.2   Poisson–Boltzmann Equation   48
            3.1.3   Debye–Hückel Theory   49

		3.1.3.1	Electrostatic Screening Length	50
		3.1.3.2	Effect of Ion Size	52
		3.1.3.3	Charge Distribution around an Ion	54
		3.1.3.4	Free Energy	56
	3.2	Charged Interfaces		57
		3.2.1	Charged Spherical Particles	58
		3.2.2	Planar Interfaces	60
		3.2.2.1	Salt-Free Solutions	61
		3.2.2.2	Salty Solutions	62
		3.2.3	Cylindrical Pore	65
		3.2.4	Charged Cylinder	69
		3.2.5	Charged Line	72
		3.2.6	Dielectric Mismatch	74
	3.3	Summary		77
4		Flexible and Semiflexible Polyelectrolytes		79
	4.1	Concepts		80
		4.1.1	Coulomb Strength	81
		4.1.2	Debye Length for Polyelectrolyte Solutions	81
		4.1.3	Chain Connectivity	82
		4.1.4	Counterion Worm	84
		4.1.5	Dielectric Mismatch	85
		4.1.6	Electrostatic Excluded Volume	87
		4.1.7	Electrostatic Persistence Length	88
		4.1.8	Hydrophobic Interaction	90
	4.2	Experimental Results		90
	4.3	Simulation Results		92
		4.3.1	Radius of Gyration	94
		4.3.2	Counterion Adsorption and Effective Charge	95
	4.4	Electrostatic Swelling with Fixed Polymer Charge		100
		4.4.1	High Salt Limit	100
		4.4.2	Low Salt Limit	102
		4.4.3	Electrostatic Blob	104
		4.4.4	Crossover Formula	106
	4.5	Self-Regularization of Polymer Charge		107
	4.6	Concentration Effects		111
	4.7	Summary		113
5		Confinement, Entropic Barrier, and Free Energy Landscape		115
	5.1	Hole in a Wall		116
	5.2	Spherical Cavities		121
		5.2.1	Gaussian Chain	122
		5.2.2	Uncharged Polymer with Excluded Volume	128
		5.2.3	Polyelectrolyte Chain	132
	5.3	Cylindrical Pores		134
		5.3.1	Flexible Chain	135
		5.3.2	Semiflexible Chain	138

| | 5.4 | Infinitely Wide Channels | 141 |
| | 5.5 | Summary | 142 |

6		Random Walks, Brownian Motion, and Drift	145
	6.1	Biased Random Walk	146
		6.1.1 Gaussian Chain Statistics	148
		6.1.2 Drift–Diffusion	149
	6.2	Brownian Motion and Langevin Equation	152
		6.2.1 Brownian Particle without External Force	153
		6.2.2 Brownian Particle under an External Force	156
		6.2.3 General Form of Langevin Equation	156
	6.3	Fokker–Planck–Smoluchowski Equation	157
	6.4	Collection of Brownian Particles	159
	6.5	Equilibrium versus Steady State	160
		6.5.1 Equilibrium	161
		6.5.2 Steady State	161
	6.6	Finite Boundaries and First Passage Time	162
		6.6.1 General Equations	163
		6.6.2 General Solutions	165
	6.7	Properties of Drift–Diffusion Process	167
		6.7.1 One Absorbing Barrier	168
		6.7.2 Two Absorbing Barriers	169
		6.7.3 Reflecting Boundary	172
		6.7.4 Radiation Boundary Condition	173
	6.8	Summary	174

7		Polyelectrolyte Dynamics	177
	7.1	Solvent Continuum and Hydrodynamic Interaction	178
	7.2	Uncharged Polymer	181
		7.2.1 General Expectations	181
		7.2.2 Rouse Model	183
		7.2.3 Zimm Dynamics	186
		7.2.4 Semidilute Solutions and Hydrodynamic Screening	189
		7.2.5 Entanglement Regime	193
	7.3	Diffusion of Polyelectrolyte Chains	194
	7.4	Electrophoretic Mobility	198
	7.5	Coil Stretch under Flow	204
	7.6	Summary	206

8		Ion Flow in Single Pores	209
	8.1	A General Scenario	209
	8.2	Equilibrium	212
	8.3	Steady State without Diffusion	213
		8.3.1 Ionic Current and Ohm's Law	214
		8.3.2 Relaxation of Charge Density	215
	8.4	Steady State with Drift and Diffusion	216
		8.4.1 Analogy to Electrical Circuits	217
		8.4.2 Solutions of PNP Equations	218

	8.4.3	Constant Field Approximation	219
		8.4.3.1 GHK Current Equation	220
		8.4.3.2 Concentration Profile	221
		8.4.3.3 Resting Potential	221
8.5	Effect of Barriers		223
8.6	Ionic Current through Protein Pores		224
	8.6.1	Gramicidin A Channel	225
	8.6.2	α-Hemolysin Pore	226
8.7	Fluctuations in Ionic Current		230
8.8	Electroosmotic Flow		234
	8.8.1	EOF near Planar Interfaces	234
	8.8.2	EOF inside Cylindrical Pores	236
8.9	Summary		238

9	Polymer Capture		241
9.1	Representative Experimental Results		241
9.2	General Considerations		244
9.3	Diffusion-Limited Capture		247
9.4	Drift-Limited Regime		250
9.5	Effect of Convective Flow		251
9.6	Polymer Capture with Electroosmotic Flow		252
9.7	Effect of Barriers on Capture Rate		256
	9.7.1	Capture Rate	259
	9.7.2	Probability of Successful Translocation	262
	9.7.3	Phenomenological Chemical Kinetics Model	265
9.8	Summary		268

10	Translocation Kinetics: Nucleation and Threading		271
10.1	Representative Experimental Results		271
	10.1.1 Translocation through α-Hemolysin Pore		272
	10.1.2 Translocation through Solid-State Nanopores		276
	10.1.3 Translocation through Channels		278
10.2	Insights from Simulations		279
10.3	Theory of Translocation Kinetics		284
	10.3.1 Nucleation Time		288
	10.3.2 Translocation Time		289
		10.3.2.1 Two Absorbing Barriers	291
		10.3.2.2 One Absorbing Barrier and One Reflecting Barrier	293
	10.3.3 Effect of Pore–Polymer Interactions		295
		10.3.3.1 Escape from a Spherical Cavity	296
		10.3.3.2 Translocation between Two Spherical Cavities	297
		10.3.3.3 Two Spherical Cavities Connected by a Pore	297
		10.3.3.4 Translocation with a Membrane Potential	300
10.4	Comparison between Experimental Data and Theory		302
	10.4.1 Translocation through α-Hemolysin Pore		302

		10.4.2 Translocation through Solid-State Nanopores	304
		10.4.3 Translocation through Channels	306
	10.5	Summary	311
11		Further Issues	313
	11.1	Nonequilibrium Conformations during Threading	313
	11.2	Amplification of Chemical Details	315
		11.2.1 Chemical Decoration of the Pore	316
		11.2.2 Secondary Structures of the Translocating Polymer	318
	11.3	Biological Examples	320
		11.3.1 Mitochondrial Transport	320
		11.3.2 Bacterial Conjugation	322
		11.3.3 Transport through Nuclear Pore	322
		11.3.4 Genome Packing in Bacteriophages	323
	11.4	Summary	325
		References	327
		Index	343

PREFACE

The process of polymer translocation occurs in many biological and biotechnological phenomena, where electrically charged polymer molecules, such as polynucleotides and proteins and their complexes, move from one region of space to another in crowded environments. Substantial research activities are currently being pursued in an effort to understand the macromolecular basis of polymer translocation. These activities are further stimulated by the societal need to sequence an enormous number of genomes immediately and inexpensively. Due to the inherent challenges of formulating the molecular basis of polymer translocation, this area has attracted a diverse set of researchers in biology, physics, chemistry, materials science, chemical engineering, and electrical engineering.

The thread that is central to polymer translocation is polyelectrolyte physics, which is perhaps one of the most challenging areas of modern research. The challenge in understanding the complex behavior of polyelectrolyte molecules arises from three long-range forces due to chain connectivity, electrostatic interactions, and hydrodynamic interactions. In addition, translocation of polyelectrolyte molecules through a protein pore or a solid-state nanopore becomes more complex by the polymer–pore interactions, confinement effects, and flow fields in the system. Unraveling the rich phenomenology of polymer translocation requires a grasp of modern concepts of polymer physics and polyelectrolyte behavior.

With this goal in mind, this book strives to present a summary of the key concepts of polyelectrolyte structures, electrolyte solutions, ionic flow, mobility of charged macromolecules, polymer capture by pores, and threading of macromolecules through pores. The main concepts and theoretical results are presented without formal derivations whereas the cited references provide adequate derivations. For situations where there is a lack of readily usable references, derivations are given. Every effort has been made to give the reader an overview of basic concepts, established experimental facts, relevance of the concepts to real systems, ongoing challenges, and strategies for applying these ideas and summarized formulas to design new experiments. An attempt has also been made to avoid heavy mathematics and an exhaustive repetition of published literature.

ACKNOWLEDGMENTS

This book would not have been possible without the collaborations with my former and current students, as should be evident from the bibliography. I am sincerely grateful to all of them for their intellectual stimulation and friendship. I especially thank Stephen Mirigian, Mithun Kumar Mitra, and Jing Hua for many valuable suggestions for improving the presentation and for help with many figures. I am indebted to S. Bezrukov, E. DiMarzio, D. Hebert, R. Kumar, J. Machta, and C. Santangelo for their comments on one or more chapters.

I also gratefully acknowledge support from the National Institutes of Health and the National Science Foundation, without which the writing of this book and some of the advances presented here would have been almost impossible. I am also indebted to many collaborators outside my laboratory. In particular, it is a pleasure to acknowledge the tremendous encouragement and support from H. Bayley, S. Bezrukov, D. Branton, D. Deamer, E. DiMarzio, J. Kasianowicz, and M. Schmidt.

Finally, it is a great pleasure to thank Luna Han for her invitation to write a book on polymer translocation and her constant encouragement throughout the writing stage. It is also a pleasure to thank Kyle Meyer and Anto Aroshini Fernandez for their tremendous support during the production of the book.

AUTHOR

Murugappan Muthukumar is Wilmer D. Barrett Distinguished Professor at the University of Massachusetts, Amherst. After his undergraduate education at the University of Madras, India, he studied chemical physics at the University of Chicago, Illinois, followed by postdoctoral research at the Cavendish Laboratory, Cambridge University, Cambridge, U.K. He has taught polymer physics and biophysics at the University of Massachusetts for over 25 years. His research interests cover many areas of polymer science and physical biology, including polymer statistics, dynamics and thermodynamics of polymer solutions, and phase transitions of polymer systems. His current major interests concern the basic mechanisms of polymer crystallization, biomineralization, polyelectrolyte physics, assembly of viruses, and polymer translocation.

Muthukumar is a fellow of the American Physical Society and has received the Alfred P. Sloan fellowship, the Dillon Medal and the Polymer Physics Prize of the American Physical Society, the Chancellor's Medal of the University of Massachusetts, a Senior Humboldt Award, and the Gutenberg Lecture Award from the University of Mainz.

GENERAL PREMISE

Polymer translocation is one of the most fundamental macromolecular processes in life. This ubiquitous phenomenon deals with how electrically charged polymer molecules, such as polynucleotides and proteins, move from one region of space to another in crowded environments. Examples of biological phenomena for which polymer translocation is a crucial fundamental step include passage of mRNA through nuclear pore complexes, injection of DNA from a virus head into a host cell, gene swapping through pili, and protein translocation across biological membranes through channels (Lodish et al. 2007, Alberts et al. 2008). In addition, primarily due to societal and technological demands on DNA sequencing, there has recently been a tremendous effort to monitor and control the translocation of single macromolecules through a single pore made of proteins or synthetic solid-state materials. Although these apparently diverse phenomena emerge from different specific chemical details that are unique to each of these phenomena, we seek to identify the most common universal features behind these phenomena. The chemical details indeed decorate the basic universal feature of the passage of long macromolecules differently and impart specific directions and targets. We will first attempt to identify the common universal aspects of translocation and then to explore ways of incorporating the specific details relevant to different contexts. After illustrating the richness of the phenomenon with a few examples, we will offer an operating definition of the process and introduce the main concept, namely the entropic barrier idea (Muthukumar 2007), behind the polymer translocation. This will be followed by a brief outline of the various significant components, which need to be put together for a molecular understanding of the polymer translocation phenomenon.

1.1 BIOLOGICAL CONTEXTS

If we were to look into any volume element inside a eukaryotic cell (Figure 1.1), we are most likely to meet charged polymer molecules (such as proteins, RNA, and DNA) and electrolyte ions. In fact, the cell is a very crowded environment, and due to the nature of electrical (Coulomb) forces mediating the structures and functions of the constituent molecules inside the cell, it can be considered to be a thick "Coulomb soup." A comprehensive fundamental understanding of the structures, dynamics, and mobilities of single macromolecules and their complexes with other molecules in the *in vivo* environments, even in

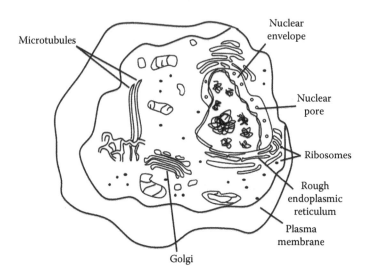

Figure 1.1 A cell is a crowded "Coulomb soup," with charged macromolecules and their assemblies moving between different compartments.

the absence of penetration through channels, is currently absent. Our expedition of trying to understand the translocation process by these molecules and their complexes, which themselves remain as poorly understood systems, becomes even more daunting. It is therefore necessary to investigate isolated translocation events before embarking on the coupled multiple translocation processes occurring simultaneously *in vivo*.

Even an isolated translocation process in *in vivo* is extremely rich in its phenomenology. As an illustration of the richness of details and complexities, consider the nuclear pore complex (Alberts et al. 2008), a crude sketch of which is given in Figure 1.2. The pore itself is apparently self-assembled by roughly a hundred different proteins with elaborate structural motifs: a basket-like cage with several openings capable of sieving different-sized molecules, a capillary in the middle of the passage with the capacity to dilate under macromolecular pressures, suspension of the passage into the double membrane of the nuclear envelope with a combination of hydrophobic and hydrophilic moieties, and charged polymer bristles protruding into the outside of the nucleus. The typical size of this assembly along the nuclear envelope is about 100 nm. It is through such an elaborate assembly, the mRNA, present as an mRNP complex of mRNA and more than 30 different carrier proteins inside the nucleus, is threaded into. In its own right, the mRNP complex is big with typical radius size of 50 nm. Thus, a structurally correlated object of about 50 nm is somehow pushed into the nuclear pore complex, and mRNA undergoes translocation. What is amazing is that this process is taking place all the time with fidelity, as instruction for synthesis of coded proteins would not occur without mRNA translocation. Indeed, we are yet to understand how this phenomenon takes

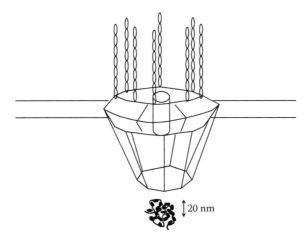

$\updownarrow$ 20 nm

Figure 1.2 A sketch of the nuclear pore complex, and the mRNP. The typical feature sizes of these structures are 20–100 nm.

place. Nevertheless, it is evident that the phenomenon is manifest at very large length and timescales in comparison with atomistic scales. It is perhaps more fruitful to borrow ideas from polymer physics (de Gennes 1979), dealing with large-scale behaviors of macromolecules, to gain the "bird's eye view" and then to reckon the specific higher resolution features.

Not all contexts of polymer translocation are as complicated as in the nuclear pore complex. The passage of dsDNA from a virus head into a host cell as a single-file threading (Figure 1.3) and the transfer of DNA molecule from one bacterium to another (Figure 1.4) are examples of less complex situations. Again, the relevant length scales in the translocation phenomenon are much larger than atomistic length scales, calling for ideas from polymer physics.

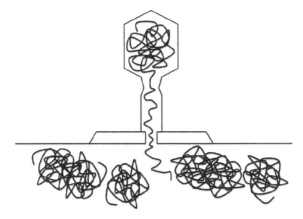

Figure 1.3 Cartoon of threading of DNA from a bacteriophage into a host cell.

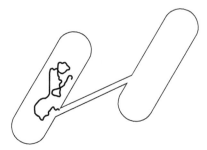

Figure 1.4 Cartoon of gene swapping between bacteria.

1.2 SINGLE-MOLECULE EXPERIMENTS

The above-mentioned *in vivo* biological phenomena are too complex to directly monitor one long macromolecule undergoing translocation in its totality. Fortunately, there have recently been many exciting single-molecule nanopore-based electrophysiology experiments, whereby the features of translocation by single polymer chains can be measured in great quantitative detail. Although these experiments were stimulated by the societal need of having to sequence enormous number of genomes immediately and inexpensively, they are serendipitously paving the way toward a fundamental molecular understanding of the phenomenon of polymer translocation.

In the single-molecule nanopore-based translocation experiments (Kasianowicz et al. 1996), a single nanopore is incorporated (either by a self-assembly of proteins or by ion-beam sculpting) into a membrane separating a donor (*cis*) chamber and an acceptor (*trans*) chamber. Each chamber contains a buffer solution with a strong electrolyte such as KCl. In many of the experiments, involving protein channels, the pore is a heptameric self-assembly of α-hemolysin (αHL) with a length of ~ 10 nm and a narrow constriction of ~ 1.4 nm, as sketched in Figure 1.5a (Song et al. 1996). In the case of solid-state nanopores (Figure 1.5b) (Chen et al. 2004a), the diameter is in the tunable range of 3–10 nm and the length is in the order of 10 nm or more. When an external voltage is applied across the membrane, the pore allows passage of small ions and the resulting ionic current is measured. When this experiment is repeated with ssDNA/RNA originally present in the *cis* chamber (with negative electrode), the measured ionic current decreases significantly whenever the polymer interferes with the pore. A typical trace of ionic current versus time for the passage of a polymer chain through αHL is given in Figure 1.6. Although every encounter with the pore is caused by identical polymer molecules, the resultant ionic response is stochastic. As marked in Figure 1.6, there are apparently three timescales. The time t_0 is the approach time between two successive events and we may define the inverse of the average t_0 as the capture rate R_c, independent of whether the polymer actually underwent the translocation process or not. Also, as indicated in Figure 1.6, there are partial blockades

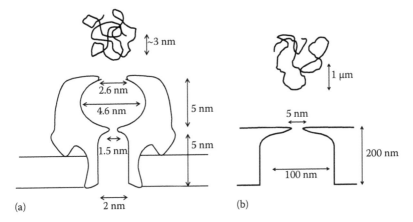

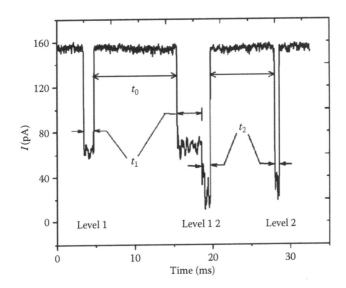

Figure 1.5 Sketches of (a) the α-hemolysin pore and (b) solid-state nanopore used in single-molecule electrophysiology experiments.

Figure 1.6 A typical ionic current associated with the encounter of polymer chains with an αHL pore.

(with duration t_1) and deep blockades (with duration t_2). Furthermore, the common feature of the experimental results is that the distributions of t_0, t_1, t_2, and the various blocked current levels are very broad.

The details of the time-dependence of the ionic current bear information on the manner in which polymer molecules attempt to translocate through a pore and the underlying molecular mechanism of polymer threading. Experiments show that the average translocation time, for single-file translocation processes, is directly proportional to the polymer length and inversely proportional to the applied voltage, in spite of the fact that the translocation time generally has a broad distribution (Kasianowicz et al. 1996). The capture rate depends on the polymer concentration, the direction of the translocation (*cis*-to-*trans* vs. *trans*-to-*cis*), and the applied voltage above a threshold value (Henrickson et al. 2000). In addition, polymer sequences and their ability to spontaneously form secondary structures influence their migration through nanopores as manifest in the corresponding ionic current traces (Akeson et al. 1999, Meller et al. 2000). In fact, such distinguishing features in the ionic current traces, associated with translocation of different polymer sequences, raised high hopes for cultivating single-molecule electrophysiology technique into a fast sequencing technology (Branton et al. 2008).

As a complement to the threading of single-stranded DNA/RNA, and to avoid potential complications from the role of the vestibule of α-hemolysin pore in interpreting the ionic current traces, much experimental effort has gone into forming solid-state nanopores with diameters large enough to thread double-stranded DNA (Chen et al. 2004a). The ionic current traces associated with the passage of dsDNA through these solid-state nanopores are apparently more complex than the corresponding results for the α-hemolysin pore. Now, the polymer can translocate in quantized configurations such as a single file, chain with one hairpin, etc. Even the seemingly simplest situation of translocation of dsDNA through a nanopore of 15 nm diameter exhibits rich puzzles. Time-resolved fluorescence studies have revealed that a depletion (capture) region of about 3 μm (much larger than the pore size) develops in front of the pore (Chen et al. 2004a). The DNA molecules were found to diffuse slowly ($\sim 4s$) until they approach the capture region. Once the molecules reach the capture region, they were found to be depleted rapidly (~ 50 ms) by active pulling through the pore.

In addition to nanopores, nanoscopic channels have also been used to investigate the translocation of DNA molecules (Han and Craighead 2000). Consider a periodic alternation of deep ($\sim 2 \mu$m) and shallow (~ 100 nm) wells, with the width of both of these wells being far wider than the size of a DNA molecule. Experimental measurement of the average time τ taken by one λDNA molecule to pass through a pair of adjacent deep and shallow wells showed that it takes shorter time for longer molecules in accordance with an empirical formula, $\tau \sim N^{-0.42}$, where N is proportional to the polymer length. This counterintuitive finding is in direct opposition to the linear increase of τ with N for single-file translocation through pores.

The general picture that emerges from the above selective description of the phenomenology of polymer translocation is that the phenomenon is quite complex even in simple experimental setups and is controlled by numerous factors. The translocation time of one macromolecule depends on the chain length, chemical sequence of the polymer, chain stiffness in terms of whether single-stranded or double-stranded, applied voltage, chemical nature of the pore, pore geometry, and flow fields in the experimental setups. In general, the

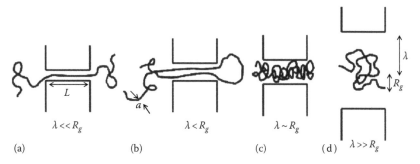

Figure 1.7 Different regimes (a–d) of polymer confinement by the pore. The process is translocation for $a \leq \lambda < R_g$.

translocation process is stochastic with broad distributions of various measures of the process, even though identical molecules are undergoing translocation.

1.3 NOMENCLATURE

It is perhaps useful to associate certain specific criteria in defining the process of polymer translocation, in order to distinguish it from the general transport of macromolecules. Consider a uniform pore with radius λ and length L, through which a chain of average radius of gyration R_g undergoes translocation. Let a be the radius of each of the monomers constituting the polymer. If λ is slightly larger than a but much smaller than R_g, the chain can undergo translocation only as a single file or as a hairpin (Figure 1.7a and b). If λ is much larger than a but smaller than R_g, the chain can be squeezed inside the pore as sketched in Figure 1.7c. On the other hand, if λ is much larger than R_g (Figure 1.7d), then the polymer undergoes transport through the capillary as in free solutions, except for the possible adsorption/depletion effects at the walls of the pore. In the nomenclature adopted here, the phenomenon of polymer translocation refers to the constrained motion of polymer chains where the size of the pore is smaller than the size of the polymer, $a \leq \lambda < R_g$.

1.4 ENTROPIC BARRIER IDEA

One of the inherent properties of an isolated polymer chain is its ability to assume a large number of conformations $\mathcal{N}$. As a result, the chain entropy ($k_B \ln \mathcal{N}$; k_B is the Boltzmann constant) can be high and its free energy F is given by

$$F = E - TS = E - k_B T \ln \mathcal{N}, \tag{1.1}$$

where E is the energy of interaction between the monomers and the surrounding solvent molecules, and T is the absolute temperature. There can be additional

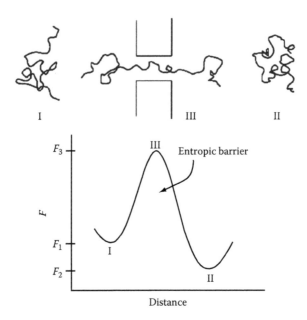

Figure 1.8 Genesis of the entropic barrier for polymer translocation.

entropic contributions to F due to a reorganization of solvent molecules accompanying conformational changes of the chain. When such a chain is exposed to a restricted environment such as a pore, the number of conformations that can otherwise be assumed by the chain is reduced, and as a result the chain entropy decreases and the chain free energy increases. This effect is depicted in Figure 1.8.

F_1, F_2, and F_3 are the free energies of the chain in regions I, II, and III, respectively. Owing to the reduction of conformations in region III, F_3 is higher than F_1 and F_2. We shall call $(F_3 - F_1)$ the entropic barrier to the passage of the chain out of region I. Although this barrier is called the entropic barrier, it is indeed a free energy barrier because additional enthalpic contributions to F_3 can arise from the interactions between the polymer and the pore. In general, the environment of the chain in region II can be different from that in region I (due to different electrochemical potentials in these regions), so that F_2 is not necessarily equal to F_1. The net driving potential for polymer translocation from region I to region II is $(F_1 - F_2)$. The polymer chain must negotiate the entropic barrier in order for it to successfully arrive at the opposite side of the pore.

It is important to recognize that the role of conformational entropy of polymer chains in various biological processes cannot be treated as only a minor factor. Since the temperature T is essentially fixed for a given physiological system and because only rather minor variations are permitted in E for a fixed T, the only way the free energy landscape can be dramatically modified must be through the entropy *S*. The ability of polymer molecules to undergo large conformational changes, without losing their topological connectivity, makes them ideal candidates for large entropic changes. No wonder that life is made of polymer strings instead of, say, cubes or spheres. With the help of such entropic considerations, we will formulate the arguments for the structure, dynamics, mobility, and translocation of polymer chains in what follows in the book.

1.5 PHYSICS OF TRANSLOCATION

Given the rich features of the translocation phenomenon, the objective is to identify the various significant contributing factors and to assess their relative contributions to translocation. Even with the modern computational technologies, it is impossible to build from atomistic details and force fields at sub-nanometer resolution and calculate the behavior of the whole macromolecular assemblies of hundreds of nanometers in size. The nature of forces among all atoms of translocating polymer molecules, enzymes, and protein pores in salty and crowded aqueous solutions with highly heterogeneous dielectric function remains as a huge challenge to be unraveled. Nevertheless, it is worthwhile to explore theoretical possibilities where local details at the spatial resolution of amino acids and nucleotides are surrogated into coarse-grained parameters at multiple nanometer resolution. This will allow implementation of well-established concepts from polymer physics. With this attitude, we will present basic concepts, arguments, predictions, and comparison with experimental results related to polymer translocation in the following chapters.

The general scope of the translocation process may be divided into several separate parts. There are three essential steps associated with the transit of a polymer through a nanopore (illustrated in Figure 1.9 for a structureless blunt pore): (1) drift–diffusion, (2) capture, and (3) translocation.

1.5.1 Drift–Diffusion

In the first step (far away from the pore), the polymer undergoes a combination of drift, due to the externally imposed force fields, and diffusion arising from collisions with solvent molecules. The drift–diffusion of the polymer is established by the structure and size of the polymer, the nature of the background fluid (such as solvent quality and ionic strength), and influences from external forces (such as electric field and pressure gradient).

1.5.2 Capture

At the pore mouth, force fields may be generated by chemical decoration of the inside surface of the pore. More importantly, steep electric potential gradients may occur at the pore mouth due to the dielectric mismatch between the layer in which the pore is embedded and the rest of the system, in the presence of an applied voltage gradient. Furthermore, strong flow currents may arise at the

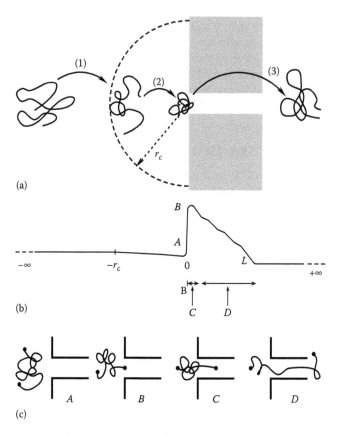

Figure 1.9 (a) Three main stages of polymer translocation process: (1) drift–diffusion, (2) capture, and (3) translocation; (b) free energy landscape; and (c) three stages in the third translocation step.

pore mouth due to the movement of water through the pore. In particular, for situations dealing with charge-bearing pores containing electrolyte solutions, this force, called the electroosmotic force, can be quite significant. Since the flux of the water flow must be continuous in the system, and since there is only a narrow passage for fluid flow inside the pore, strong velocity gradients may develop at the pore mouth. All of these effects generate an effective sucking force at the pore entrance, which in turn tries to capture the polymer to facilitate the translocation. Thus, near the pore, the flow field and the electric field can be significantly influenced by electroosmotic forces, dielectric mismatch between the pore wall and the aqueous medium, ionic strength gradients, and pressure gradients. Depending on the details of these contributing factors, the nature of the flow field within a range of r_c from the pore can be qualitatively different from that outside this range. r_c can vary between subnanometers to microns. Within the range of r_c, the polymer may undergo conformational deformation.

By experiencing such forces within r_c, the polymer approaches the pore mouth, designated as step (2) in Figure 1.9a. The capture of the polymer at the pore mouth is controlled by the strength of the sucking force at the pore entrance and by the range of the flow field in front of the pore where the velocity gradients are strong.

1.5.3 Translocation

In general, when the polymer is caught at the pore mouth, it is in a jammed state without any initial correlation between the chain ends and the pore mouth. The chain needs to unravel itself to place one of its ends at the pore mouth for the single-file translocation to occur, and then to thread through the pore. This step of translocation consists of three stages: (a) chain-end localization, (b) nucleation, and (c) threading. An entropic barrier must be overcome in placing one of the chain ends at the pore mouth from a jammed coil state (designated as $A \rightarrow B$ in Figure 1.9b and c), in order to enable the eventual single-file translocation. This is due to the requirement that one end must be at a specific spacial location, instead of all possible locations, whereby the chain end is losing its translational entropy. After the localization of one chain end, there is an additional entropic barrier for reducing the conformational degrees of freedom for the chain in order to be squeezed into the pore. The polymer chain is thus hung across the entropic barrier. As will be seen later, only if a sufficient number of monomers crosses this "nucleation barrier" can the chain undergo further translocation. The nucleation stage is $B \rightarrow C$ in Figure 1.9b and c. The final stage of translocation, $C \rightarrow D$, is a downhill threading process, which is in its own right a drift–diffusion process. The chain is finally kicked out of the pore into the receiver compartment as the ultimate step.

The shape of the pore can lead to additional complexities. As an example, the free energy landscape for a protein pore, such as αHL, containing a trap-like vestibule in front of the pore, is qualitatively different (Figure 1.10) from that for a blunt pore. Here, the jammed coil at the pore mouth is separated by two barriers, one for successful translocation into the *trans* side and the other for the return to the *cis* side.

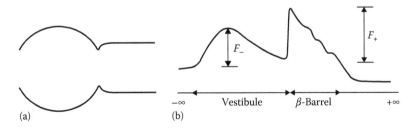

(a) (b)

Figure 1.10 The polymer can get jammed in the metastable state at the pore mouth, for αHL, with two barriers for forward and backward movements: (a) sketch of αHL and (b) free energy profile.

1.6 OUTLOOK

The above description highlights only the generic physical aspects of the translocation process. In the translocation step, specific to a particular polymer and a particular pore, the fine details of the electrostatic and hydrophobic properties of the amino acids constituting the protein pores, charge decoration of the inside wall of the solid-state nanopores, geometry of the pore, and the polymer sequence contribute significantly. An accounting of these effects manifest at both subnanometer level and microscopic level is essential for a fundamental understanding of polymer translocation. In view of our approach to adopt a coarse-grained methodology, we will implement concepts from polymer physics cultivated over the past seven decades to explore this phenomenon. We shall first devote several chapters to discuss size, shape, and structure of isolated polymers in equilibrium and under flow, and their dynamics and mobility in free solutions containing a certain amount of strong electrolytes. We will address the various origins of the capture zone and the process of polymer capture. Quantitative descriptions of the entropic barrier and the free energy landscape associated with the translocation, and the kinetics of polymer threading into pores will be presented next. We will then put all of these components together in order to understand the various experimental results on translocation.

2

SIZE, SHAPE, AND STRUCTURE OF MACROMOLECULES

Polymer molecules are monomers contiguously connected by covalent bonds in a chain-like fashion. The monomers themselves are groups of atoms, and can be either identical repeat units (as in polyethylene or polyuridylic acid) or chemically different units (as in a protein molecule or a deoxyribonucleic acid (DNA) containing different bases). Depending on the chemical nature of the repeat units of the polymer, the number of monomers per chain, and the nature of the solvent in which the polymer is dispersed, the molecule can assume different sizes and shapes such as globular, coil-like, and rod-like. It might seem at the outset that it is necessary to treat each polymer in a given solvent condition as a unique case by accounting for the specific chemical nature of the polymer and solvent. However, it turns out that there are certain universal laws that can describe average polymer conformations. It is possible to surrogate the local degrees of freedom of chemical specificity into a few parameters and obtain useful coarse-grained models in order to understand the universal properties of polymer chains.

In this chapter, we shall introduce various measures of polymer conformations and some of their universal laws. We will give a summary of coarse-grained models of polymer conformations and discuss how local details are parametrized in these models. The basic vocabulary of polymer statistics, including concepts like persistence length, radius of gyration, hydrodynamic radius, size exponent, structure factor, fractal dimension, excluded volume parameter, and coil–globule transition, will be introduced. In experiments exploring the translocation phenomenon discussed in this book, the polymer is electrically charged and the solvent medium is an aqueous electrolyte solution. When electrical charges are present in a polar dielectric medium, there are additional significant concepts that are required to describe polymer statistics. This in itself constitutes a separate field of study and still remains to be fully understood. In view of this, we relegate the discussion of charged polymers and electrolyte solutions to the following chapters. In the present chapter, we shall consider only uncharged polymers.

2.1 MEASURES OF POLYMER CONFORMATIONS

When a polymer is dispersed in a solvent, there are generally three kinds of pairwise interactions at the local level: monomer–monomer, monomer–solvent,

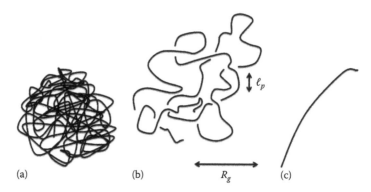

Figure 2.1 Major conformations of isolated polymer chains: (a) globule, (b) coil, and (c) rod-like. R_g is the radius of gyration and ℓ_p is the persistence length.

and solvent–solvent. If the hydrophobic interaction, due to van der Waals attractive forces, between the monomer units were to be dominant over the monomer–solvent interaction, then the monomers would aggregate together to form a globular structure (Figure 2.1a) by excluding the solvent molecules out of the globule. On the other hand, if the monomer–solvent interaction is preferable over the monomer–monomer attractive interaction, the solvent becomes a good solvent for the swelling of the polymer, which then adopts a swollen coil-like conformation (Figure 2.1b). In an average sense, the coil would look like a rough porous ball of wool, carving out a rough sphere of revolution with a radius R_g, called the radius of gyration (defined below). For some polymers, the chemical details associated with adjacent monomer units along the chain backbone are such that rotation of these monomers around the connecting chemical bond can be severely restricted. Furthermore, conformations of adjacent monomers could be locked together by hydrogen bonding, as in the helical conformations of polypeptides and double-stranded DNA molecules. As a result, the chain backbone can be locally stiff. If the contour length of the polymer is short enough, then it would look rod-like (Figure 2.1c), with the obvious shape anisotropy. If the contour length of the polymer is very long, the chain would be rod-like locally (for distances less than or comparable to the persistence length ℓ_p, defined below) but would bend and curve at longer distances appearing overall as a coil. Such polymers are called semiflexible polymers. Indeed, the chain can undergo conformational transitions between the coil and globular states when experimental conditions alter the relative weights of the monomer–monomer, monomer–solvent, and solvent–solvent energies. Also, the same polymer in identical experimental conditions could be either rod-like or coil-like, depending on its molecular length.

We shall now define some quantities, which are either measured experimentally or computed theoretically, to describe polymer conformations. These definitions are general, independent of the particular conformations taken by the polymer. As an example, consider a specified conformation of a polyethylene

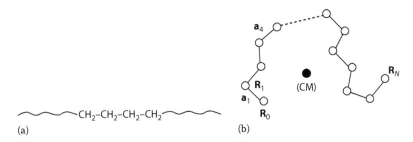

Figure 2.2 (a) Backbone structure of a polyethylene chain. (b) A typical conformation of a skeletal chain. $\mathbf{R}_i$ is the distance vector of the ith united atom from the center of mass (CM) of the conformation, and $\mathbf{a}_i$ is the bond vector connecting $(i-1)$th and ith united atoms.

chain (Figure 2.2a) with $(N+1)$ methylene monomers. Denoting each repeat unit (a methylene group in this case) as a united atom (skeletal atom), the skeletal structure of a conformation can be drawn as in Figure 2.2b. Here $\mathbf{R}_i$ is the position coordinate of the ith skeletal atom from the center of gravity (CM) of the specified chain conformation, and $\mathbf{a}_i$ is the bond vector of the ith skeletal bond. For each of such conformations, quantities like the end-to-end distance $\mathbf{R}$ and radius of gyration R_g can be defined as follows. Since the chain can adopt many conformations during the typical measurement times, we construct averages of these quantities over all possible conformations. These averages constructed in equilibrium are time-independent. We also assume that the repeat units are identical with identical bond lengths connecting them.

1. *Mean square end-to-end distance,* $\langle R^2 \rangle$:

$$\langle R^2 \rangle = \langle (\mathbf{R}_N - \mathbf{R}_0)^2 \rangle = \sum_{i=1}^{N} \sum_{j=1}^{N} \langle \mathbf{a}_i \cdot \mathbf{a}_j \rangle, \qquad (2.1)$$

where the angular brackets denote the averages over allowed conformations at a given experimental condition.

2. *Radius of gyration,* R_g: This is defined as the root mean square radius of gyration, where the mean square radius of gyration is given by

$$R_g^2 = \frac{1}{(N+1)} \sum_{i=0}^{N} \langle R_i^2 \rangle. \qquad (2.2)$$

This quantity is measured in static scattering techniques using light, x-rays, and neutrons.

3. *Hydrodynamic radius,* R_h:

$$R_h = \left(\frac{1}{(N+1)^2} \sum_{i=0}^{N} \sum_{j>i} \left\langle \frac{1}{|\mathbf{R}_i - \mathbf{R}_j|} \right\rangle \right)^{-1}. \qquad (2.3)$$

This quantity is measured in dynamic light-scattering technique, and its origin lies in the hydrodynamic interactions in the solution.

4. *Size exponent, ν*: It is evident from the above definitions that after taking the conformational averages, $\langle R^2 \rangle$, R_g, and R_h depend only on the number of monomers $(N + 1)$ in the chain and the bond length (a). As will be derived in the following sections, it can be shown that each of these three quantities is directly proportional to an exponent ν of the number of monomers per chain. The only difference between the three relations of $\langle R^2 \rangle^{1/2}$, R_g, and R_h is the numerical prefactor of order unity. Also, the difference between the number of monomers $(N + 1)$ and the number of bonds (N) can be omitted for the large values of N for polymers typically dealt with in translocation experiments. By suppressing the proportionality constant for the root mean square end-to-end distance, average radius of gyration and the hydrodynamic radius, and using the generic symbol R to represent any of these three quantities, we write

$$R \sim aN^\nu, \tag{2.4}$$

where R is taken to represent the average "size" of the polymer, and ν is called the size exponent. The bond length a is used in the above equation to make both sides of the equation to have the same dimension of length. Furthermore, it is sometimes useful to think of the polymer coils as statistically fractal objects with their own fractal dimensions embedded in the space of three dimensions of the solution (or two dimensions corresponding to a membrane). Note that N is directly proportional to the mass of the polymer and that a compact object (with its own dimension being three) obeys the relation, $R^d \sim N$, where d is the space dimension ($d = 3, 2$, and 1, for a sphere, disk, and line, respectively). Analogous to this geometric relation, we define a dimension for the average polymer conformation by rewriting the above equation as

$$R^{d_f} \sim N, \tag{2.5}$$

where d_f is called the fractal dimension of the polymer and is defined according to the above two equations as

$$d_f \equiv \frac{1}{\nu}. \tag{2.6}$$

The fractal dimension of the polymer is different from the space dimension d in which the polymer is present.

5. *Shape factor, R_g/R_h*: The ratio of R_g to R_h is sometimes used to remark on the anisotropy of the shape of the molecule. Since the dependence of R_g and R_h on N is the same with identical size exponent, the ratio is only a numerical factor reflecting the different values of the proportionality factors in their relations to N. For example, the shape factor is 0.77 for a compact sphere and increases to values of about 4 for rod-like conformations.

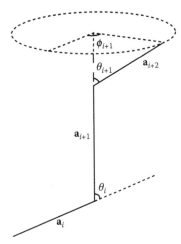

Figure 2.3 Sketch of local conformations defining the bond angle θ and the dihedral angle ϕ.

6. *Persistence length, ℓ_p:* The bond angles between the contiguous chemical bonds along the chain backbone cannot be arbitrary and are restricted by quantum mechanical properties. This feature results in the persistence of the direction of a bond over a certain distance along the chain contour. There can be additional reasons for this orientational persistence, due to hydrogen bonding among consecutive monomers as well. The relative orientation of a bond next to the preceding two bonds is defined by two angles, namely the bond angle θ and the dihedral angle ϕ, as illustrated in Figure 2.3. Here, the bond angle between the ith and $(i + 1)$th bond vectors is defined as $180 - \theta_i$. The dihedral angle ϕ_{i+1} is the angle between the plane of the bond vectors $\mathbf{a}_{i+2}$ and $\mathbf{a}_{i+1}$ and the plane of the bond vectors $\mathbf{a}_{i+1}$ and $\mathbf{a}_i$. The chemical nature of the atoms constituting the united atoms influences the allowed values of the dihedral angles, which then are manifest as the persistence length of the polymer. There are two ways of defining the persistence length. In one way, the average of the product of the orientation (that is the bond vector) of the ith bond and that of the jth bond is monitored as a function of the distance along the chain backbone $|\, i - j\,|$. This correlation function obeys the typical formula,

$$\langle \mathbf{a}_i \cdot \mathbf{a}_j \rangle = a^2 \exp\left(-\frac{|\, i - j\,|}{\ell_p} \right), \tag{2.7}$$

where ℓ_p is defined as the persistence length. For distances $|\, i - j\,|$ less than ℓ_p, the bond orientations are correlated and hence the conformation is rod-like. For distances $|\, i - j\,|$ larger than ℓ_p, the bond orientations are uncorrelated and hence the conformation can become coil-like. Another way to define the persistence length of the polymer is by projecting the end-to-end distance vector of the chain on the first bond in the limit of large values of n. We shall later introduce a model in order to extract the persistence lengths of polymers from experimental data.

7. *Monomer density distribution, $\rho(r)$:* For coil-like conformations, as sketched in Figure 2.1b, the density of monomers at the center is expected to be high and progressively decreasing with the radial distance until reaching the coil's boundary. This is to be contrasted with a compact object where the density is uniform until the radius at which it discontinuously becomes zero. Let $\rho(r)$ be the number of monomers inside a spherical volume of radius r around a tagged monomer inside a large polymer coil. It can be shown (de Gennes 1979), based on general arguments for self-similar fractal objects, that the number density of monomers depends on the radial distance according to

$$\rho(r) \sim \frac{1}{r^{3-d_f}}, \tag{2.8}$$

where d_f (reciprocal of the size exponent v) is the fractal dimension of the coil. This result is valid as long as the radial distance is not too close to the monomeric dimensions or not larger or comparable to the polymer radius. Basically, this relation is obtained by constructing the ratio of number of monomers in a volume of radius r to this volume. If the coil can be assumed to be self-similar inside the polymer coil, then the numerator of this ratio is proportional to r^{d_f} (see Equation 2.5) and the denominator is proportional to r^3, thus leading to the above equation. Since d_f is less than three (except for solid-like conformations), $\rho(r)$ decreases algebraically with the radial distance (Figure 2.4a). This result is contrasted with the corresponding result for a solid object in Figure 2.4b. Therefore, the topological correlation arising from chain connectivity leads directly to long-ranged correlation between monomer densities for polymer chains that assume noncompact and ramified conformations.

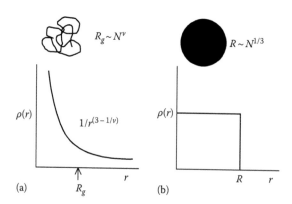

Figure 2.4 (a) Algebraic decay of monomer density correlation with radial distance. The determining factor is the fractal dimension (reciprocal of the size exponent v). (b) The density profile is a step function for solid objects.

8. *Form factor, s(**k**)*: The easiest method to determine the molar mass, size, and the size exponent of a macromolecule for a given experimental condition is to use scattering techniques with light, x-rays, or neutrons. The choice of the scattering beam in these techniques depends on the contrast between the polymer and the solution background: refractive index for light, electron density for x-rays, and neutron-scattering length for neutron scattering. The key quantity that is measured in these experiments is the form factor of the polymer chain, defined as

$$s(\mathbf{k}) \equiv \frac{1}{N} \sum_i \sum_j \langle \exp(i\mathbf{k} \cdot (\mathbf{R}_i - \mathbf{R}_j)) \rangle, \qquad (2.9)$$

where $\mathbf{R}_i$ is the position vector of the ith scattering element of the macromolecule. The magnitude of the scattering wave vector $\mathbf{k}$ depends on the wave length λ_s of the incident beam and the scattering angle θ_s (Figure 2.5),

$$k = \frac{4\pi}{\lambda_s} \sin \frac{\theta_s}{2}. \qquad (2.10)$$

The expression for the form factor connects theoretical predictions on the right-hand side to experimental measurements of $s(\mathbf{k})$ on the left-hand side. k is the reciprocal of the probe length that is being explored in a particular scattering experiment. In measurements of R_g, it is necessary to satisfy the condition $kR_g < 1$ whereby the characteristic length probed in the experiment is larger than R_g. On the other hand, if we are interested in exploring the internal structure within the polymer coil, the probe length must be smaller than R_g so that $kR_g > 1$. In general, the form factor has the following limiting behaviors:

$$s(\mathbf{k}) = \begin{cases} N\left(1 - \frac{k^2 R_g^2}{3} + \cdots\right), & kR_g < 1 \\ \sim \frac{1}{k^{1/\nu}}, & kR_g > 1. \end{cases} \qquad (2.11)$$

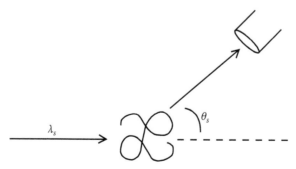

Figure 2.5 Sketch of scattering experiments. λ_s is the wavelength of the incident beam and θ_s is the scattering angle.

These limiting laws are valid for all fractal-like objects. The above equation allows us to determine N, R_g, and ν in one set of scattering experiments by judiciously tuning the scattering wave vector k, namely λ_s and θ_s.

9. *Total number of conformations*, $\mathcal{Z}$: The total number of conformations of a chain at any given experimental condition depends on its number of monomers N as

$$\mathcal{Z}(N) = \bar{z}^N N^{\gamma-1}, \tag{2.12}$$

where $\bar{z}$ is essentially the effective coordination number for the orientation of the adjacent bonds, and γ is a critical exponent. The $\bar{z}$ can be alternatively written as $\exp(-\mu/k_B T)$, where μ is the chemical potential per segment. The exponent γ depends on the nature of the polymer and the background fluid, and also on any spatial restriction on the polymer. In writing the above result for the allowed number of conformations, namely, the partition sum in the canonical ensemble (McQuarrie 1976), potential energies of interaction among segments and the boundaries are accounted for. The Helmholtz free energy F_N of the chain is related to the partition sum according to $F_N = -k_B T \ln \mathcal{Z}(N)$ (McQuarrie 1976),

$$\frac{F_N}{k_B T} = \frac{\mu N}{k_B T} - (\gamma - 1) \ln N, \tag{2.13}$$

where Equation 2.12 is used. The free energy of the chain has an extensive part proportional to N and a logarithmic part proportional to $(\gamma-1) \ln N$. As we shall see in later chapters, the logarithmic part plays a crucial role in constructing the free energy landscape for polymer translocation. An additional factor of volume V of the system, corresponding to the number of possible locations for the placement of one end of the chain, is needed on the right-hand side of Equation 2.12, if we were to be interested in the total free energy of the system.

10. *Free energy of a constrained chain*: It is of considerable interest to count the number of conformations of a polymer chain with some constraints imposed on it. For example, by fixing one end of the chain, the other end is allowed to take an end-to-end distance $\mathbf{R}$. Now, the number of chain conformations $\mathcal{N}(\mathbf{R})$ depends on $\mathbf{R}$, and the chain entropy for a given $\mathbf{R}$ is

$$S(\mathbf{R}) = k_B \ln \mathcal{N}(\mathbf{R}), \tag{2.14}$$

so that the free energy now is

$$F(\mathbf{R}) = E(\mathbf{R}) - TS(\mathbf{R}), \tag{2.15}$$

with E corresponding to the total internal energy of the chain with the end-to-end distance fixed at $\mathbf{R}$. The result of $F(\mathbf{R})$ depends on the particular model of the chain that will be used. From this result, the tensile force $\mathbf{f}$ required to maintain an end-to-end distance $\mathbf{R}$ can be obtained by using the relation,

$$\mathbf{f} = -\frac{\partial F(\mathbf{R})}{\partial \mathbf{R}}. \tag{2.16}$$

2.2 UNIVERSAL BEHAVIOR

It would seem that the above measures would be unique to a particular polymer for a given experimental condition. However, in spite of the diversity of monomeric details, macromolecules obey certain universal laws. As an example, consider the following collection of dilute polymer solutions at 25°C: polystyrene (PS) in benzene (Miyaki et al. 1978), poly(D-β-hydroxybutyrate) (PHB) in trifluoroethanol (Miyaki et al. 1977), poly(α-methylstyrene) (PαMS) in toluene (Kato et al. 1970), and poly(isobutylene) (PIB) in cyclohexane (Matsumoto et al. 1972). Clearly the chemical details are different in these systems. Nevertheless, when the ratio of the square of the radius of gyration to molar mass (M, proportional to N) is plotted against $M^{1/5}$, a straight line is observed (Fujita 1990), showing that R_g is proportional to $N^{3/5}$ for each of these systems (Figure 2.6). Even more remarkably, many denatured proteins (such as

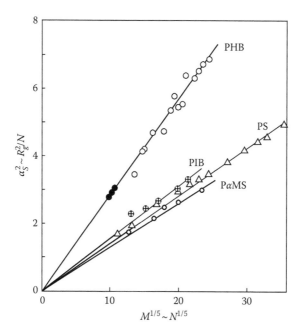

Figure 2.6 Different polymers exhibit the same universal law, $R_g \sim N^{3/5}$. Only the prefactors of this law are specific to the polymer–solvent combinations. (Adapted from Fujita, H., *Polymer Solutions*, Elsevier, Amsterdam, the Netherlands, 1990.)

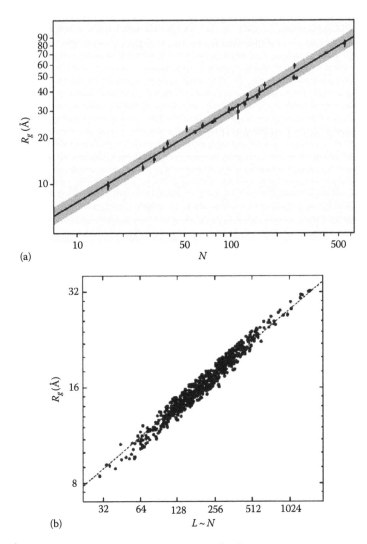

Figure 2.7 (a) Coil-like universal behavior of unfolded proteins. Many proteins follow the universal law: $R_g = R_0 N^v$, with $R_0 = 0.193$ nm and $v = 0.598$. (Adapted from Kohn, J.E. et al., *Proc. Natl. Acad. Sci. USA*, 101, 12491, 2004.) (b) Many proteins in their native states exhibit the universal law, $R_g \sim N^{1/3}$, corresponding to globular statistics. (Adapted from Banavar, J.R. et al., *J. Chem. Phys.*, 122, 234910, 2005. With permission.)

lysozyme, RNase A, ubiquitin, cytochrome C, and carbonic anhydrase) follow the same power law (Figure 2.7a) with the size exponent being 3/5 (Kohn et al. 2004, Fitzkee and Rose 2004). In their native states, hundreds of proteins follow a similar power law (Banavar et al. 2005), now the size exponent (given by the slope of the double logarithmic plot) being 1/3, as shown in Figure 2.7b. In

Figure 2.7b, R_g is plotted against the contour length L (which is proportional to the number N of amino acid residues) of the protein molecules. The graphs in Figures 2.6 and 2.7 clearly show that polymer chains exhibit certain universal features, despite their different chemical specificities.

In view of the universal nature of certain properties of polymer chains, such as the relation between polymer size and its length, it is desirable to integrate the local degrees of freedom that reflect the chemical specificity into a few parameters and build coarse-grained models to capture the essentials of the global properties of polymer statistics. In order to accomplish this purpose, there are two major features of polymer chains that require coarse graining. One deals with the chain connectivity and the other deals with the potential interaction among various monomers that are not neighbors along the chain backbone. The latter is generically called the excluded volume interaction. We shall first introduce a model for the treatment of the excluded volume interaction in the next section, followed by a summary of a few coarse-grained models for chain connectivity. We shall later combine these two together to describe conformational changes of isolated polymer chains.

2.3 EXCLUDED VOLUME INTERACTION

Consider a real chain dispersed in a solvent. Following the discussion of Figure 2.2b, a conformation of the skeletal part of the chain is represented in Figure 2.8a, where the monomers i and j, separated by a distance $\mathbf{r}_{ij}$, are marked. In reality, every monomer interacts with the nearby solvent molecules and other monomers, in addition to its neighbors along the chain backbone. As a result, there is an effective potential interaction energy $u(\mathbf{r}_{ij})$ between the ith and jth monomers, mediated by solvent molecules, in addition to the connectivity forces operating on i and j from their backbone neighbors. For any polymer–solvent system, *ab initio* calculation of $u(\mathbf{r}_{ij})$ is a very difficult task. Nevertheless, we expect this interaction to become infinitely repulsive as the two monomers approach too close to each other, due to the fact that one monomer will exclude a certain volume for the placement of another monomer.

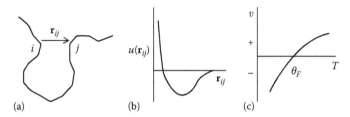

Figure 2.8 (a) Excluded volume interaction between a pair of monomers. Details of the interaction energy, sketched in (b), unique to a particular combination of the polymer and solvent are parametrized by the parameter v. (c) v appears to be zero at a special temperature θ_F.

Of course, $u(\mathbf{r}_{ij})$ is zero if the monomers are far away from each other. As in the case of a pair of atoms or molecules in a gaseous state, we expect a van der Waals–type attractive interaction between the monomers at some intermediate distances between the monomers, as sketched in Figure 2.8b. The depth and range of the attractive well depend on the specificities of the monomer and the solvent. We call the effective pairwise interaction energy $u(\mathbf{r}_{ij})$ between the monomers as the excluded volume interaction.

Since the precise details of the pairwise potential energy between monomers are very difficult to determine, we define a parameter, called the excluded-volume parameter v, that is specific to a particular combination of the polymer and solvent,

$$v = \int d\mathbf{r}_{ij} \left[1 - \exp\left(-\frac{1}{k_B T} u(\mathbf{r}_{ij}) \right) \right]. \qquad (2.17)$$

This parameter is actually the second virial coefficient for the monomers in the solvent medium. Clearly, v is temperature-dependent. At higher temperatures, as the monomers collide against each other, the interaction energy is repulsive (positive) so that the first term inside the square bracket of the above equation dominates over the second term. Therefore, v is positive at higher temperatures. At lower temperatures, the monomers tend to attract (u is negative), and now the second term dominates over the first term so that v is negative. The temperature dependence of v is sketched in Figure 2.8c.

It is obvious from Figure 2.8c that there is a special temperature that is unique to a particular combination of the polymer and solvent, at which the net effective pairwise interaction energy between the monomers would appear to vanish due to a cancelation of the repulsive and attractive contributions. This can happen only at one special temperature for a given system. This special temperature is called the ideal temperature, or the theta temperature, or the Flory temperature (θ_F). At this temperature, the excluded volume parameter v is identically zero.

In addressing the coarse-grained models of polymer chains, we first describe the skeletal models without any excluded volume effects, by considering the special case of $v = 0$. We shall address the role of the excluded volume effect in Sections 2.5 and 2.6.

2.4 COARSE-GRAINED MODELS OF CHAIN CONNECTIVITY

2.4.1 Freely Jointed Chain

Let us briefly return to the skeletal conformation given in Figures 2.2b and 2.3, where the bond angles θ_i and the dihedral angles ϕ_i are depicted. Let us also imagine a hypothetical situation where there are no restrictions on θ_i and ϕ_i. Of course, this is a violation of fundamental laws of chemical bonds. Nevertheless, this imagination allows us to map a polymer conformation to a trajectory of a random walk and to import the familiar properties of random walks to polymer chain statistics. When there are no restrictions on θ_i and ϕ_i, the skeletal bonds can rotate freely and bend back on themselves. Also, the bond vectors are completely uncorrelated in their directions. This hypothetical model is called the freely jointed chain.

The mean square end-to-end distance of a freely jointed chain follows from the second equality of Equation 2.1 as

$$\langle R^2 \rangle = N_0 a^2, \tag{2.18}$$

where N_0 is now the number of the skeletal bonds in the chain, and a is the bond length. The change to the symbol N_0 is to avoid confusion with the symbol for the number of segments N defined below. In obtaining the above result, $\langle \mathbf{a}_i \cdot \mathbf{a}_j \rangle = 0$, unless $i = j$. This result is exactly the same as the famous Einsteinian law that the mean square displacement of a particle undergoing Brownian motion is directly proportional to the number of steps taken by the particle. The implication of Equation 2.18 is that the size exponent for a freely jointed chain is half,

$$\nu = \frac{1}{2}. \tag{2.19}$$

It is a straightforward, but lengthy, exercise to calculate $\langle R^2 \rangle$ by constraining the bond angle to be fixed (freely rotating chain). Also, much effort (Flory 1969) has gone into calculating the consequences of some preferred dihedral angles on $\langle R^2 \rangle$. The net result for large values of N_0, in the absence of any excluded volume effect, is

$$\langle R^2 \rangle = C_\infty N_0 a^2, \tag{2.20}$$

where C_∞ is a temperature-dependent numerical prefactor independent of N_0. It is a measure of the average number of bonds beyond which the bond orientations are uncorrelated. The value of C_∞ is specific to the chemical details of the polymer backbone and is the ratio by which $\langle R^2 \rangle$ is expanded over a freely jointed chain by restrictions on bond angle and dihedral angle. Therefore, it is called the characteristic ratio of the polymer. The characteristic ratio is a measure of the local stiffness of the chain backbone. The nonuniversal value of C_∞ is related to the persistence length of the polymer (Section 2.4.4).

2.4.2 Kuhn Chain Model

As seen above, the key result of the freely jointed chain statistics and incorporation of local conformational details such as restricted bond angles and dihedral angles is that $\langle R^2 \rangle$ is always proportional to the chain length, that is, the size exponent $\nu = 1/2$, as long as the excluded volume effect is absent. The only

Figure 2.9 Kuhn chain model: a real chain conformation of N_0 bonds with bond length a is mapped into a freely jointed chain of N steps, each of length ℓ.

place where the chemical identity of the polymer backbone appears is in the prefactor of the relation,

$$\langle R^2 \rangle \sim N_0. \tag{2.21}$$

This conclusion therefore suggests that we could build an equivalent freely jointed chain, without compromising the N_0-dependence of the polymer size given by Equation 2.20, by defining an effective bond length, which is renormalized by incorporating the consequences of C_∞. Such a model is the Kuhn chain model and is the basis of all coarse-grained models of polymer chains.

Let ℓ be the distance along the chain contour beyond which local conformations are uncorrelated. We imagine (Figure 2.9) that the real chain (with N_0 skeletal bonds), in the absence of excluded volume interactions, can be replaced by an equivalent chain consisting of N hypothetical bonds (or segments) each of length ℓ connected by free joints. In other words, the chain can be taken to be a string of segments each of length ℓ. Clearly, each segment may contain many actual monomers. This model chain is referred to as the Kuhn chain. ℓ is called the Kuhn statistical segment length or simply Kuhn length, and N is called the number of Kuhn steps.

The Kuhn parameters N and ℓ are related to the number of actual skeletal bonds and bond length by requiring that both Kuhn and real chains have the same chain length L at full extension. For the Kuhn chain, $L = N\ell$ since the model chain is freely jointed. For the real chain of the type discussed in Section 2.1, $L = N_0 a \cos(\theta/2)$, where $(180 - \theta)$ is the bond angle. Therefore,

$$L = N\ell = N_0 a \cos\left(\frac{\theta}{2}\right). \tag{2.22}$$

Also the mean square end-to-end distance for the Kuhn chain (being a freely jointed chain) is

$$\langle R^2 \rangle = N\ell^2. \tag{2.23}$$

Equating this with the result of Equation 2.20, we get

$$\langle R^2 \rangle = N\ell^2 = C_\infty N_0 a^2. \tag{2.24}$$

TABLE 2.1

Some Examples of Relationship between
Kuhn Length and Bond Length

Polymer	a (nm)	ℓ (nm)
Polyethylene	0.154	1.24
Poly-L-alanine	0.38	3.6
dsDNA (0.2 M salt)	0.35	30

It follows from the above two equations that

$$N = \frac{L^2}{\langle R^2 \rangle} = \frac{N_0 \cos^2(\theta/2)}{C_\infty}, \tag{2.25}$$

$$\ell = \frac{\langle R^2 \rangle}{L} = \frac{C_\infty a}{\cos(\theta/2)}. \tag{2.26}$$

The two Kuhn parameters, N and ℓ, are related to N_0 and a of the real chain according to the above equations. Since C_∞ depends on temperature, both N and ℓ are in principle temperature-dependent. Nevertheless, N is directly proportional to N_0. Some examples (Cantor and Schimmel 1980) of the relationship between ℓ and a are given in Table 2.1.

In the rest of the book, we shall use N to denote interchangeably the number of Kuhn segments, the degree of polymerization of the chain, and chain length. By using the same symbol R to denote the various measures of the polymer radius, one of the main results of the Kuhn chain model is the universal law,

$$R \sim N^\nu, \tag{2.27}$$

with the size exponent,

$$\nu = \frac{1}{2}. \tag{2.28}$$

In addition to the above result for the size exponent, several quantities such as the probability distribution function for finding a particular end-to-end distance can be derived for the Kuhn model. In particular, the results become simple if the end-to-end distance is smaller than the chain contour length $N\ell$. In this limit, for example, the probability distribution function for the end-to-end distance is a Gaussian function. In view of this, a Kuhn model chain with large enough N is called a Gaussian chain. The major properties of a Gaussian chain are now summarized.

2.4.3 Gaussian Chain

Since a Gaussian chain is a freely jointed chain of large number of Kuhn steps without excluded volume interactions, it does not describe any chain statistics in almost all experiments pertinent to translocation experiments. However, the merit of this model chain is the amenability for explicit closed-form analytical expressions for various quantities of interest. Therefore, we give below a summary of key results for the various measures of conformations introduced in Section 2.1, based on Gaussian chain statistics. These results should be taken as a qualitative guidance for describing the consequences of chain connectivity. Revisions of these results due to excluded volume interactions will be addressed in Sections 2.5 and 2.6.

a. *Probability of finding the end-to-end distance at* **R**:

$$P(\mathbf{R}) = \left(\frac{3}{2\pi N\ell^2}\right)^{3/2} \exp\left(-\frac{3R^2}{2N\ell^2}\right) \tag{2.29}$$

The shape of this Gaussian function is given in Figure 2.10a. Here, $R_e^3 P(R)$ is plotted against R/R_e, where $R_e = \sqrt{N}\ell$ is the root mean square end-to-end distance for a Gaussian chain. The radius of gyration, root mean square end-to-end distance, and the hydrodynamic radius (in units of R_e) are also marked.

b. *Mean-square end-to-end distance*:

$$\langle R^2 \rangle = N\ell^2 \tag{2.30}$$

c. *Mean square radius of gyration*:

$$R_g^2 = \frac{N\ell^2}{6} \tag{2.31}$$

d. *Hydrodynamic radius*:

$$R_h = \frac{3}{4\sqrt{6}}\sqrt{N}\ell \tag{2.32}$$

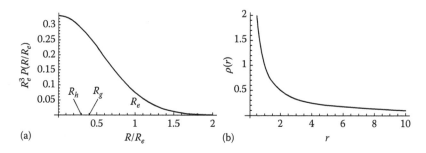

(a)

(b)

Figure 2.10 (a) The normalized end-to-end distance probability distribution function for a Gaussian chain. The radius of gyration R_g and the hydrodynamic radius R_h are shown in units of the root mean square end-to-end distance R_e. (b) Sketch of the long-ranged segment density profile falling inversely with the radial distance from the center of the chain.

e. *Size exponent*:

$$R \sim N^{\nu}, \quad \nu = \frac{1}{2} \qquad (2.33)$$

f. *Fractal dimension*:

$$d_f = 2 \qquad (2.34)$$

g. *Shape factor*:

$$\frac{R_g}{R_h} = \frac{4}{3} \qquad (2.35)$$

h. *Segmental density profile*:

$$\rho(r) \sim \frac{1}{r} \qquad (2.36)$$

The long-range correlation of segment density arising from chain connectivity is sketched in Figure 2.10b.

i. *Form factor*:

$$s(k) = \frac{2N}{k^4 R_g^4} \left[k^2 R_g^2 - 1 + \exp\left(-k^2 R_g^2 \right) \right] \qquad (2.37)$$

This form factor for a Gaussian chain is known as the Debye structure factor. The asymptotic limits of this equation are as in Equation 2.11 with $\nu = 1/2$.

j. *Total number of conformations*: Upon integrating $P(\mathbf{R})$ over all possible values of $\mathbf{R}$, the total number of conformations of a Gaussian chain becomes

$$\mathcal{Z}(N) = \bar{z}^N, \qquad (2.38)$$

with the critical exponent (defined through Equation 2.12) given by

$$\gamma = 1. \qquad (2.39)$$

k. *Chain entropy for a fixed end-to-end distance*: Since the entropy is $k_B \ln P(\mathbf{R})$, we get from Equation 2.29,

$$S(\mathbf{R}) = \frac{3k_B}{2} \ln \left(\frac{3}{2\pi N \ell^2} \right) - \frac{3k_B R^2}{2N \ell^2}. \qquad (2.40)$$

The first term on the right-hand side is a constant independent of $\mathbf{R}$. The second term shows that the chain entropy decreases quadratically with the end-to-end distance due to the reduction in the number of conformations. Therefore, coil-like conformations are entropically favorable, and it would cost energy to expand the chain to rod-like conformations.

l. *Free energy for a fixed end-to-end distance*: Substituting the expression for the chain entropy from the above equation, the free energy of a Gaussian chain with end-to-end-distance at $\mathbf{R}$ follows from $F = E - TS$ as

$$F(\mathbf{R}) = E - \frac{3k_BT}{2} \ln\left(\frac{3}{2\pi N\ell^2}\right) + \frac{3k_BTR^2}{2N\ell^2}, \tag{2.41}$$

with the total energy being simply the number of segments times the energy ϵ_s of each segment (because of absence of any intersegment excluded volume interactions),

$$E = N\epsilon_s. \tag{2.42}$$

By combining the above two equations, we see that the free energy of a chain with the end-to-end distance at $\mathbf{R}$ is given by

$$F(\mathbf{R}) = \text{constant} + \frac{3k_BTR^2}{2N\ell^2}. \tag{2.43}$$

The constant depends on N and T, but not on $\mathbf{R}$. The above equation shows that the free energy of the Gaussian chain depends quadratically on the end-to-end-distance. This behavior is analogous to that of a Hookean spring. Therefore, it is sometimes imagined that a Gaussian chain is like a Hookean spring, or a dumbbell, with a force constant k_G given by $3k_BT/(2N\ell^2)$,

$$F(\mathbf{R}) = F(0) + k_G R^2; \qquad k_G = \frac{3k_BT}{2N\ell^2}. \tag{2.44}$$

Since the spring constant (proportional to T/N) can be very small, with N being very large, the chain is highly elastic resulting in many familiar societal benefits from polymers.

The above result is not valid for large chain extensions. A more careful analysis (des Cloizeaux and Jannink 1990) of the Gaussian chain statistics, by allowing fluctuations in each Kuhn segment length, shows that the free energy of the chain is given by

$$F(\mathbf{R}) = F(0) + \frac{3k_BT}{2}\left(\frac{R^2}{N\ell^2} - 1 - \ln\frac{R^2}{N\ell^2}\right). \tag{2.45}$$

This equation reduces to Equation 2.43 for large values of $R^2 \gg N\ell^2$.

m. *Force to pull*: The tensile force f required to maintain one end of a Gaussian chain at a radial distance R from the location of the other end follows from $\mathbf{f} = -\partial F(\mathbf{R})/\partial \mathbf{R}$ and Equation 2.45 as

$$f = \frac{3k_BT}{\ell}\left(\frac{R}{N\ell} - \frac{\ell}{R}\right). \tag{2.46}$$

Naturally, the tensile force is zero when the end-to-end distance is equal to its root mean square value, $\sqrt{N}\ell$. For large $R \gg N\ell$, we get

$$\frac{f\ell}{k_B T} = \frac{3R}{N\ell}. \tag{2.47}$$

n. *Finite extensibility*: According to the above formulas derived for Gaussian chain statistics, the tensile force is linear with the end-to-end distance R for large values of R. In fact, it is theoretically possible for this model chain to extend the chain ends to infinite distances, which is physically unrealistic. This result is a mathematical consequence of the assumption $R \ll N\ell$ in obtaining the Gaussian statistics. This assumption is of course not valid at large chain extensions. A general formula for the force–distance relation can be derived (Flory 1953), without making this assumption. For a freely jointed Kuhn chain, with the tensile force acting uniformly on all segments, the relation between the end-to-end distance R and the tensile force f is given by

$$\frac{R}{N\ell} = \coth\left(\frac{f\ell}{k_B T}\right) - \frac{k_B T}{f\ell} \equiv \mathcal{L}\left(\frac{f\ell}{k_B T}\right). \tag{2.48}$$

where the Langevin function $\mathcal{L}$ is defined. By inverting the above equation, the tensile force can be written in terms of the end-to-end distance,

$$\frac{f\ell}{k_B T} = \mathcal{L}^{-1}\left(\frac{R}{N\ell}\right) \tag{2.49}$$

where $\mathcal{L}^{-1}$ is the inverse of the Langevin function $\mathcal{L}$ defined above. The difference between the results for the Gaussian chain (Equation 2.47) and the Kuhn chain with finite extensibility (Equation 2.49) is shown in Figure 2.11.

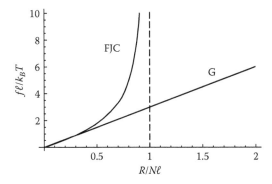

Figure 2.11 Relation between tensile force and chain extension for the freely jointed Kuhn chain (FJC) (Equation 2.49) and the Gaussian chain (G) (Equation 2.47). The vertical dashed line represents the fully extended state.

2.4.4 Wormlike Chain

Many polymer chains are not completely flexible under the usual experimental conditions of interest. In order to incorporate the local chain stiffness, the Kuhn model is modified slightly by introducing a "bond angle" $180 - \theta$ between the consecutive Kuhn steps, as sketched in Figure 2.12a. Obviously, this angle is a parameter to capture the backbone stiffness of the chain. Further, let us assume that the Kuhn steps are freely rotating, and now the model is called the Kratky–Porod or wormlike chain model.

Analogous to Equation 2.1, the mean square end-to-end distance of the Kuhn chain is

$$\langle R^2 \rangle = \sum_i \sum_j \langle \boldsymbol{\ell}_i \cdot \boldsymbol{\ell}_j \rangle. \tag{2.50}$$

Since the adjacent Kuhn steps are freely rotating, the projection of a Kuhn step onto the direction of the preceding step is $\cos \theta$, so that the correlation function for the segmental orientations becomes

$$\langle \boldsymbol{\ell}_i \cdot \boldsymbol{\ell}_j \rangle = \ell^2 \cos^{|i-j|} \theta. \tag{2.51}$$

By using this result and taking the definition of the persistence length ℓ_p to be the projection of the end-to-end distance vector on the first step (Cantor and Schimmel 1980, Yamakawa 1997),

$$\ell_p \equiv \lim_{N \to \infty} \frac{\boldsymbol{\ell}_1}{\ell} \cdot \sum_{j=1}^{N} \boldsymbol{\ell}_j, \tag{2.52}$$

we get

$$\ell_p = \frac{\ell}{(1 - \cos \theta)}. \tag{2.53}$$

Clearly, as the parameter θ approaches zero, the persistence length becomes very large. For fixed values of ℓ_p and the contour length $L \equiv N\ell$ for a specific

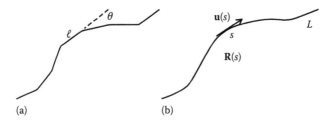

(a)   (b)

Figure 2.12 Sketches of a wormlike chain. (a) Freely rotating Kuhn chain. (b) Space curve representation: $\mathbf{R}(s)$ is the position vector of the segment at the contour variable s; $\mathbf{u}(s)$ and $\partial \mathbf{u}(s)/\partial s$ are the local tangent and curvature, respectively.

polymer, we make a continuous representation of the freely rotating Kuhn chain by taking ℓ smaller and smaller so that the chain curves continuously. Now, θ between the adjacent small steps is necessarily very small. Since $(1 - \cos\theta) \simeq -\ln\cos\theta$ for $\theta \sim 0$,

$$\cos\theta = \exp\left(-\frac{\ell}{\ell_p}\right). \tag{2.54}$$

Therefore, the orientational correlation function follows from Equations 2.51 and 2.54 as

$$\langle \boldsymbol{\ell}_i \cdot \boldsymbol{\ell}_j \rangle = \ell^2 e^{-\ell|i-j|/\ell_p}. \tag{2.55}$$

Thus, ℓ_p is a measure of the distance along the chain contour over which segmental orientations are correlated, as introduced in Section 2.1. Substituting this result in Equation 2.50, the mean square end-to-end distance is given by

$$\langle R^2 \rangle = 2\ell_p L \left[1 - \frac{\ell_p}{L}\left(1 - e^{-L/\ell_p}\right) \right]. \tag{2.56}$$

This is the full expression for the mean square end-to-end distance in the Kratky–Porod model. The limits of small persistence length with respect to the contour length and vice versa follow from this formula as

$$\langle R^2 \rangle = \begin{cases} 2\ell_p L \equiv \ell L, & \ell_p \ll L \\ L^2, & \ell_p \gg L. \end{cases} \tag{2.57}$$

Therefore, the Kratky–Porod model reduces to the rod conformation for $\ell_p \gg L$, and the freely-jointed random-walk conformation for $\ell_p \ll L$. As noted in the above equation, the Kuhn length ℓ is equivalent to twice the persistence length ℓ_p,

$$\ell = 2\ell_p, \tag{2.58}$$

and hence the Kuhn length is also a measure of chain stiffness. Similar to Equation 2.57, the two limits for the radius of gyration of a wormlike chain are

$$R_g^2 = \begin{cases} L\ell/6, & \ell_p \ll L \\ L^2/12, & \ell_p \gg L. \end{cases} \tag{2.59}$$

Therefore, depending on the ratio of persistence length to the chain length, the same polymer can have different effective size exponents.

An alternate interpretation of the parameter θ that enters in the definition of ℓ_p can be reached by considering a continuous representation of the chain backbone. Let us consider a space curve (Figure 2.12b) of contour length L to represent a wormlike chain. Here, the arc length variable s represents a segment along the contour ($0 \leq s \leq L$), whose position, local tangent, and local curvature are $\mathbf{R}(s), \mathbf{u}(s) = \partial \mathbf{R}(s)/\partial s$, and $\partial^2 \mathbf{R}(s)/\partial s^2 = \partial \mathbf{u}(s)/\partial s$, respectively. Based on the elasticity theory of rods, the energy to bend a rod, per unit length, is half the bending force constant (ϵ_b) times the inverse square of the local radius of curvature R_c,

$$\frac{U_b}{L} = \frac{\epsilon_b}{2} \left(\frac{1}{R_c} \right)^2. \tag{2.60}$$

By integrating over the whole contour length of the chain, the total energy U_b of the chain due to all local bends along the contour is

$$U_b = \frac{\epsilon_b}{2} \int_0^L \left(\frac{\partial \mathbf{u}}{\partial s} \right)^2 ds. \tag{2.61}$$

The correlation function for the segmental orientations of the wormlike chain follows from an analogy with a quantum mechanical free particle (Yamakawa 1997) as

$$\langle \mathbf{u}(s_1) \cdot \mathbf{u}(s_2) \rangle = e^{-|s_1 - s_2|/\ell_p}, \tag{2.62}$$

where the persistence length ℓ_p is related to the bending energy,

$$\ell_p = \frac{\epsilon_b}{k_B T}. \tag{2.63}$$

2.5 CHAIN SWELLING BY EXCLUDED VOLUME EFFECT

In addition to effects arising from local stiffness along the chain, monomers of real chains undergo excluded volume interactions among themselves mediated by solvent molecules as pointed out in Section 2.3. Since the chain connectivity has been parametrized in terms of Kuhn steps, we define the second virial coefficient (v) for a pair of Kuhn segments, in an equivalent way to Equation 2.17, as

$$v = \int d\mathbf{r}_{ij} \left[1 - \exp \left(-\frac{1}{k_B T} u(\mathbf{r}_{ij}) \right) \right], \tag{2.64}$$

where i and j denote the Kuhn segments, and $u(\mathbf{r}_{ij})$ is the effective interaction energy between the two segments. Since the right-hand side of the above equation has the dimension of volume, the excluded volume parameter v can be rewritten as

$$v \equiv w\ell^3 \equiv \left(\frac{1}{2} - \chi \right) \ell^3, \tag{2.65}$$

by expressing the unit of volume as the cube of the Kuhn length. The parameter χ defined above is called the Flory–Huggins χ parameter. The chemical specificity of a particular combination of polymer and solvent at a given temperature appears in the value of this parameter. We shall use w and χ interchangeably to address the effect of excluded volume interactions among polymer segments. The temperature dependence of w is as sketched in Figure 2.8c for v. Also, since the intersegment pairwise potential $u(\mathbf{r}_{ij})$ is short-ranged, in the absence of charged moieties on the polymer, we parametrize this as the pseudo-potential,

$$\frac{u(\mathbf{r}_{ij})}{k_B T} \equiv v\delta(\mathbf{r}_{ij}), \tag{2.66}$$

where $\delta(\mathbf{r}_{ij})$ is the Dirac delta function. In the usual sense of high-temperature phenomena, the above definition is consistent with Equation 2.64. Basically, this means that whenever two segments are in contact, the cost of energy is w in units of $k_B T$.

Let us consider the consequences of including the excluded volume effect in the Gaussian chain model (same as the freely jointed Kuhn chain model with large number of Kuhn steps). To begin with, let us consider repulsive excluded volume interaction, that is, $w > 0$, corresponding to temperatures higher than the Flory temperature. We expect a swelling of the chain due to excluded volume effect under these conditions. The free energy F (in units of $k_B T$) of the chain is the sum of contributions from chain connectivity $F_{connectivity}$ and excluded volume effect F_{eve},

$$F = F_{connectivity} + F_{eve}. \tag{2.67}$$

$F_{connectivity}$ is given by Equation 2.43 for a Gaussian chain. F_{eve} is the product of energy cost per contact in volume element ℓ^3, probability of finding two segments to make the contact at any space location within the coil and the volume of the coil. Since the probability of finding two segments at a location is proportional to the square of the segment density, we get

$$F_{eve} \sim \frac{w\ell^3}{2} \left(\frac{N}{R^3}\right)^2 R^3, \tag{2.68}$$

where we have left out some numerical prefactors. R is the coil radius and the segment density is N/R^3. By combining Equations 2.43, 2.67, and 2.68,

$$F = \frac{3R^2}{2N\ell^2} + \frac{w\ell^3}{2}\frac{N^2}{R^3}. \tag{2.69}$$

Therefore, the repulsive excluded volume interaction favors larger values of R (by lowering F_{eve}), and the entropic part due to chain connectivity favors smaller values of R (by lowering $F_{connectivity}$). As a result, an optimum is attained. This is obtained by minimizing F with respect to R. Leaving out the coefficients, the result is

$$\frac{\partial F\ell}{\partial R} = \frac{R}{N\ell} - \frac{w\ell^4 N^2}{R^4} = 0, \tag{2.70}$$

so that the optimum coil radius, called the Flory radius, follows as

$$R_F \sim \ell w^{1/5} N^{3/5}. \tag{2.71}$$

For strong repulsive excluded volume interactions, corresponding to "good" polymer solutions $(T > \theta_F)$, the size exponent is 3/5, called the Flory exponent ν_F

$$\nu_F = \frac{3}{5}. \tag{2.72}$$

The size exponent given by the above equation is universal for all polymer chains in their good solvents. In fact, data given in Figure 2.6 for different polymer systems obey this universal value of the size exponent. The place where the specificity of the polymer–solvent pair appears is in the slope of the lines in Figure 2.6, which is proportional to $w^{2/5}$ as given in Equation 2.71. The excluded volume parameter w is of course nonuniversal, depending on the specificity of the intersegment potential interactions for a particular polymer–solvent system.

The result of Equation 2.71 is valid only asymptotically for very large values of the excluded volume parameter. The actual variable that determines the strength of the excluded volume effect is $w\sqrt{N}$, which is known as the Fixman parameter. Both the chain length and the thermodynamic parameter w, appearing together in the combination of $w\sqrt{N}$, determine the extent of the excluded volume effect. If $w\sqrt{N}$ is small, then the Gaussian chain size exponent of $1/2$ is the result. In general, depending on the magnitude of the excluded volume parameter, there is a crossover between the two limits of Gaussian chain (random-walk statistics) and the excluded volume chain (self-avoiding-walk statistics). A formula that is useful in describing such a crossover is

$$\left(\frac{R^2}{N\ell^2}\right)^{5/2} - \left(\frac{R^2}{N\ell^2}\right)^{3/2} = \frac{4}{3}\left(\frac{3}{2\pi}\right)^{3/2} w\sqrt{N}, \tag{2.73}$$

where R is the root mean square end-to-end distance. This equation is derived by taking the correct form (Equation 2.45) for the entropic contribution to the total free energy of the chain and writing the numerical coefficient (that is consistent with the perturbation theory [Muthukumar and Nickel 1984]) for the excluded volume part in Equation 2.68. The free energy $F(R)$ of the chain with its end-to-end distance at R, which leads to the above formula, is

$$\frac{F(R)}{k_B T} = \frac{3}{2}\left(\frac{R^2}{N\ell^2} - 1 - \ln\frac{R^2}{N\ell^2}\right) + \frac{4}{3}\left(\frac{3}{2\pi}\right)^{3/2} w\frac{N^2\ell^2}{R^3}, \tag{2.74}$$

where Equations 2.45 and 2.68, with correct numerical coefficients, are used. R-independent constant terms (except $-3/2$) are ignored.

The above result for the coil size (Equation 2.73) crosses over smoothly between the Gaussian and Flory limits,

$$\frac{R}{\ell} = \begin{cases} \sqrt{N}, & w = 0 \\ \left(\frac{4}{3}\right)^{1/5} \left(\frac{3}{2\pi}\right)^{3/10} w^{1/5} N^{3/5}, & w\sqrt{N} \gg 1. \end{cases} \tag{2.75}$$

A similar formula to the above can be derived for the radius of gyration. The qualitative features and the exponents in the asymptotic regimes are the same, the only difference being a slight variation in the numerical coefficients. The factor 4/3 on the right-hand side of Equations 2.73 and 2.75 must be replaced by 134/105 (Yamakawa 1971).

When a flexible chain can swell maximally due to repulsive excluded volume interactions among its segments, the size exponent is 3/5 and the fractal dimension of the chain is 5/3. In this asymptotic regime,

$$\nu = \frac{3}{5}, \quad d_f = \frac{5}{3}. \tag{2.76}$$

Rigorous calculations (Muthukumar and Nickel 1987, des Cloizeaux and Jannink 1990) based on renormalization group theory give a more accurate value for the size exponent as $\nu = 0.5886$. The difference between this value and the Flory value is indiscernible experimentally, and, as a result, we will use the more transparent Flory value in this book.

The critical exponent γ determining the total number of chain conformations, and hence the nonextensive part of the N-dependence of the chain free energy, is

$$\gamma = \frac{6}{5}, \tag{2.77}$$

as can be readily seen by substituting the N-dependence of the optimum value of R in the free energy expression (Equation 2.74). On the other hand, the renormalization group calculations (Muthukumar and Nickel 1987, des Cloizeaux and Jannink 1990) give

$$\gamma \sim 1.16. \tag{2.78}$$

For γ, the Flory's calculation as derived in this section is not as accurate.

There are no closed-form expressions (Muthukumar and Nickel 1987, des Cloizeaux and Jannink 1990) for other measures of polymer conformations of a swollen chain, such as the form factor and monomer density profiles. Nevertheless, the general laws discussed in Section 2.2 for statistical fractals are valid, with the approximate value of 5/3 for d_f. The monomer density decays with the radial distance r as

$$\rho(r) \sim \frac{1}{r^{4/3}}, \tag{2.79}$$

and the form factor depends on the scattering wave vector as

$$s(k) \sim \frac{1}{k^{5/3}}, \tag{2.80}$$

for large angles of scattering, $kR_g > 1$.

It is noteworthy that the probability distribution function for finding the end-to-end distance is qualitatively different for swollen chains in good solvents from that for a Gaussian chain. Since the segments are avoiding each other due to repulsive excluded volume effect, the probability that the two ends of a chain are close by is very small. This feature is sketched in Figure 2.13 for a swollen chain in the asymptotic limit. Here, $R_e^3 P(R)$ is plotted against R/R_e, where $R_e = \sqrt{N}\ell$. The Gaussian chain result is included in Figure 2.13 for comparison.

As we shall see later, effects from chain confinement in channels and pores can force the chain to be essentially in space dimensions of two or one. In the space dimension of d, F_{eve} of Equation 2.68 is $\sim w\ell^d N^2/R^d$. By repeating the above procedure of free energy minimization, the Flory radius of a swollen coil in d-dimensions is

$$R_F \sim \ell N^{\nu_d}, \tag{2.81}$$

with the size exponent given by

$$\nu_d = \frac{3}{d+2}. \tag{2.82}$$

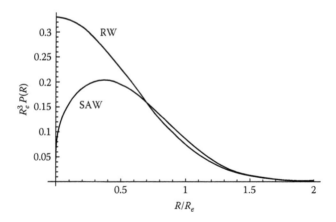

Figure 2.13 Normalized probability distribution of the end-to-end distance. The top curve is for the Gaussian chain (random walk, RW) and the bottom curve is for a fully swollen coil (self-avoiding-walk, SAW).

For the three space dimensions of interest,

$$
v = \begin{cases} 1, & d = 1 \\ 3/4, & d = 2 \\ 3/5, & d = 3. \end{cases} \tag{2.83}
$$

The values of the Flory exponent for $d = 1$ and 2 are exact, whereas the value of 3/5 in three dimensions is only approximate, but very close to the rigorous result of 0.5886. It is experimentally impossible to measure differences arising from such a small discrepancy in the exponent. As a result, the mean field theories of the type derived above have been very useful in organizing the conceptual framework for polymer physics.

2.6 COIL–GLOBULE TRANSITION

When the intersegment excluded volume interaction becomes attractive, either by the choice of a poorer solvent, or by a change in temperature, or by an increase in salt concentration (to be addressed in the next chapter), the polymer chain tends to contract in its average size. For the excluded volume parameter $w < 0$, that is, $T < \theta_F$, the analysis of the free energy given in Section 2.5 is inadequate. According to Equation 2.69, for $w < 0$, the free energy minimum of the chain would occur at $R \to 0$ with F approaching $-\infty$. Such a catastrophe is of course unphysical. As the chain contracts, the monomer density increases and a description of a chain with only two-body interactions becomes inadequate. Based on a systematic theoretical analysis (des Cloizeaux and Jannink 1990) of many-body interactions in a polymer chain, it has been shown that it is sufficient to include only the three-body interactions between the various segments, which turn out to be repulsive. These repulsive interactions counter the precipitous collapse of the chain due to the two-body attractive interactions and stabilize the chain size.

In addition to the two terms in Equation 2.69, we take the three-body contribution into account when $w < 0$. The energy per chain due to three-body interactions is proportional to the product of the triple-contact energy (w_3, in units of $k_B T$), probability of finding three monomers in a volume of ℓ^3 (proportional to the cube of monomer density, $(N/R^3)^3$) and the volume of the coil ($\sim R^3$). Adding this term to Equation 2.69 and ignoring the numerical factors, we get

$$
F = \frac{R^2}{N\ell^2} + w\ell^3 \frac{N^2}{R^3} + w_3 \ell^6 \frac{N^3}{R^6}. \tag{2.84}
$$

By minimizing the free energy with respect to R and writing w as $-|w|$, we obtain

$$
\frac{R}{N\ell} + \frac{|w|\ell^4 N^2}{R^4} - \frac{w_3 \ell^7 N^3}{R^7} = 0. \tag{2.85}
$$

Again, the numerical coefficients are ignored. When the chain collapses such that $R < \sqrt{N}\ell$ (the Gaussian chain value), we expect that the free energy contribution arising from the conformational entropy to be weaker in comparison with the two-body and three-body interaction terms. Therefore, by ignoring the first term in the above equation, we get

$$\frac{R}{\ell} = \left(\frac{w_3}{|w|}\right)^{1/3} N^{1/3}. \tag{2.86}$$

Substitution of Equation 2.86 into Equation 2.85 for R self-consistently justifies the omission of the first (entropic) term in arriving at Equation 2.86.

The main conclusion is that when w becomes negative, for temperatures below the Flory temperature, the chain attains the globular state in the asymptotic regime of large N,

$$R \sim N^{1/3}, \tag{2.87}$$

with the size exponent being $1/3$,

$$\nu = \frac{1}{3}, \quad d_f = 3. \tag{2.88}$$

It must be noted however that the globule is not fully compact, despite its dimension being three, and there is enough room for rearrangement of various monomers inside the globule. A simple crossover formula that connects the globule and the swollen coil is (Grosberg and Khokhlov 1994)

$$\left(\frac{R^2}{N\ell^2}\right)^{5/2} - \left(\frac{R^2}{N\ell^2}\right)^{3/2} = \frac{4}{3}\left(\frac{3}{2\pi}\right)^{3/2} w\sqrt{N} + w_3\left(\frac{N\ell^2}{R^2}\right)^{3/2}. \tag{2.89}$$

This expression allows the calculation of the polymer size in terms of the chain length for different values of the two parameters w and w_3. Such calculations (Grosberg and Khokhlov 1994) show that the coil–globule transition is a conformational phase transition and is similar to the liquid–gas phase transition of a van der Waals fluid.

A typical set of experimental data for the contraction of a coil as the temperature decreases is illustrated in Figure 2.14 for polystyrene in cyclohexane (Slagowski et al. 1976). For the molecular weight of 44×10^6 g/mol, the radius of gyration decreases from 348 nm at 55 °C to 121 nm at 34 °C. The Flory temperature is 35.4 °C, at which $R_g = 207$ nm. Chains of rather very large molecular weights were needed in order to experimentally observe significant coil shrinkage. However, in most of the experiments, the coil size is smaller with $R_g \leq 100$ nm.

2.7 CONCENTRATION EFFECTS

So far, we have considered only isolated chains where the average distance between any two chains in the solution far exceeds their radius of gyration.

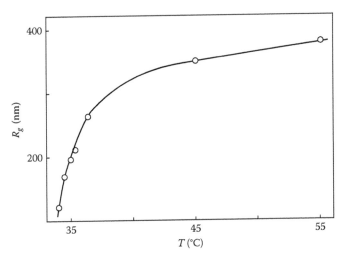

Figure 2.14 Shrinkage of R_g with lowering temperature. (From Slagowski, E. et al., *Macromolecules*, 9, 687, 1976.)

As the polymer concentration increases, the polymer size and thermodynamic properties of the solution are significantly modified in ways unique to the string-like nature of the polymers. Let us consider a solution of volume V, containing n flexible chains with N Kuhn segments per chain. The monomer concentration c is defined as

$$c = \frac{nN}{V}. \tag{2.90}$$

A useful marker in addressing the polymer concentration variable (usually defined as the total number of monomers per volume, c) is the overlap concentration c^*, at which the polymer coils are about just touching each other. If there are n chains in the polymer solution, then the total volume is roughly the volume of each coil times n. Since the chains have not yet penetrated on average into each other at this concentration, the volume of coil is R_g^3, which in turn is proportional to $N^{3\nu}$. Therefore, the monomer concentration at the chain overlap condition is

$$c^* \sim \frac{nN}{nR_g^3} \sim N^{(1-3\nu)}. \tag{2.91}$$

As a result, for a good solution, c^* is proportional to $N^{-4/5}$. The longer the chain, the lower the monomer concentration at which the chains would interpenetrate.

When $c < c^*$, we label the solution as dilute (Figure 2.15a). All of the above results in this chapter are applicable in this limit. As the concentration is increased above the overlap concentration, we have either the semidilute

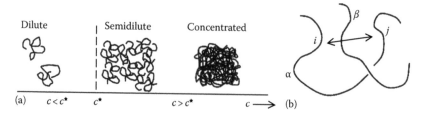

Figure 2.15 (a) Different concentration regimes. c^* is the overlap concentration. (b) Interpenetration of chains results in the screening of intrachain excluded volume interaction.

solutions (where the monomer density fluctuations can be substantial) or the concentrated solutions (with only weak monomer density fluctuations).

One of the remarkable consequences of the string-like nature of polymer chains in crowded environments is the ability for the chains to interpenetrate and attain significant entropy instead of merely colliding against each other. As a result, for concentrations higher than the overlap concentration, the intrachain excluded volume interaction is substantially screened (Edwards 1966, de Gennes 1979, Doi and Edwards 1986) by other interpenetrating chains (Figure 2.15b). This is known as the Edwards screening. Since the intrachain excluded volume effect is diminished by this screening phenomenon progressively with increasing concentration, the effective value of the size exponent of a labeled chain in good solvents decreases with polymer concentration and ultimately reaches the Gaussian chain value of $1/2$. In fact, the radius of gyration of a labeled chain in semidilute solutions depends on the polymer concentration and N according to (de Gennes 1979, Muthukumar and Edwards 1982a)

$$R_g \sim \frac{\sqrt{N}}{c^{1/8}}. \tag{2.92}$$

Also, remarkably, the radius of gyration of a labeled chain in a melt obeys the Gaussian chain scaling law (Cotton et al. 1974, Wignall et al. 1974),

$$R_g \sim \sqrt{N}, \quad \text{melt.} \tag{2.93}$$

The correlation length ξ that characterizes the monomer density fluctuations depends on the polymer concentration. For dilute solutions with $c < c^*$, the monomers are correlated over the coil radius, and therefore, $\xi \sim R_g$. For $c > c^*$, the concentration dependence of ξ depends on whether c corresponds to the semidilute regime or concentrated regime. In the semidilute regime, where the monomer density fluctuations are substantial, $\xi \sim c^{-\nu/(3\nu-1)}$, where the size exponent ν depends on the solvent quality for the polymer swelling (de Gennes 1979). On the other hand, for concentrated solutions where monomer density fluctuations are weak, $\xi \sim c^{-1/2}$, and in this regime, ξ is known as the Edwards screening length. The above results are summarized in Table 2.2 for good solutions ($\nu = 3/5$).

TABLE 2.2
R_g and ξ in Dilute, Semidilute, and Concentrated
Solutions of Flexible Chains in Good Solvents

Quantity	Dilute	Semidilute	Concentrated
R_g	$N^{3/5}$	$c^{-1/8}\sqrt{N}$	$\sqrt{N}$
ξ	R_g	$c^{-3/4}$	$c^{-1/2}$

Source: Muthukumar, M. and Edwards, S.F.,
J. Chem. Phys., 76, 2720, 1982a.

The free energy of a concentrated polymer solution has two parts. The first
part is the mean field contribution arising from the translational entropy of the
chains and solvent molecules, and the enthalpic interactions among the polymer
segments and solvent molecules. The second part is due to monomer density
fluctuations. Assuming that the components are randomly mixed without any
topological correlations associated with chain connectivity and that the solution
is incompressible, the mean field part is given by the Flory–Huggins theory
(Flory 1953) as

$$\frac{\Delta F_0}{k_B TV} = \frac{\phi}{N}\ln\phi + (1-\phi)\ln(1-\phi) + \chi\phi(1-\phi), \qquad (2.94)$$

where ϕ is the volume fraction of the polymer $nN\ell^3/V$, and the Flory–Huggins
parameter χ is defined in Equation 2.65. The first two terms on the right-hand
side of Equation 2.94 are due to the entropy of random mixing of the polymer segments and solvent molecules. The third term is the enthalpy of mixing
with the assumption of random mixing. In deriving the above equation, the
volumes of a solvent molecule and a segment are assumed to be identical for
convenience, and the reference state of pure polymer and solvent components
is taken. With this reference state in deriving the free energy of mixing, ΔF_0,
the Flory–Huggins χ parameter is a measure of the relative pairwise interaction
energy between a polymer segment and a solvent molecule (w_{ps}) with respect to
the pairwise energies for the pure components (w_{pp} for the polymer segments,
and w_{ss} for the solvent molecules),

$$\chi \sim \frac{1}{k_B T}\left[w_{ps} - \frac{1}{2}(w_{pp} + w_{ss})\right]. \qquad (2.95)$$

For good solutions where the polymer–solvent interactions are more favorable
($w_{ps} < 0$) in comparison with polymer–polymer and solvent–solvent interactions, χ is negative. It can be shown that the polymer solution is miscible and
homogeneous if χ is less than about $1/2$ (Flory 1953).

The free energy part, $\Delta F_{fl,p}$, arising from monomer density fluctuations can be derived to be (Muthukumar and Edwards 1982a)

$$\frac{\Delta F_{fl,p}}{k_B TV} = \frac{1}{24\pi \xi^3}, \qquad (2.96)$$

where ξ depends on the polymer concentration as a crossover function between the regimes of Table 2.2. By adding the above two parts, the free energy density of a crowded polymer solution follows as

$$\frac{\Delta F}{k_B TV} = \frac{\phi}{N} \ln \phi + (1 - \phi) \ln(1 - \phi) + \chi\phi(1 - \phi) + \frac{1}{24\pi \xi^3}. \qquad (2.97)$$

This equation can be used to obtain an expression for the osmotic pressure of a polymer solution inside a confining region.

We do not dwell more on the thermodynamic consequences of chain inter-penetration in the crowded semidilute and concentrated polymer solutions, as we will mainly focus on the translocation phenomenon involving single chains in this book.

2.8 SUMMARY

Although the polymer chains must possess chemical specificity in order to express their unique functions in various macromolecular processes, they exhibit certain universal behavior at larger length scales. By parametrizing the chemical details at the monomeric level, we have described various coarse-grained models, namely the Kuhn chain, Gaussian chain, and the wormlike Kratky–Porod chain. Chain stiffness is captured by the persistence length parameter.

We introduced several measures of polymer conformations, the most important being the radius of gyration R_g of the chain. It depends on the number of monomers N in the chain, according to a power law,

$$R_g \sim N^\nu, \qquad (2.98)$$

where the size exponent can take universal values. The fractal dimension d_f of the polymer is the reciprocal of ν.

Excluded volume interactions among segments mediated by the solvent lead to different scaling behaviors. In good solvents, the excluded volume interactions are repulsive and the coils are swollen ($\nu \simeq 3/5$). In poor solvents, the excluded volume interactions are attractive and the chain contracts to become globules ($\nu \simeq 1/3$).

If the chain stiffness is strong and if the contour length is short, then the polymer is rod-like ($\nu \simeq 1$). Even a stiff chain can look coil-like with random-walk statistics ($\nu = 1/2$), if the contour length is too long in comparison with the persistence length and if the excluded volume interactions are absent. At short distances along the backbone, the chain is rod-like, while at larger length scales it appears coil-like.

3

ELECTROLYTE SOLUTIONS, INTERFACES, AND GEOMETRIC OBJECTS

What happens when a pinch of table salt is added to a cup of water? The dissociated cations (sodium ions) and anions (chloride ions) in the polar solvent distribute themselves in a correlated manner due to their electrical charges. Like charges repel each other and the opposite charges attract, and the pairwise electrostatic interactions can be long-ranged. As a result, the long-ranged correlations among the dissociated electrolyte ions lead to thermodynamic properties distinctly different from those for solutions containing uncharged solute molecules, although the electrolyte solution is overall neutral. When a foreign body such as a colloidal particle or a cylindrical pore is present in an electrolyte solution, the interfaces tend to be charged with one sign, and the oppositely charged ions (counterions) hover over near the interfaces. The amount of counterions near the interfaces is dictated by the balance between the strength of the attractive interaction between opposite charges and the loss of translational degrees of freedom of free counterions. Polyelectrolytes are no exception to this phenomenon. However, the conformational degrees of freedom of the charged macromolecules play an additional role in the way the various ions are correlated. Since an adequate description of ions surrounding a fluctuating macromolecule is quite complex, approximations are necessary. On the other hand, some rigorous results are known for the behavior of electrolyte solutions near rigid geometries such as planar interfaces, spheres, cylinders, pores, and lines. Therefore, it is useful to first gain an understanding of the behavior of electrolyte solutions around rigid geometric objects, before we address suitable models and approximations for charged macromolecules. The primary focus of this chapter is to collect the key concepts of electrostatic interactions in electrolyte solutions containing rigid geometric objects.

3.1 ELECTROLYTE SOLUTIONS

Let us consider a dilute solution of a completely dissociated simple electrolyte. The charges of the ions are denoted by $z_a e$, where the subscript a refers to the different kinds (cations and anions) of ions. e is the absolute value of the unit electronic charge. z_a is a positive or a negative integer. Let n_{a0} be the number density of ions of the a-type, namely the number of ions of the a-type per unit

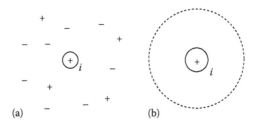

Figure 3.1 Sketches of (a) distribution of ions and (b) the cloud around the reference ion i.

volume of the solution. Since the solution as a whole is electrically neutral,

$$\sum_a n_{a0} z_a = 0. \tag{3.1}$$

Due to the electrostatic interaction being not short-ranged, the ions are not randomly distributed in the solution. Although the distance between any two nearby ions is about the same (which is set by the electrolyte concentration), the distribution of the ion charge is not uniform. The immediate neighborhood of an ion of positive charge is likely to have more of negative ions than the average number that would be at distances sufficiently far away from the reference ion. As a result, we can imagine that every ion in the solution is surrounded by a cloud of ions such that the net charge of the cloud is exactly the same as that of the reference ion but of opposite sign (Figure 3.1). Before we address the essential features of the counterion cloud around a labeled ion and the spatial variations of the ion distributions and electrical potentials in the system, let us first consider only one pair of ions.

3.1.1 Coulomb Interaction and Bjerrum Length

The electric potential $\psi(\mathbf{r})$ due to a point-charge ez_i at a radial distance $r = |\mathbf{r}|$ from the center of the ion is given by the Coulomb law,

$$\psi(\mathbf{r}) = \frac{ez_i}{4\pi\epsilon_0\epsilon} \frac{1}{r}, \tag{3.2}$$

where ϵ is the dielectric constant of the medium where the ion is placed, and ϵ_0 is the permittivity of the vacuum. The electric field due to the ion is the negative gradient of the potential,

$$\mathbf{E}(\mathbf{r}) = -\nabla\psi(\mathbf{r}). \tag{3.3}$$

Therefore, the electric potential ($\psi \sim 1/r$) and the electric field ($E \sim 1/r^2$) due to a charge are long-ranged. This is to be contrasted with the hydrophobic forces that operate only at monomeric distances and hence short-ranged. When another ion of charge ez_j is placed at a distance r from the first ion, the electrostatic energy of the second ion in the field created by the first ion is given by

$$u_{ij}(\mathbf{r}) = ez_j\psi(\mathbf{r}). \tag{3.4}$$

Substituting Equation 3.2 for $\psi(\mathbf{r})$ in the above equation, the electrostatic energy of a pair of ions of charges $z_i e$ and $z_j e$, separated by a distance r, is given by the Coulomb energy,

$$u_{ij}(\mathbf{r}) = \frac{z_i z_j e^2}{4\pi\epsilon_0\epsilon}\frac{1}{r}. \tag{3.5}$$

The strength of the interaction depends on the valencies of the ions and the dielectric constant of the medium, in addition to the fundamental constants e and ϵ_0. A convenient measure of the strength of the electrostatic interaction between two ions in a medium with uniform dielectric constant ϵ is the Bjerrum length ℓ_B defined as

$$\ell_B \equiv \frac{e^2}{4\pi\epsilon_0\epsilon k_B T}. \tag{3.6}$$

The Bjerrum length is the distance at which the electrostatic energy of bringing two charges is equal to the thermal energy $k_B T$. If the distance between two monovalent ions is shorter than ℓ_B, the ions would interact stronger; for $r > \ell_B$, the interaction would be weaker. The value of ℓ_B depends importantly on the dielectric constant of the medium. For example, for aqueous solutions at 25°C, the Bjerrum length is

$$\ell_B \simeq 0.7\,\text{nm}, \qquad \epsilon = 80 \quad \text{and} \quad T = 25°\text{C}, \tag{3.7}$$

where the dielectric constant is taken to be 80 along with $e = 1.602 \times 10^{-19}\,\text{C}$, $\epsilon_0 = 8.854 \times 10^{-12}\,\text{C}^2/\text{N}\cdot\text{m}^2$, and $k_B = 1.381 \times 10^{-23}\,\text{J/K}$. On the other hand, for oil-like media such as alkanes ($\epsilon = 2.25$), ℓ_B is

$$\ell_B \simeq 24.9\,\text{nm}, \qquad \epsilon = 2.25 \quad \text{and} \quad T = 25°\text{C}. \tag{3.8}$$

As the interior of a folded protein or the neighborhood of the backbone of a polyelectrolyte has lower dielectric constant, the counterions are bound more strongly to the charged monomers in the interior, in comparison with the monomers at the outside boundary of the molecule. Also, as the experimental conditions change, the chain conformations can alter substantially. For example, a folded protein or a globule-like polyelectrolyte can open up. Simultaneously to this change, the previously bound counterions would become unbound as the local dielectric constant becomes higher (i.e., ℓ_B becomes smaller). The binding of counterions to the polymer is self-regulated with the changes in polymer conformations in a self-consistent manner, as we shall see in the next chapter. It also must be recognized that the dielectric constant of a solvent is temperature dependent, and as a result, ℓ_B depends on temperature through both ϵ and T in Equation 3.6. A special feature of aqueous solutions is that ℓ_B is essentially insensitive to the temperature. The temperature dependence of ϵ for water is such that the product ϵT is insensitive to the temperature.

3.1.2 Poisson–Boltzmann Equation

The electric potential around a labeled ion in an electrolyte solution is determined by the correlated distribution of other ions in the system. The electric potential $\psi(\mathbf{r})$ at any spatial location $\mathbf{r}$ is given by the Poisson equation (Griffiths 1999),

$$\nabla^2 \psi(\mathbf{r}) = -\frac{\rho(\mathbf{r})}{\epsilon_0 \epsilon}, \tag{3.9}$$

where $\rho(\mathbf{r})$ is the local charge density of ions at $\mathbf{r}$. The total charge density due to all ions at $\mathbf{r}$ is given by

$$\rho(\mathbf{r}) = \sum_a e z_a n_a(\mathbf{r}) \tag{3.10}$$

where $n_a(\mathbf{r})$ is the local number density of a-type ions (cations or anions). The potential energy of an ion of the a-type in the electric field created by $\psi(\mathbf{r})$ around the ion is $e z_a \psi(\mathbf{r})$. We assume that the local number density of each kind of ions follows the Boltzmann distribution,

$$n_a(\mathbf{r}) \sim \exp\left[-\frac{e z_a \psi(\mathbf{r})}{k_B T}\right]. \tag{3.11}$$

The potential energy from nonuniform distribution of ions in the ion cloud vanishes at very large distances from the reference ion, because the ion distribution is uniform (n_{a0}) with respect to the reference ion at such large distances. Therefore, the proportionality factor in the above equation is the average number density n_{a0} so that

$$n_a(\mathbf{r}) = n_{a0} \exp\left[-\frac{e z_a \psi(\mathbf{r})}{k_B T}\right]. \tag{3.12}$$

Combining the Poisson equation (Equation 3.9) and the Boltzmann equation (Equation 3.12), and using Equation 3.10, we get the Poisson–Boltzmann equation,

$$\nabla^2 \psi(\mathbf{r}) = -\frac{e}{\epsilon_0 \epsilon} \sum_a z_a n_{a0} \exp\left[-\frac{e z_a}{k_B T} \psi(\mathbf{r})\right]. \tag{3.13}$$

We must be aware that the Boltzmann formula for the distribution of ions is not exact, and the potential of mean force from all ions should be used in the argument of the exponential instead of the average potential. As a result, the Poisson–Boltzmann equation is known to lead to some incorrect results (McQuarrie 1976, Barthel et al. 1998). Nevertheless, this formalism is adequate in explaining many experimental situations. In particular, we shall implement the results of the Poisson–Boltzmann equation in the treatment of ionic current through nanopores.

3.1.3 Debye–Hückel Theory

The Poisson–Boltzmann equation for the electric potential, or equivalently distribution of ions, is nonlinear, and the solution of the equation requires numerical methods, except for a few special situations. However, tremendous simplification arises, without losing the major concepts of ion correlations, by linearizing the above nonlinear equation, Equation 3.13.

Let us consider experimental conditions where the electrostatic interaction of the ions is relatively weak such that the electrical potential energy of an ion interacting with its cloud, $ez_a\psi$, is small in comparison with the thermal energy $k_B T$. Under these conditions, the exponential of Equation 3.13 can be expanded as a series,

$$\nabla^2\psi(\mathbf{r}) = -\frac{e}{\epsilon_0\epsilon}\sum_a z_a n_{a0}\left[1 - \frac{ez_a}{k_B T}\psi(\mathbf{r}) + \cdots\right].\tag{3.14}$$

Due to the electroneutrality condition of Equation 3.1, the first term on the right-hand side of the above equation vanishes. By ignoring all the higher order terms inside the square brackets except the linear term in ψ, we get the linearized Poisson–Boltzmann equation,

$$\nabla^2\psi(\mathbf{r}) = \kappa^2\psi(\mathbf{r}),\tag{3.15}$$

where

$$\kappa^2 \equiv \frac{e^2}{\epsilon_0\epsilon k_B T}\sum_a z_a^2 n_{a0}.\tag{3.16}$$

The linearized Poisson–Boltzmann equation is referred to as the Debye–Hückel equation. The constant coefficient κ^2 appearing in the above equation is one of the important parameters in the discussion of electrolyte solutions and polyelectrolytes. We shall describe its physical interpretation and experimental relevance below and in Section 3.1.3.1.

Let us consider the electric potential around a reference ion i with charge ez_i. Since the ion cloud is spherically symmetric on average, Equation 3.15 can be rewritten as

$$\frac{1}{r^2}\frac{d}{dr}\left(r^2\frac{d\psi}{dr}\right) = \kappa^2\psi,\tag{3.17}$$

where r is the radial distance from the reference ion i. Assuming that all ions are point-like and using the boundary conditions that the electric potential vanishes as $r \to \infty$ and it is given by Equation 3.2 for very small values of r, the solution of the above equation is

$$\psi(r) = \frac{ez_i}{4\pi\epsilon_0\epsilon} \frac{e^{-\kappa r}}{r}. \tag{3.18}$$

Thus, the potential is Coulomb-like at distances shorter than κ^{-1}, and it becomes small at distances longer than κ^{-1}. The net effect of correlations among all ions is to screen the potential from an ion and make it less long-ranged than the Coulomb potential. The potential given by Equation 3.18 is called the Debye–Hückel potential, screened Coulomb potential or the Yukawa potential. κ^{-1} is called the Debye length or the electrostatic screening length.

The electrostatic energy between two charges ez_i and ez_j separated by a distance r_{ij} is the product of ez_j and the potential due to the charge ez_i,

$$u_{ij}(r_{ij}) = \frac{e^2 z_i z_j}{4\pi\epsilon_0\epsilon} \frac{e^{-\kappa r_{ij}}}{r_{ij}}. \tag{3.19}$$

Using the definition of the Bjerrum length, Equation 3.6, Equation 3.19 can be rewritten as

$$u_{ij}(r_{ij}) = z_i z_j \ell_B k_B T \frac{e^{-\kappa r_{ij}}}{r_{ij}}. \tag{3.20}$$

The electrostatic potential energy between two ions given by Equations 3.19 and 3.20 is called the Debye–Hückel potential energy. The collective effect of the ions in the solution is to screen the Coulomb interaction between a pair of ions given by Equation 3.5 resulting in the screened electrostatic interaction given by Equation 3.19.

3.1.3.1 Electrostatic Screening Length

The electrostatic screening length, also called the Debye length ξ_D, is the key measure of the range of electrostatic interaction among charges in a medium. It is defined as the reciprocal of κ, as mentioned above

$$\xi_D \equiv \kappa^{-1}. \tag{3.21}$$

Substitution of Equation 3.16 in the above definition gives

$$\xi_D = \left(\frac{e^2}{\epsilon_0\epsilon k_B T} \sum_a z_a^2 n_{a0} \right)^{-1/2}. \tag{3.22}$$

This result can be expressed in terms of the electrostatic strength parameter (Bjerrum length ℓ_B), by using Equation 3.6 in the above equation, as

$$\xi_D = \left(4\pi \ell_B \sum_a z_a^2 n_{a0}\right)^{-1/2}. \tag{3.23}$$

Thus, ξ_D is inversely proportional to the square root of the Bjerrum length for a fixed concentration of the electrolyte,

$$\xi_D \sim \frac{1}{\sqrt{\ell_B}}. \tag{3.24}$$

Whereas the Bjerrum length is a property of the solvent, the Debye length is a property of the solution. The dependence of ξ_D on the electrolyte concentration can be written in terms of the experimentally convenient concentration unit (molarity, i.e., number of moles per liter, with liter being dm^3). The number density n_{a0} of ions of a-type is

$$n_{a0} = 10^3 N_A c_{a0}, \tag{3.25}$$

where c_{a0} is the concentration of ions of a-type in molarity, N_A is the Avogadro number, 6.023×10^{23}, and the factor of 10^3 arises from the relation $1000 \, dm^3 = 1 \, m^3$. Utilizing the above conversion factor, κ^2 becomes

$$\kappa^2 = 4000\pi \ell_B N_A \sum_a z_a^2 c_{a0}, \tag{3.26}$$

where c_{a0} is in molarity, and κ^{-1} and ℓ_B are in the SI unit of m. It is also convenient to group the valencies of ions and their concentrations together by defining the ionic strength I of the electrolyte solution as

$$I \equiv \frac{1}{2} \sum_a z_a^2 c_{a0}. \tag{3.27}$$

In terms of the ionic strength, κ^2 becomes

$$\kappa^2 = 8000\pi \ell_B N_A I, \tag{3.28}$$

or more explicitly

$$\kappa^2 = \frac{2000 e^2 N_A}{\epsilon_0 \epsilon k_B T} I. \tag{3.29}$$

For an aqueous electrolyte solution at 25°C (with $\epsilon = 78.54$),

$$\kappa = 2.32 \times 10^9 \sqrt{2I} \ m^{-1}, \tag{3.30}$$

and the Debye length is

$$\xi_D = \frac{0.43}{\sqrt{2I}} \text{ nm.} \tag{3.31}$$

For monovalent salts, ξ_D is conveniently given by

$$\xi_D \simeq \frac{0.3}{\sqrt{c_s}} \text{ nm,} \tag{3.32}$$

where c_s is the salt concentration in units of moles per liter.

The typical values of the Debye length are given in Table 3.1 for (1:1) (sodium chloride type) and (2:1) (calcium chloride type) salts at concentration c_s. The dependence of ξ_D on c_s, for (1:1) and (2:1) type salts, is presented in Figure 3.2. In general, ξ_D decreases with salt concentration, and more sharply for ions with higher valencies. Specifically, the Debye length is 0.78 nm (which is comparable to the value of the Bjerrum length) at 150 mM monovalent salt in the solution. It progressively decreases with c_s and becomes merely 0.3 nm (which is less or comparable to monomer sizes) at 1 M of monovalent salt. These values for the electrostatic range become even smaller if multivalent ions such as magnesium and calcium are present in the solution. Thus, for higher salt concentrations typically used in the translocation experiments, the electrostatic correlation is not long-ranged, and it is possible to combine the electrostatic interaction among monomers with the short-ranged excluded volume interaction in describing equilibrium properties. We shall return to this issue in the next chapter.

3.1.3.2 Effect of Ion Size
Instead of taking the ions as point charges, as done above, the Debye–Hückel theory can be extended to ions with finite sizes. Let each of the ions (cations and anions) be modeled as a hard sphere of diameter a with its charge located

TABLE 3.1
Debye Length in Nanometer for Different Electrolyte Concentrations and Valencies (Aqueous Solutions at 25°C)

c_s (mol/L)	(1:1)	(2:1)
10^{-4}	30.4	17.6
10^{-3}	9.6	5.55
10^{-2}	3.04	1.76
10^{-1}	0.96	0.555
0.150	0.78	0.453
1.0	0.304	0.176

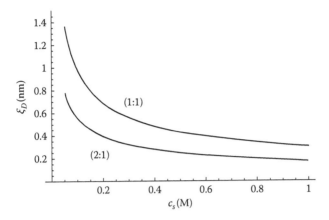

Figure 3.2 Dependence of the Debye length on salt concentration for (1:1) and (2:1) types.

at the origin of the hard sphere (Figure 3.3a). Further, let us assume that each of the ions is made of a material with the same dielectric constant as the solvent. This model is known as the restricted primitive model (McQuarrie 1976) of electrolytes. The electric potential around a reference ion i of diameter a and charge ez_i can be readily derived (McQuarrie 1976), by extending the above Debye–Hückel theory to finite ion sizes, to be

$$\psi(r) = \frac{ez_i}{4\pi\epsilon_0\epsilon} \frac{e^{-\kappa(r-a)}}{r(1+\kappa a)}, \quad r > a, \tag{3.33}$$

and

$$\psi(r) = \frac{ez_i}{4\pi\epsilon_0\epsilon r} - \frac{ez_i\kappa}{4\pi\epsilon_0\epsilon(1+\kappa a)}, \quad 0 < r < a. \tag{3.34}$$

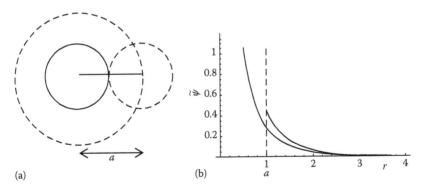

(a)

(b)

Figure 3.3 (a) Ions with diameter a. (b) Electric potential around a charge of ez_i with diameter a (Equation 3.33) for $a = 1$ nm and $\kappa^{-1} = 0.78$ nm. The lower curve is (Equation 3.18) for point charges. The ordinate is $4\pi\epsilon_0\epsilon\psi/ez_i$.

The electric potential given by Equation 3.33 is called the extended Debye–Hückel potential and is plotted in Figure 3.3b. For comparison, the result of Equation 3.18 is included in the figure. The finite size of the ion modifies the prefactor by a factor of $\exp(\kappa a)/(1+\kappa a)$, and the beginning of the decay of the potential is shifted to the ion diameter. The distance dependence is the same as the screened Coulomb potential for $r > a$. For $a \to 0$, Equation 3.33 reduces to the Debye–Hückel potential.

The first term on the right-hand side of Equation 3.34 is obviously the contribution from the reference ion itself (according to the Coulomb law, Equation 3.2). As a result, the second term must be the contribution of all ions outside the reference ion. Hence, the second term is identified as the electric potential ψ_{cloud} from the ion cloud acting on the reference ion,

$$\psi_{cloud} = -\frac{ez_i\kappa}{4\pi \epsilon_0 \epsilon(1 + \kappa a)}. \tag{3.35}$$

We now consider the charge distribution inside the cloud that leads to the potential given by Equation 3.35.

3.1.3.3 Charge Distribution around an Ion

The net charge density distribution of the ion cloud around a reference ion i of charge ez_i follows from Equations 3.10 and 3.12 as

$$\rho_{cloud}(\mathbf{r}) = \sum_a ez_a n_{a0} \exp\left[-\frac{ez_a}{k_BT}\psi(\mathbf{r})\right]. \tag{3.36}$$

As the cloud is outside the reference ion, corrections of order $1/n$, with n being the total number of ions in the solution, are ignored. As in the Debye–Hückel theory, by expanding the exponential term and keeping up to the linear term in ψ, we obtain

$$\rho_{cloud}(\mathbf{r}) = -\frac{e^2}{k_BT}\left(\sum_a z_a^2 n_{a0}\right)\psi(\mathbf{r}). \tag{3.37}$$

With the definition of the inverse Debye length κ (Equation 3.16), the charge density in the cloud becomes

$$\rho_{cloud}(\mathbf{r}) = -\epsilon_0 \epsilon \kappa^2 \psi(\mathbf{r}). \tag{3.38}$$

Substituting Equation 3.33 for the electric potential, we get

$$\rho_{cloud}(\mathbf{r}) = -\frac{ez_i\kappa^2}{4\pi(1 + \kappa a)}\frac{e^{-\kappa(r-a)}}{r}. \tag{3.39}$$

The charge density around an ion has a sign opposite to that of the reference ion and is proportional to the potential in the Debye–Hückel approximation. The total charge surrounding the reference ion is

$$ez_{cloud} = \int_a^\infty \rho_{cloud}(r)4\pi r^2 dr, \qquad (3.40)$$

where radial symmetry is used and the lower limit of the integral reflects that the centers of ions in the cloud cannot approach the reference ion within the diameter a of the ion (Figure 3.3a). Also, the evaluation of the integral of Equation 3.40 leads to the expected result that the total charge around the reference ion i is equal and opposite to the charge of the reference ion,

$$ez_{cloud} = -ez_i. \qquad (3.41)$$

The fraction of net charge, dez_{cloud}, between r and $r + dr$ in the cloud is $4\pi r^2 \rho_{cloud}$, proportional to $\kappa r \exp(-\kappa r)$, as seen from Equations 3.39 and 3.40. Rewriting this result,

$$\rho_{cloud}(r) \equiv \left| -\frac{(1+\kappa a)(dez_{cloud})\exp(-\kappa a)}{ez_i\kappa} \right| = \kappa r e^{-\kappa r}. \qquad (3.42)$$

The form and the extent of the cloud around an ion are illustrated in Figure 3.4, where $\rho_{cloud}(r)$ is plotted against κr. The maximum occurs at the Debye length $\xi_D \equiv \kappa^{-1}$, and the most important region of the ion cloud is in the neighborhood of $r \sim \kappa^{-1}$. As seen in Table 3.1 and Figure 3.2, the location of this region moves to larger values as the solution becomes more dilute in salt concentration.

For distances near the reference ion, larger than the ion radius but shorter than the distance of contact between two ions, the above results of Equation 3.41 and Figure 3.4 lead to a simple geometric representation of the cloud.

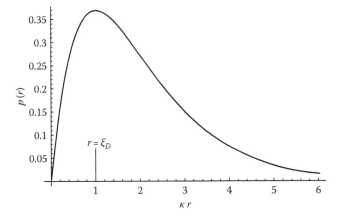

Figure 3.4 Fraction of net charge of the cloud versus the radial distance in units of the Debye length ξ_D.

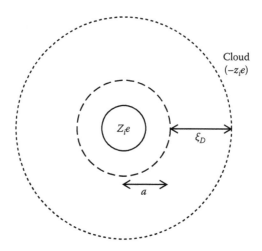

Figure 3.5 Geometrical interpretation of a cloud as a thin spherical shell of radius $(a + \xi_D)$ with net charge being the negative of the charge of the reference ion.

The cloud around the reference ion i with charge ez_i can be imagined equivalently as a thin spherical shell with a net charge of $-ez_i$ at a radial distance of $a + \kappa^{-1}$ from the center of the ion i. The electric potential from this spherical layer on the ion i is precisely as given in Equation 3.35, by substituting $a + \kappa^{-1}$ for r in Equation 3.2. Thus, the net potential acting at r from the center of the ion i is the sum of the self-potential arising from itself and the potential arising from the cloud taken as a thin shell of opposite charge, located at the distance of the Debye length away from the diameter of the ion. This result, as given by Equation 3.34, is sketched in Figure 3.5. Of course, for distances larger than the distance of pairwise contact, namely the diameter of the ion, the potential is given by Equation 3.33 and Figure 3.3b.

As we shall see later in dealing with charged macromolecules, the clouds of counterions around each of the charged monomers dominate the behavior of these macromolecules. In fact, the physics of polyelectrolytes can be termed as counterion physics. It is thus essential to have a good grasp of the results presented in Section 3.1.3 to understand the structure and mobility of charged macromolecules.

3.1.3.4 Free Energy

The electrostatic correlations of the ions in the electrolyte solution contribute to the Helmholtz free energy of the system, depending on the Debye length and ion size. Based on the Debye–Hückel theory for the restricted primitive model (Figure 3.3a), the result (McQuarrie 1976) is

$$\frac{F_{el}}{k_B T} = -\frac{V}{4\pi a^3}\left[\ln(1 + \kappa a) - \kappa a + \frac{1}{2}\kappa^2 a^2\right], \qquad (3.43)$$

where V is the total volume of the system. For low salt concentrations (i.e., $\xi_D \equiv \kappa^{-1}$ is large), κa may become so small that the above equation simplifies to

$$\frac{F_{el}}{k_B T} = -\frac{V\kappa^3}{12\pi}, \quad \kappa a \to 0. \tag{3.44}$$

The above equations are valid also for nonneutral plasma containing only counterions of a polyelectrolyte (Muthukumar 2002a), without considering the polymer charge in the treatment of ionic clouds. As the degree of ionization of a macromolecule changes, due to changes in experimental conditions, the counterion concentration changes, and consequently the free energy due to correlations of counterions changes. Such effects must be accounted for in constructing the free energy landscape for the translocation process.

3.2 CHARGED INTERFACES

All electrolyte solutions, pertinent to experimental systems, contain interfaces such as the walls of the container, internal surfaces of pores, surfaces of lipid bilayers, and surfaces of macroions present in solutions. Due to either chemical reactions, ionic equilibria, or adsorption of specific ions at the interfaces, all interfaces usually bare charges. Also, charges can be deliberately injected into the interfaces by using electrodes with an externally applied voltage. The charge density at the interface determines the electrical potential and density distribution of the electrolyte near and away from the interfaces. In general, the behavior of electrolyte solutions near charged interfaces can be quite complex. However, the salient conceptual issues associated with charged interfaces can be gleaned by considering suitably tailored theoretical models with simplifying assumptions.

As examples, we shall consider planar interfaces, cylindrical pores, large spherical and cylindrical macroions, and even an infinitely thin line of charges. Since the ideas presented in the previous section can be readily implemented to spherical macroions (Figure 3.6a), this example will be dealt with first. Next, planar-charged interfaces (Figure 3.6b) will be discussed in some detail, due to the common occurrence of this geometry in experiments. The situation of rolling the planar interface into a pore (Figure 3.6c) will follow next. Finally, a cylinder bearing charges on the outside (Figure 3.6d), along with the asymptotic limit of an infinitely thin line of charges (Figure 3.6e), will be discussed.

In realistic situations, the dielectric constants of the media across the interfaces are different. Consideration of the dielectric mismatch is quite important in assessing the quantitative aspects of forces among ions in the neighborhood of interfaces. A full treatment of dielectric heterogeneity (Frohlich 1958, Verwey and Overbeek 1999) in electrolyte solutions is beyond the scope of this book. Nevertheless, we shall consider one example in Section 3.2.6 to illustrate the effects due to dielectric mismatch across interfaces. Our primary

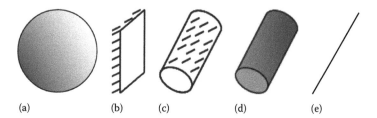

Figure 3.6 Interfaces of (a) spherical particles, (b) planar surfaces, (c) cylindrical pores, (d) cylindrical particles, and (e) infinitely thin lines.

focus is only on the major concepts related to electric potential, electric field, and ion distributions near the interfaces, by ignoring the quantitative details arising from the dielectric mismatch at interfaces.

3.2.1 Charged Spherical Particles

Consider a spherical macroion of radius R with its total charge Q distributed uniformly on its surface, present in an electrolyte solution. Since the total system is electrically neutral, the solution contains the counterions of the macroion and the dissociated salt ions. One kind of the dissociated ions, say the anions, has the same sign as the charge of the macroion. The ions of this kind are called coions. The other kind, now the cations, has the same sign as that of the counterions of the macroion. Without regard to the possible difference in the specificity of the counterions from the macroion and the salt, we shall call both of these as simply counterions. The electric potential and distributions of counterions and coions around the macroion can be obtained by solving the Poisson–Boltzmann equation (Equation 3.13). No analytical solutions are possible for this nonlinear equation with this geometry. Just as we did in Section 3.1.3, we make the linearization approximation. The Debye–Hückel theory derived for small ions can be immediately applied to the macroion case as well. Let a be the radius of the effective sphere that excludes the small ions (Figure 3.7), where

$$a = R + a_i, \tag{3.45}$$

with a_i being the mean radius of small ions. Copying the result of Equation 3.33, the electric potential outside the effective sphere of the macroion is given by the Debye–Hückel potential,

$$\psi(r) = \frac{Q}{4\pi\epsilon_0\epsilon} \frac{e^{-\kappa(r-a)}}{r(1+\kappa a)}, \quad r > a. \tag{3.46}$$

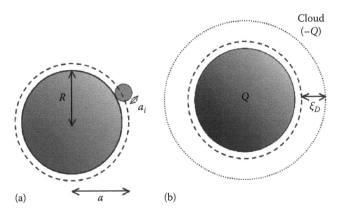

Figure 3.7 Effective radius of a spherical macroion, $a = R + a_i$. The ion cloud is at the Debye length from a. For electrolyte ions much smaller than the macroion, $a \simeq R$.

For distances shorter than the sum of particle radius and small ion radius, Equation 3.34 gives

$$\psi = \frac{Q}{4\pi \epsilon_0 \epsilon} \left[\frac{1}{r} - \frac{\kappa}{(1 + \kappa a)} \right], \quad r < a. \tag{3.47}$$

If the radius of the small ion is negligible in comparison with the radius R of the macroion (or a spherical colloidal particle) of net charge Q, the electric potential from the macroion is

$$\psi(r) = \frac{Q}{4\pi \epsilon_0 \epsilon} \frac{e^{-\kappa(r-R)}}{r(1 + \kappa R)}, \quad r > R. \tag{3.48}$$

The properties of $\psi(r)$ and the counterion cloud are exactly the same as described in Section 3.1.3. The ionic cloud around the macroion contributes to the surface potential of the macroion, in addition to the potential arising from the charge of macroion itself. Equation 3.47 yields for $r = R$,

$$\psi = \frac{Q}{4\pi \epsilon_0 \epsilon} \left[\frac{1}{R} - \frac{\kappa}{(1 + \kappa R)} \right]. \tag{3.49}$$

A slight rearrangement of this equation gives

$$\psi = \frac{Q}{4\pi \epsilon_0 \epsilon} \left[\frac{1}{R} - \frac{1}{R + \kappa^{-1}} \right]. \tag{3.50}$$

Therefore, the potential on the surface of a spherical-charged particle is the sum of the potential from its charge and that from its ion cloud, as a spherical shell with the opposite charge of the particle, located at the Debye length (κ^{-1}) away from the surface of the macroion. We shall use this result in describing the mobility of macroions in electrolyte solutions.

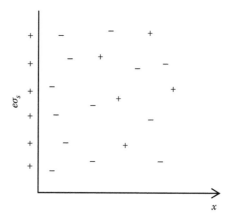

Figure 3.8 An electrolyte solution near an interface.

3.2.2 Planar Interfaces

Let us consider an electrolyte solution at a planar interface with uniform charge density $e\sigma_s$ (Figure 3.8). Let the ions in the solution be point-charges and the solution has a uniform dielectric constant ϵ. The image charges due to the dielectric mismatch at the interface are ignored. The spatial variation of the electric potential away from the interface and the density distributions of the counterions and coions can be obtained exactly from the Poisson–Boltzmann equation for this model. This theory is known as the Gouy–Chapman theory. In fact, this situation is one of the few cases where the nonlinear Poisson–Boltzmann equation can be solved exactly, and helps to have confidence in using the Debye–Hückel theory under appropriate experimental conditions. We shall give below only the important results without derivations (Evans and Wennerstrom 1999).

Since the planar interface is taken as infinitely wide, the spatial variations of the electric potential and ion distributions occur in only one dimension along the x-axis normal to the interface (Figure 3.8). Therefore, the Poisson–Boltzmann equation (Equation 3.13), becomes

$$\frac{d^2\psi(x)}{dx^2} = -\frac{e}{\epsilon_0\epsilon}\sum_a z_a n_{a0}\exp\left[-\frac{ez_a}{k_BT}\psi(x)\right], \tag{3.51}$$

where n_{a0} is the average number density of the a-type ions in the solution. The electric field due to the interface vanishes at distances far away from the interface. Also, let ψ_0 be the surface potential. These are used as boundary conditions in solving Equation 3.51,

$$\psi(x=0) = \psi_0, \tag{3.52}$$

and

$$\frac{d\psi}{dx}\bigg|_{x\to\infty} = 0. \tag{3.53}$$

We now give the results of this calculation for the two cases of (a) salt-free solutions and (b) salty solutions.

3.2.2.1 Salt-Free Solutions

Here, the ions in the solution are only the counterions to the charges on the surface in order to maintain the overall electroneutrality of the system. Let the valency of the counterion be z. The electric potential depends on the distance from the interface logarithmically,

$$\frac{ze\psi(x)}{k_BT} = 2\ln(x+\lambda) + C, \tag{3.54}$$

where C is a constant term independent of x, and λ (called the Gouy–Chapman length) is

$$\lambda \equiv \frac{1}{2\pi\ell_B \mid z\sigma_s \mid}. \tag{3.55}$$

Here ℓ_B is the Bjerrum length defined in Equation 3.6. Since $\ln(x+\lambda)$ in Equation 3.54 can be written as $\ln(1+x/\lambda)$ by adjusting the x-independent C term, λ defines a characteristic distance in the system. The Gouy–Chapman length λ is a measure of the range of attraction for the counterion cloud from the interface. The fall of the electric potential with distance from the interface is illustrated in Figure 3.9 for monovalent counterions and three surface charge densities ($\sigma_s = 0.1, 1$, and 10 per nm^2) and $\ell_B = 0.7$ nm. The values of the Gouy–Chapman length are also included in these figures to indicate the characteristic length for the decay of the potential.

The electric field follows from Equations 3.3 and 3.54 as

$$E(x) = -\frac{k_BT}{ze}\frac{2}{(x+\lambda)}. \tag{3.56}$$

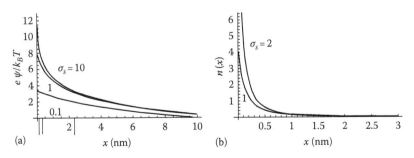

Figure 3.9 (a) Decay of electric potential with distance from the interface. (b) Counterion crowding near a charged interface.

For positively charged interfaces, the counterions are negatively charged, and the electric field is normal to the surface pointing into the electrolyte solution. The opposite is true for negatively charged interfaces. The electric field at the interface E_s in salt-free solutions follows from the above equation as

$$E_s = \frac{e\sigma_s}{\epsilon_0\epsilon},$$ (3.57)

a general result known as the contact theorem (Israelachvili 1992).

The density profile of the counterions can be derived by substituting Equation 3.54 in the Boltzmann law (Equation 3.12) as

$$n(x) = \frac{2\pi\ell_B\sigma_s^2}{\left(1+\frac{x}{\lambda}\right)^2}.$$ (3.58)

The number density of counterions at the interface ($x = 0$) is

$$n(x = 0) = 2\pi\ell_B\sigma_s^2.$$ (3.59)

Thus, the counterions pile up at the interface. The counterion density, due only to the electrostatic attraction and thermal motion, at the interface is proportional to the square of the surface charge density and to the Bjerrum length.

3.2.2.2 Salty Solutions

When a strongly dissociating salt is present in the solution in addition to the counterions, the general behavior of the electric potential is the same as in salt-free solutions. In terms of the density distributions of ions, we need to consider the coions as well. For simplicity, let us assume that the added salt is of the symmetric type ($z{:}z$) with the valencies of the cations and anions of the salt and the counterions being the same as z. Only the signs are different depending on whether they are positively charged or negatively charged. According to the Gouy–Chapman theory (Evans and Wennerstrom 1999), the electric potential $\psi(x)$ is given by

$$\tanh\left(\frac{ze\psi(x)}{4k_BT}\right) = e^{-\kappa x}\tanh\left(\frac{ze\psi_s}{4k_BT}\right),$$ (3.60)

where ψ_s is the surface potential, and κ is the inverse Debye length. According to the Gouy–Chapman theory, $\tanh(ze\psi/4k_BT)$ decreases exponentially with x.

Since k_BT/e sets the scale for the electric potential in terms of the temperature, we give a special symbol to it,

$$\psi_\theta \equiv \frac{k_BT}{e}.$$ (3.61)

At 25°C, $\psi_\theta = 25.7\,\text{mV}$.

For weak potentials such that $z\psi \ll \psi_\theta$, Equation 3.60 reduces to the result of linearized Poisson–Boltzmann equation,

$$\psi(x) = \psi_s e^{-\kappa x}. \tag{3.62}$$

This is the Debye–Hückel law for the one-dimensional variation along the x-axis, normal to the two-dimensional interface.

The extent of error arising from the Debye–Hückel approximation to the full Gouy–Chapman theory is presented in Figure 3.10a, by comparing Equations 3.60 and 3.62. Here the electric potential ψ is plotted against x for a 0.01 M solution of a (1:1) electrolyte and at the constant surface potential of 77.1 mV (i.e., three times the room temperature value, $3\psi_\theta$). For this electrolyte solution, the Debye length is 3 nm and the potential is significant even at 10 nm. The discrepancy between the Gouy–Chapman (nonlinear Poisson–Boltzmann) and Debye–Hückel (linearized Poisson–Boltzmann) is rather mild for surface potentials that are even three times $k_B T/e$. This offers confidence in utilizing the Debye–Hückel approximation to extract the main qualitative features of more complex situations of polyelectrolyte–electrolyte solution interfaces.

The surface charge density σ_s, namely the number of unit charges per nm^2, is uniquely related to the surface potential ψ_s as a function of the salt concentration (Evans and Wennerstrom 1999). The result is

$$\sigma_s = \frac{\kappa}{2\pi z \ell_B} \sinh\left(\frac{z\psi_s}{2\psi_\theta}\right). \tag{3.63}$$

Substituting the values of the Bjerrum length and the Debye length (in terms of monovalent salt concentration c_s in molarity) for aqueous solutions at room

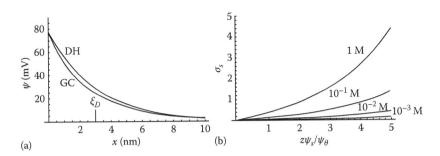

Figure 3.10 (a) Comparison of Gouy–Chapman (GC) and Debye–Hückel (DH) results for the electric potential. $\kappa^{-1} = 3$ nm (0.01 M monovalent salt) and the surface potential is 77.1 mV corresponding to three times the equivalent of room temperature. (b) Relation between the number of charges on the surface per nm^2, σ_s, and the surface potential ψ_s for different concentrations of a monovalent salt.

temperatures, a more convenient form is

$$\sigma_s = 0.731\sqrt{c_s}\sinh\left(\frac{\psi_s}{2\psi_\theta}\right),\qquad (3.64)$$

with $\psi_\theta = 25.7\,\mathrm{mV}$. The dependence of this relation between the surface charge density and the surface potential on c_s is given in Figure 3.10b. The dependence is progressively steeper as c_s increases. For a fixed surface potential, the surface charge density increases as the electrolyte concentration is increased. For example, the number of surface charges per nm^2 triples roughly as the electrolyte concentration is increased from 0.1 to 1 M, at ψ_s being four times ψ_θ. Equivalently, the surface potential decreases with an increase in c_s for fixed surface charge density.

As seen above (Figure 3.10a), the electric potential falls off roughly exponentially with the distance from the interface for a constant surface potential. A more realistic experimental scenario occurs when the surface maintains a constant surface charge density. By combining Equations 3.60 and 3.63, the electric potential can be obtained in terms of σ_s. The typical result for $\psi(x)$ is illustrated in Figure 3.11a for $c_s = 0.1$ and 1 M monovalent salt in water at $T = 25°C$, by keeping σ_s at the constant value of 1 per nm^2. The electric potential decreases roughly exponentially with x with the characteristic distance being the Debye length. Also, the electric field at the interface is independent of the salt concentration when the surface charge is fixed (as seen from the contact theorem [Equation 3.57]).

The distribution of the ions in the solution from the interface can be readily obtained from the Boltzmann law (Equation 3.12) by using the result for the potential. The distributions of counterions and coions are given in Figure 3.11b for 0.1 and 1 M monovalent electrolyte solutions in water at $T = 25°C$, by keeping the surface charge density σ_s at the constant value of 1 per nm^2. For each electrolyte concentration, the upper curve represents the counterion

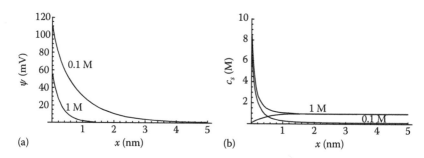

Figure 3.11 (a) Dependence of electric potential on electrolyte concentration for fixed surface charge density $\sigma_s = 1\ \mathrm{nm}^{-2}$. (b) Concentration profiles of counterions and coions in molarity for bulk concentrations of monovalent electrolyte at 0.1 and 1 M. Surface charge density is fixed at 1 nm^{-2}.

distribution describing the excess counterions near the surface. There is naturally a depletion of coions near the surface given by the lower curve. There
exists a threshold distance beyond which the interface has little influence on
the counterion excess. The value of this threshold distance increases as the
electrolyte concentration decreases. The area enveloped by the curves of counterions and coions for each electrolyte concentration is the net charge of the
solution and must be exactly the opposite of the surface charge.

The key result is that the counterions prefer to be closer to highly charged
interfaces, although they are not permanently adsorbed. It can be shown
(Israelachvili 1992, Evans and Wennerstrom 1999) that the total number
density of all ions at the interface is given by

$$n(x = 0) = n_\infty + \frac{\epsilon_0 \epsilon}{2k_B T} E_s^2, \qquad (3.65)$$

where n_∞ is the number density of all ions in the bulk solution far away from
the interface, and E_s is the electric field at the interface, given by Equation 3.57.
The above equation is a general result, known as the Grahame equation, and
can be used to assess the accumulation of various ions at a charged interface
essentially as a monolayer.

Since the excess of counterions and depletion of coions occur only within a
distance comparable to the Debye length, it is useful to construct a geometrical representation of the ion cloud near the interface, as in Section 3.2.1. The
capacitance $C_{interface}$ of the interface is given by $d\sigma_s e/d\psi_s$ and it follows from
Equation 3.63 that

$$C_{interface} \equiv \frac{de\sigma_s}{d\psi_s} = \frac{\epsilon_0 \epsilon}{\kappa^{-1}} \cosh\left(\frac{ze\psi_s}{2k_B T}\right). \qquad (3.66)$$

For $z\psi_s \ll \psi_\theta$, the capacitance of the interface assumes the simple form,

$$C_{interface} = \frac{\epsilon_0 \epsilon}{\kappa^{-1}}. \qquad (3.67)$$

This expression is exactly the capacitance of a parallel-plate capacitor with two
parallel plates with a separation distance of the Debye length κ^{-1} confining
a dielectric medium of permittivity $\epsilon_0 \epsilon$. In view of this equivalence between
the capacitance of a charged interface (for sufficiently weak surface potentials)
and that of a parallel-plate capacitor, a charged interface is called an "electric
double layer." Combining the above-described concepts of strong accumulation
(adsorption) of counterions near the interface, finite range for the excess of
counterions and depletion of coions, and the electric double layer, a cartoon of
a charged interface can be depicted as in Figure 3.12.

3.2.3 Cylindrical Pore

Consider a cylindrical pore of length L and radius a containing a symmetric
(z:z) type electrolyte at the number concentration n_0. Let the internal wall of

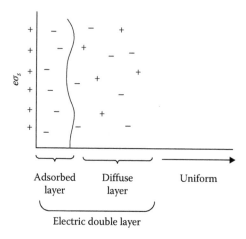

Figure 3.12 Cartoon of a charged interface as an electric double layer.

the pore be charged with a uniform charge density $e\sigma_s$, or equivalently let the surface potential be ψ_s. Let the axis of the cylinder be along the z-direction, and the radial direction r is perpendicular to the z-axis (Figure 3.13a).

The electric potential inside the pore is calculated by following exactly the same procedure as in the above cases. The potential is given by the Poisson–Boltzmann equation (Equation 3.13). For symmetric electrolytes, the valencies z_i are the same for both the anions and cations, except for the sign. Also, it turns out that, for both types of ions, $n_{i0} = n_0$ is a good approximation.

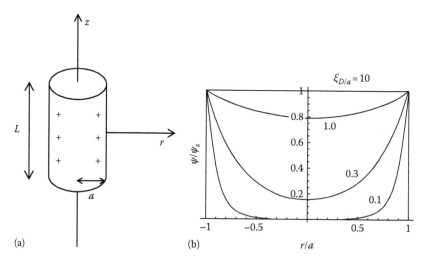

(a)

(b)

Figure 3.13 (a) Cylindrical pore of radius a and length L with surface potential ψ_s. (b) The variation of potential inside a cylindrical pore of radius a for $\xi_D/a = 0.1, 0.3, 1.0,$ and 10.

As a result, Equation 3.13 becomes

$$\nabla^2 \psi(\mathbf{r}) = -\frac{ezn_0}{\epsilon_0 \epsilon} \left[\exp\left(\frac{-ez\psi(\mathbf{r})}{k_B T}\right) - \exp\left(\frac{ez\psi(\mathbf{r})}{k_B T}\right) \right]. \tag{3.68}$$

Using the relation, $\sinh(x) = (e^x - e^{-x})/2$, we get

$$\nabla^2 \psi(\mathbf{r}) = \frac{2ezn_0}{\epsilon_0 \epsilon} \sinh\left(\frac{ez\psi(\mathbf{r})}{k_B T}\right). \tag{3.69}$$

If we consider long enough pores to ignore end effects in the interior of the pore, there is a cylindrical symmetry about the axis of the pore. For this two-dimensional problem, the above Poisson–Boltzmann equation is solved in the cylindrical coordinate system of Figure 3.13, where r is the radial distance from the pore axis. Rewriting the Laplacian ∇^2 in the cylindrical coordinate system,

$$\frac{1}{r}\frac{d}{dr}\left(r\frac{d\psi(r)}{dr}\right) = \frac{2ezn_0}{\epsilon_0 \epsilon} \sinh\left(\frac{ez\psi(r)}{k_B T}\right). \tag{3.70}$$

In general, this equation needs to be solved numerically to get the spatial variations of the electric potential and charges (Gross and Osterle 1968). Therefore, as usual, we make the Debye–Hückel approximation by linearizing the above equation,

$$\frac{1}{r}\frac{d}{dr}\left(r\frac{d\psi(r)}{dr}\right) = \kappa^2 \psi(r), \tag{3.71}$$

where

$$\kappa^2 = \frac{2e^2 z^2 n_0}{\epsilon_0 \epsilon k_B T}. \tag{3.72}$$

κ^{-1} is the Debye length and its properties are as in Section 3.1.3. Due to the radial symmetry, the electric field (the negative gradient of the potential) at the center of the axis must be zero. Using this as a boundary condition and taking the potential at the surface as $\psi(r = a) = \psi_s$, the Debye–Hückel equation for the present geometry can be solved. The result is

$$\psi(r) = \psi_s \frac{I_0(\kappa r)}{I_0(\kappa a)}, \tag{3.73}$$

where $I_n(x)$ is the nth-order modified Bessel function of the first kind. This function is an increasing function of the argument and is conveniently tabulated in handbooks of mathematical functions (Abramowitz and Stegun 1965). The result of the above equation is presented in Figure 3.13b, where $\psi(r)/\psi_s$ is plotted against r/a for different values of the ratio of the Debye length to the pore radius. For high salt concentrations such that the Debye length is much shorter than the pore radius, the electric potential is zero over most of the interior of the pore. On the other hand, the electric potential is nearly constant across the pore for larger Debye lengths than the pore radius, possibly occurring at low salt concentrations. As an example, for a pore of 1 nm radius containing monovalent salts inside a pore, 1 and 0.1 M salt correspond to the curves labeled with 0.3 and 1.0, respectively, for the ratio of the Debye length to the pore radius (see Table 3.1). These results can be correct only qualitatively, due to the linearization approximation used in solving the Poisson–Boltzmann equation. However, there are no significant differences between the results in Figure 3.14 and the numerical results from the nonlinear Poisson–Boltzmann equation (Gross and Osterle 1968). A more serious problem might be that even the Poisson–Boltzmann formalism breaks down for such crowded environments as the inside of nanopores. In fact, the electrostatics inside a nanopore remains as a major challenging problem due to the lack of understanding of structure of water inside the confined region and the role of image charges arising from dielectric mismatch at the pore walls.

The net charge density $\rho(r)$ inside the pore follows from Equations 3.12 and 3.71. With the Debye–Hückel approximation used in getting Equation 3.73 (i.e., linearization of the exponentials in Equation 3.12), $\rho(r)$ is $-\epsilon_0\epsilon\kappa^2\psi(r)$,

$$\rho(r) = -\epsilon_0\epsilon\kappa^2\psi_s\frac{I_0(\kappa r)}{I_0(\kappa a)}. \tag{3.74}$$

The radial dependence of the net charge density (due to the excess of counterions and depletion of coions) from the wall of the pore into the solution is similar to the results of Figure 3.13b, except for the multiplicative factor of κ^2.

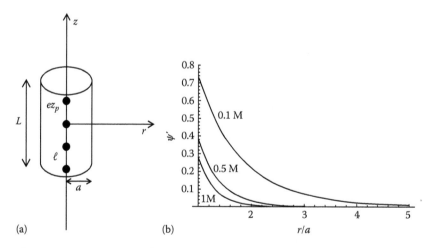

(a)   (b)

Figure 3.14   (a) A cylinder with its charges along its central axis. (b) Dependence of potential on r/a for monovalent salt concentrations at 0.1 M ($\xi_D \simeq 0.98$ nm), 0.5 M ($\xi_D \simeq 0.424$ nm), and 1 M ($\xi_D \simeq 0.3$ nm) ($\psi' \equiv 2\pi\epsilon_0\epsilon\ell\psi/ez_p$).

The value of the ratio of the Debye length to the pore radius, pertinent to an experiment, is obtained from the radius of the nanopore and the electrolyte concentration (Table 3.1). For the typical salt concentrations used in single-molecule electrophysiology experiments, the ion cloud is close to the wall of the pore, its thickness being given by the Debye length. The interior of the pore near the pore axis has no net charge due to the uniform distribution of counterions and coions.

3.2.4 Charged Cylinder

In view of the appearance of a short dsDNA molecule roughly as a charged rod, the electrostatics around a rod can be used to get some insight into the electric potential and ion distributions around some models of rod-like polymers. Consider a cylinder of radius a and length L with its axis along the z-direction and the radial distance perpendicular to the z-axis (Figure 3.14a). Let the identical charges in the cylinder be distributed uniformly along the cylinder axis with ℓ being the charge separation. The charge density is ez_p/ℓ and the total charge is $Q = ez_pL/\ell$, where z_p represents the valency and sign of the individual charges. Outside the cylinder, the solution has the symmetric $(z_c:z_c)$ type electrolyte, and the counterions from the salt and the cylinder are taken to be identical. Analogous to the restricted primitive model of simple electrolytes discussed in Section 3.1.3.2, the volume of the cylinder of radius a excludes counterions and coions. Again, the simplifying assumption that there is no dielectric mismatch between the inside and outside of the cylinder is made.

The electric potential $\psi(r)$ at the radial distance from the axis of the cylinder is given by the same Poisson–Boltzmann equation (Equation 3.70) as for the case of cylindrical pores, due to the cylindrical symmetry for sufficiently long cylinders. The only difference now is in the boundary conditions to be used in solving Equation 3.70. Using the Gauss law (Young and Freedman 2000), the electric field $E(r)$ at a radial distance r from the cylinder axis, without any regard to the presence of ions in the electrolyte solution, is given by

$$E(r) = \frac{ez_p}{2\pi\epsilon_0\epsilon\ell r}, \tag{3.75}$$

so that the electric field at the surface of the cylinder is

$$E_s = \frac{ez_p}{2\pi\epsilon_0\epsilon\ell a}. \tag{3.76}$$

The electric potential due to the charge distribution inside the cylinder follows from Equations 3.3 and 3.76 as

$$\psi(r) = -\frac{ez_p}{2\pi\epsilon_0\epsilon\ell} \ln\left(\frac{r}{a}\right). \tag{3.77}$$

The electric potential is an arbitrary constant at the cylinder surface ($r = a$), and it must be emphasized that there is no mathematical divergence in the potential for the physically relevant values of r at the surface of the cylinder and its neighborhood in the solution, for cylinders of finite radius. For large distances from the cylinder, the potential varies logarithmically, reflecting the two-dimensional spatial symmetry around an infinite cylinder.

The ion cloud in the solution modifies the result of Equation 3.77. Exact solution of the Poisson–Boltzmann equation (Equation 3.70) is known for the salt-free solutions containing only the counterions (Alfrey et al. 1951). As expected, the electric potential falls off smoothly with the radial distance, and there exists a counterion cloud near the cylinder. In order to get insight into the basic nature of the electrostatics in salty electrolyte solutions around a charged thin cylinder, we linearize Equation 3.70 to get the Debye–Hückel theory (Equation 3.71). Solving this equation with the boundary conditions that the electric field vanishes far away from the cylinder and that it is given by Equation 3.76 at the surface of the cylinder, the result is

$$\frac{ez_c\psi(r)}{k_BT} = 2z_pz_c\frac{\ell_B}{\ell}\frac{1}{\kappa a}\frac{K_0(\kappa r)}{K_1(\kappa a)}, \tag{3.78}$$

where $K_n(r)$ is the modified Bessel function of nth order (Abramowitz and Stegun 1965). Noting that the strength of electrostatic interaction compared to the thermal energy is given by the Bjerrum length, we define a dimensionless Coulomb strength parameter Γ as the ratio of the Bjerrum length to the charge separation distance ℓ along the cylinder axis,

$$\Gamma \equiv \frac{\ell_B}{\ell} = \frac{e^2}{4\pi\epsilon_0\epsilon k_BT\ell}. \tag{3.79}$$

We shall call Γ as the charge density parameter or Coulomb strength parameter interchangeably. The electric potential follows from Equations 3.78 and 3.79 as

$$\frac{ez_c\psi(r)}{k_BT} = \frac{2z_pz_c\Gamma}{\kappa a}\frac{K_0(\kappa r)}{K_1(\kappa a)}, \tag{3.80}$$

depending on two dimensionless parameters κa and Γ.

The spatial variation of the potential given by the above equation is presented in Figure 3.14b, where $e\psi(r)/2z_p\Gamma k_BT$ is plotted against r/a for different values of the ratio of the Debye length to the radius of the cylinder. As an illustration, let us assume that a dsDNA molecule can be modeled as a cylinder of radius $a = 0.9$ nm and charge separation length $\ell = 0.17$ nm, so that $\Gamma = 4.2$ at 20 °C. If the concentration of monovalent salt is 0.1, 0.5, and 1 M, the corresponding Debye lengths are 0.98, 0.42, and 0.3 nm, respectively. The potential variations for these three concentrations are given in Figure 3.14b. For low salt concentrations, the range of the potential can be substantially longer in comparison with the radius of the cylinder. However, as the salt concentration increases, the electric potential away from the cylinder axis dies quickly.

If the Debye length is very large as in the case of extremely dilute electrolyte solutions, and if the radius of the cylinder is very small, the arguments of the modified Bessel functions in the above equations can become small. For very small values of the argument, the limiting behaviors of the modified Bessel functions (Abramowitz and Stegun 1965) are

$$K_0(\kappa r) \simeq -\ln(\kappa r), \tag{3.81}$$

and

$$K_1(\kappa a) \simeq \frac{1}{\kappa a}. \tag{3.82}$$

Therefore, in the limit of $\kappa a \to 0$, the electric potential is given by

$$\psi(r) \simeq \frac{2k_B T z_p \Gamma}{e}[-\ln(\kappa r)]. \tag{3.83}$$

Substitution of Equation 3.79 in the above equation yields

$$\psi(r) = -\frac{e z_p}{2\pi \epsilon_0 \epsilon \ell}(\ln r - \ln \xi_D). \tag{3.84}$$

The first term $(\ln r)$ in the above equation corresponds to the potential due to the charges inside the cylinder, and the second term $(-\ln \xi_D)$ is the potential due to the ion cloud essentially located at the Debye length ξ_D.

The above formula (Equation 3.84) is valid only for large Debye lengths. For the usual electrolyte concentrations used in experiments, Equation 3.80 should be used. In general, the electric potential around a cylinder is directly proportional to the charge density parameter Γ, defined in Equation 3.79. If Γ is large, the potential can become so large that the assumption of linearization of the Poisson–Boltzmann equation $(e z_c \psi/k_B T \ll 1)$ breaks down. Therefore, for large Γ values, the Debye-Hückel description is valid only for large distances away from the cylinder where the potential becomes weak enough to satisfy the assumption made in the derivation of Equation 3.80. Also, similar to the case of planar interfaces, for very short distances from the surface, the counterions adsorb essentially into a monolayer. Thus, we can imagine three regimes (Hiemenz and Rajagopalan 1997) for the variation of the electric potential as a function of the radial distance from the cylinder axis: (a) the Debye–Hückel regime at large distances, (b) Gouy regime at closer distances from the cylinder, where the nonlinear Poisson-Boltzmann description is needed, and (c) a monolayer of adsorbed counterions, called the Stern regime. Although the quantitative details of the crossover behaviors between these regimes are rather intricate, the basic feature is that the counterions accumulate near the charged surface and the electric potential decays into the solution with a characteristic distance for the decay roughly given by the Debye length. The dependencies of the electric potential and ion distribution on the radial distance are smooth as evident from the exact solutions for salt-free solutions around thin-charged cylinders (Alfrey et al. 1951).

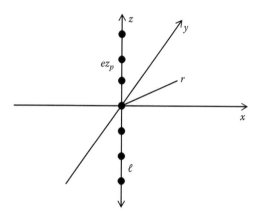

Figure 3.15 Geometry of a line charge used in deriving Manning condensation.

3.2.5 Charged Line

There is extensive theoretical description in the literature for an electrolyte solution containing an infinitely thin and infinitely long line of charges, as this model is the simplest that might mimic a long rigid polyelectrolyte. Using this model, an argument, called the Manning condensation, can be made to derive a simple relation for the accumulation of counterions near the line charge. In view of the simplicity of the Manning relation for the condensation of counterions on the line charge, we give details of its derivation. The geometry and the parameters of the model are given in Figure 3.15. Let the separation distance between the consecutive uniform charges be ℓ and each charge be ez_p, as in the preceding section. The difference now is that the line charge does not have any radius. The line charge is along the z-direction and r is the radial distance from the line charge in a plane perpendicular to the line charge. The electric field at r follows from the Gauss law (Section 3.2.4), as

$$E(r) = \frac{ez_p}{2\pi\epsilon_0\epsilon\ell r}, \qquad (3.85)$$

and (because $E = -d\psi/dr$)

$$\psi(r) = A - \frac{ez_p}{2\pi\epsilon_0\epsilon\ell}\ln r, \qquad (3.86)$$

where A is an integration constant. The potential energy W for bringing a counterion of valency z_c to the location at r is

$$W(r) = (-ez_c)\psi(r). \qquad (3.87)$$

Therefore, the number of counterions at r is proportional to

$$\exp\left(-\frac{W(r)}{k_BT}\right) = \exp\left(\frac{ez_cA}{k_BT}\right)\exp\left(-\frac{e^2z_pz_c}{2\pi\epsilon_0\epsilon\ell k_BT}\ln r\right). \qquad (3.88)$$

By combining the exponential and the logarithm in the last factor, we get

$$\exp\left(-\frac{W(r)}{k_BT}\right) = \frac{1}{r^{2\Gamma z_p z_c}}\exp\left(\frac{ez_cA}{k_BT}\right),\tag{3.89}$$

where Γ is the charge density parameter defined in Equation 3.79.

The number of counterions inside a cylinder of radius r_0 (of length ℓ) is

$$n_{r_0} \sim \int_0^{r_0}(2\pi r)e^{-W(r)/k_BT}dr.\tag{3.90}$$

Combining Equations 3.89 and 3.90 yields

$$n_{r_0} \sim \int_0^{r_0} r\frac{1}{r^{2\Gamma z_p z_c}}dr,\tag{3.91}$$

which upon integration results in

$$n_{r_0} \sim \left.\frac{1}{r^{2(\Gamma z_p z_c - 1)}}\right|_0^{r_0}.\tag{3.92}$$

This shows that n_{r_0} diverges at $r \to 0$ for $z_p z_c \Gamma > 1$. This apparent divergence can be mathematically avoided by assuming that $z_p z_c \Gamma$ is never allowed to be greater than unity. In other words, we imagine that enough counterions condense on the line charge and reduce the charge density parameter to be Γ_{eff} so as to make $z_p z_c \Gamma_{eff}$ to become unity. Hence, the effective charge separation ℓ_{eff} is imagined to be greater than ℓ.

The above argument, constructed to avoid the divergence in the number of counterions on an infinitely thin and infinitely long line charge, is the Manning condensation. According to this hypothesis, a plot of Γ_{eff} against Γ is given in Figure 3.16a. By accounting for the counterion condensation, the charge fraction α of the line charge is defined as the ratio of ℓ to ℓ_{eff},

$$\alpha \equiv \frac{\ell}{\ell_{eff}} = \frac{\Gamma_{eff}}{\Gamma}.\tag{3.93}$$

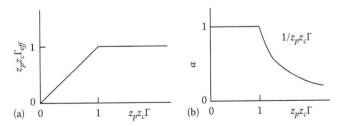

Figure 3.16 Counterions condense on the line charge for $z_p z_c \Gamma > 1$. (a) Effective charge density parameter. (b) Fraction of uncondensed charge density, $\alpha = \ell/\ell_{eff}$.

For $\Gamma z_p z_c < 1$, $\ell_{eff} = \ell$ so that $\alpha = 1$. For $\Gamma z_p z_c > 1$, $\Gamma_{eff} z_p z_c = 1$ so that $\alpha = 1/\Gamma z_p z_c$,

$$\alpha = \begin{cases} 1, & \Gamma z_p z_c < 1 \\ \frac{1}{\Gamma z_p z_c}, & \Gamma z_p z_c > 1 \end{cases} \tag{3.94}$$

as sketched in Figure 3.16b.

The simplicity and implication of the above results can be illustrated by considering a line charge model of dsDNA. Let the charge separation length ℓ be 0.17 nm, so that $\Gamma = 4.2$ at 20 °C. Let $z_p = 1$ and $z_c = 1$. Therefore, for conditions where counterions would condense ($\Gamma z_p z_c > 1$), the fraction of condensed counterions, $1 - \alpha$, follows from Equation 3.94 as

$$1 - \frac{1}{\Gamma} = 0.76. \tag{3.95}$$

Therefore, the effective charge fraction is only about a quarter. The effective charge of each phosphate group depends on the valency of the counterion,

$$(ez_p)_{eff} = \frac{e}{z_c \Gamma}. \tag{3.96}$$

However, it must be emphasized that these results are only mathematical artifacts, as the exact solutions for the potential and counterion distribution, for the line charge model with a finite radius, do not show any such phase-transition-like discontinuities (Alfrey et al. 1951). The three major objections to the line-charge model of flexible polyelectrolyte are zero thickness, infinite length, and no-chain flexibility.

There has been extensive discussion about the validity and applicability of the Manning condensation (Manning 1969, 1978). While the above argument is exact for the particular model of an infinitely thin and infinitely long one-dimensional line charge, the results of Equation 3.94 cannot be applied to experimental systems involving flexible and semiflexible polyelectrolyte molecules (Holm et al. 2004b, Muthukumar 2004). The discontinuity of α shown in Figure 3.16b is also not to be expected for these experimental systems (Alfrey et al. 1951). However, the condition

$$\Gamma z_p z_c = 1 \tag{3.97}$$

has been used as a qualitative measure for binding of counterions around a polyelectrolyte. This interpretation of $\Gamma z_p z_c = 1$ is equivalent to the criterion for the formation of a Bjerrum ion-pair at a separation distance of ℓ. This makes Equation 3.97 attractive in spite of the above objections.

3.2.6 Dielectric Mismatch

Consider a point charge Q inside a semi-infinite medium of dielectric constant ϵ_1 at a distance d away from a planar interface that separates the first medium

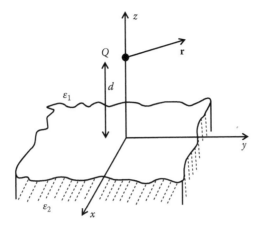

Figure 3.17 Planar surface separating two media of dielectric constants ϵ_1 and ϵ_2. A point-charge Q inside the medium of ϵ_1 at a distance d from the interface exerts an electric potential at $\mathbf{r}$ from its location.

from another semi-infinite medium of dielectric constant ϵ_2. Let the interface be the plane at $z = 0$ in the coordinate system of Figure 3.17. The electric potential at $\mathbf{r}$ due to the charge Q is no longer given by Equation 3.2, which is valid for a homogeneous infinite medium. The dielectric mismatch alters the potential. The standard arguments (Griffiths 1999) in electrostatics show that the effect of dielectric mismatch is equivalent to placing an image charge Q' at a distance d below the interface and considering the effects of Q and Q' in a medium of uniform dielectric constant ϵ_1. The result is

$$\psi = \begin{cases} \dfrac{1}{4\pi\epsilon_0\epsilon_1}\left(\dfrac{Q}{r} + \dfrac{Q'}{r'}\right), & z > 0 \\[2mm] \dfrac{1}{4\pi\epsilon_0}\dfrac{2}{(\epsilon_1 + \epsilon_2)}\dfrac{Q}{r}, & z < 0 \end{cases} \tag{3.98}$$

as given in Figure 3.18, where the image charge Q' is

$$Q' = \frac{\epsilon_1 - \epsilon_2}{\epsilon_1 + \epsilon_2}Q. \tag{3.99}$$

The distances r and r' are, respectively, $(x^2 + y^2 + (z - d)^2)^{1/2}$ and $(x^2 + y^2 + (z + d)^2)^{1/2}$, as marked in Figure 3.18.

Depending on the ratio ϵ_1/ϵ_2, the image charge may attract or repel the charge Q. For $\epsilon_1 \gg \epsilon_2$, the image charge has the same sign and magnitude as Q, and the charge Q is repelled by the interface. On the other hand, if $\epsilon_2 \gg \epsilon_1$, the charge in a medium of dielectric constant ϵ_1 is attracted by the medium of dielectric constant ϵ_2. As an example of the net effect from the charge Q and its image charge Q', let us consider the electric potential at a distance r much

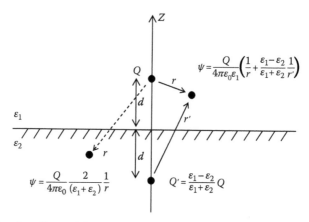

Figure 3.18 Effects of dielectric mismatch. The image charge is Q'. The potential due to Q at a distance r depends on whether r is in medium of ϵ_1 or medium of ϵ_2.

larger than the distance d of the charge from the interface. This follows from the above equations as

$$\psi = \frac{Q}{4\pi\,\epsilon_0}\,\frac{2}{(\epsilon_1 + \epsilon_2)}\,\frac{1}{r}. \tag{3.100}$$

Therefore, an effective dielectric constant ϵ_{eff} may be identified in describing the electric potential at distances far away from Q, by writing the above equation as

$$\psi = \frac{Q}{4\pi\,\epsilon_0\epsilon_{eff}r}, \tag{3.101}$$

where

$$\epsilon_{eff} = \frac{\epsilon_1 + \epsilon_2}{2}. \tag{3.102}$$

If the dielectric constants of an aqueous medium and an oily cavity, such as the interior of a protein or a hydrophobic polymer, are taken to be 80 and 3, respectively, then the effective dielectric constant at large distances is roughly 40. Thus, the apparent dielectric constant can be different by a factor of two. The complementary problem of how a charge buried in an oily enclosure is subjected to an attractive force by its image charge, which is now present in the surrounding aqueous medium with higher dielectric constant, can be readily addressed from the above equations.

Analogous calculations of the effects of dielectric mismatch due to curved interfaces become highly technical. Nevertheless, if one is interested in very accurate estimates of various forces near interfaces with dielectric mismatch, such as solid-state nanopores or protein pores embedded in a lipid membrane, such computations as the above must be resorted to. We shall avoid these details in this book and focus mainly on general concepts and qualitative trends.

3.3 SUMMARY

The electrolyte solutions are characterized by two length scales: the Bjerrum length and the Debye length. The Bjerrum length is a property of the solvent and depends on the dielectric constant and temperature. It is the separation distance between two unit charges at which the electrostatic energy is comparable to the thermal energy. The Debye length is a property of the solution depending on the concentration and valencies of dissociated ions as well as the dielectric constant and temperature. It is the range over which ions are correlated.

Charged interfaces in electrolyte solutions cause counterions to adsorb. A cloud with net opposite charge hovers around the interface, with a characteristic thickness comparable to the Debye length. At distances much larger than the Debye length, the effect of charged interfaces is essentially absent.

The Poisson–Boltzmann formalism is used to compute the electric potentials and charge distributions. In general, this requires numerical work. The Debye–Hückel theory assumes a weak electrical energy compared to the thermal energy allowing closed form analytical formulas for various quantities of interest in electrolyte solutions.

Formulas for the electric potential and ion distributions are given for various geometries: planar interfaces, spheres, cylinders, cylindrical pores, and lines.

4

FLEXIBLE AND SEMIFLEXIBLE POLYELECTROLYTES

Let us throw a tiny amount of a polyelectrolyte salt, such as deoxyribonucleic acid sodium salt or sodium polystyrene sulfonate, into an aqueous solution containing a known amount of a simple electrolyte such as potassium chloride. In due course, the polymer salt dissolves into a homogeneous solution consisting of the charged polymer molecules, their dissociated counterions, and the cations and anions of the simple electrolyte. As in the cases of rigid charged particles in electrolyte solutions, an ion cloud would surround a polyelectrolyte chain, due to an optimization between the attractive interaction among opposite charges and the loss of translational freedom of free ions. A generic picture of a polyelectrolyte is drawn in Figure 4.1a. For one particular conformation of a polymer molecule, some of the counterions hover around the chain backbone. However, unlike rigid bodies, the polymer has the intrinsic capacity to assume enormous number of conformations due to the chain flexibility. The polymer conformations are influenced by the electric forces arising from the charges of counterions and salt ions, and the charges on the chain backbone itself. The polymer conformations, in turn, influence the spatial distribution of the small ions. After an elapse of a certain short time, the polymer conformations in equilibrium would have changed (Figure 4.1b). Now the ion cloud surrounding the polymer skeleton will contain different configurations of the counterions, with some of the original counterions replaced by new ones at new locations. Thus, on an average, we imagine a counterion worm around a polymer chain. The counterion worm is dynamic, with the counterions continuously binding and unbinding with the polymer backbone at random locations, but maintaining an average number inside the worm. As a result, the effective charge of the polyelectrolyte is not the same as its chemical charge that could be estimated by assuming that all ionizable groups of the polymer fully dissociate. The effective charge is unique to the average polymer conformation, and it self-regulates with changes in polymer conformations accompanying changes in experimental conditions. The repulsion among the effective charges on the polymer backbone is manifest as electrostatic swelling for flexible polymers, or equivalently as chain stiffening for semiflexible polymers. The primary focus of this chapter is to combine the key concepts of electrostatic interactions (Chapter 3) and the various polymer models (Chapter 2), toward a description of the equilibrium properties of polyelectrolyte molecules in dilute solutions. After

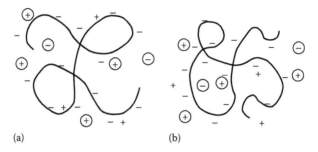

Figure 4.1 Cartoon of a polyelectrolyte chain (negatively charged as an example). + is counterion; circled + and − represent ions from the salt. Oppositely charged ions may adsorb on the polymer. (a) and (b) are different conformations in equilibrium.

presenting major concepts and some key experimental and simulation results, we shall describe simple treatments of electrostatic effects endowed on polymer conformations.

4.1 CONCEPTS

Firstly, before describing the key concepts, we consider the parameters that reflect the various experimental variables such as the length, charge, and concentration of polymer molecules, and the concentration of simple salt in a polyelectrolyte solution. Let the polyelectrolyte be a homopolymer with each repeat unit carrying one ionizable group. Upon ionization in a polar solvent, each repeat unit is assumed to carry one monovalent charge e. If the charge separation between any two consecutive monomeric charges along the chain backbone is ℓ_0 (comparable to a couple of bond lengths), the chemical charge density q_ℓ along the chain contour is e/ℓ_0. Since the basic model for chain connectivity is the Kuhn model with segmental length ℓ, we write the chemical charge density as ez_p/ℓ,

$$q_\ell = \frac{e}{\ell_0} = \frac{ez_p}{\ell}, \tag{4.1}$$

where z_p is the average number of ionizable monomers per one Kuhn length, according to the chemical formula of the polyelectrolyte. The total chemical charge of one polymer molecule is eN_0 (with N_0 being the number of ionizable monomers per chain), or equivalently ez_pN, where N is the number of Kuhn segments per chain. Let the charge of each counterion be ez_c. Although all repeat units of the chain are ionizable, it turns out that only a certain fraction of these dissociate and consequently bear the monomer charge. As we shall see below, this fraction depends on the experimental conditions. Let α be the average degree of ionization. The number of counterions n_c in a solution containing

n polyelectrolyte chains of uniform length with an average degree of ionization α is given by the electroneutrality condition as

$$n_c = \left| \frac{z_p}{z_c} \right| \alpha n N. \tag{4.2}$$

In addition, there are cations and anions from the simple salt, with their charges and numbers represented by $e z_a$ and n_{a0} for the a-type ions. Let V be the total volume of the solution, so that the number densities of salt ions and counterions are n_{a0}/V and n_c/V, respectively. The average segment number density of the polymer in the solution is nN/V. We now enumerate the key concepts pertinent to the equilibrium properties of polyelectrolyte chains.

4.1.1 Coulomb Strength

As discussed in Chapter 3, there are two basic length scales describing the electrostatic interactions among ions, namely the Bjerrum length and the Debye length. The Bjerrum length is the distance between two ions at which their electrostatic energy equals the thermal energy. Dividing the Bjerrum length by the Kuhn length, and absorbing the magnitudes of the charges of the segment and the counterion, we define a parameter,

$$\Gamma = |\, z_p z_c \,| \, \frac{\ell_B}{\ell} = |\, z_p z_c \,| \, \frac{e^2}{4\pi \epsilon_0 \epsilon k_B T \ell}, \tag{4.3}$$

where the definition of ℓ_B from Equation 3.6 is used. We shall call Γ as the Coulomb strength parameter. This dimensionless parameter is a measure of the strength of the attractive interaction between the counterion and the monomer. This is also the charge density parameter given in Equation 3.79, for monovalent charges.

4.1.2 Debye Length for Polyelectrolyte Solutions

The range of electrostatic interaction between two ions in an electrolyte solution is given by the Debye length (Section 3.1.3.1). For electrolyte solutions, all types of ions contribute to the Debye length ξ_D, as given by Equation 3.22. We make a slight modification for polyelectrolyte solutions. When an electrolyte solution contains long polymers, which are gigantic in size in comparison with the small ions from the simple electrolyte, we expect the timescales for the rearrangements of the small ions and polymer conformations to be widely separated. Therefore, it is convenient to imagine (de Gennes et al. 1976, Muthukumar 1996a) that the polymer molecules are present in an effective medium where the degrees of freedom of all small ions are integrated out. The effective medium is assumed to obey the Debye–Hückel description, with the Debye length arising from only the counterions and the ions from the simple electrolyte,

$$\xi_D \equiv \kappa^{-1} \equiv \left[4\pi \ell_B \left(z_c^2 \frac{n_c}{V} + \sum_a z_a^2 \frac{n_{a0}}{V} \right) \right]^{-1/2}. \tag{4.4}$$

Using the conversion factor discussed in Section 3.1.3.1, the Debye length can be expressed in terms of the molarities of simple salt ions and counterions as

$$\xi_D^{-2} \equiv \kappa^2 = 4000\pi \ell_B N_A \left(z_c^2 c_{c0} + \sum_a z_a^2 c_{a0} \right), \tag{4.5}$$

where N_A is the Avogadro number, c_{a0} is the concentration of ions of a-type in molarity, and c_{c0} is the counterion concentration in molarity and is related to the polymer concentration c_{p0} in molarity according to

$$c_{c0} = \left| \frac{z_p}{z_c} \right| \alpha N c_{p0}. \tag{4.6}$$

The polyelectrolyte chains exist in the effective medium created by the counterions and ions from the simple salt. The charges on the polymer chains interact among themselves mediated by the neutralizing plasma that makes up the system except the polymer molecules. We further assume that the mediation by the background is adequately described by the Debye–Hückel theory. This assumption allows us to deduce the key features, without resorting to heavy numerical work that will be needed to solve the very complex Poisson–Boltzmann equations for such topologically correlated objects as flexible polymers.

As in the case of rigid charged objects in an electrolyte solution, an ion cloud is expected to surround the backbone of a polyelectrolyte chain. The nature of this ion cloud must be dictated by the correlations among the polymer segments and various ions. These correlations arise from the long-range electrostatic interactions, short-range hydrophobic effects, and the chain connectivity manifest as a long-ranged monomer density distribution. The net result of the ion cloud around the polymer is that the effective polymer charge is different from the nominal chemical charge of the polymer. The extent of counterion adsorption is affected by many factors including the nature of the counterion (size and valency), concentration of added salt, polymer concentration, solvent dielectric constant, and temperature. Naturally, the net charge of the polymer is one of the fundamental features of the polyelectrolyte as it controls all properties of the molecule in terms of its structure, dynamics, and mobility. Next, we consider the key conceptual factors contributing to the effective polymer charge.

4.1.3 Chain Connectivity

The electric potential in and around a flexible polyelectrolyte chain is directly dependent on how the monomers are spatially correlated. Unlike the situations in Chapter 3, a flexible polymer molecule cannot be simply assumed to be a sphere or cylinder with all its charges on the outer surface. Instead, the polymer is essentially a statistical fractal, with its monomer density falling off algebraically with the radial distance from its center (Figure 2.4a). Indeed, the monomer density profile is set up self-consistently by the mutual correlations between the chain connectivity and electrostatic interactions among all charged monomers and small ions. As an example, a snapshot from a numerical simulation of an equilibrated single flexible polyelectrolyte in salt-free conditions is given in Figure 4.2a. The values of the parameters used in getting this snapshot are $N = 60$ and $\Gamma = 2.8$ (corresponding to aqueous solutions containing sodium polystyrene sulfonate type polymers), and the volume of the system corresponds to the monomer density of $5.63 \times 10^{-6}\ell^{-3}$. Many counterions are seen to hover over the chain backbone. The identities of the counterions

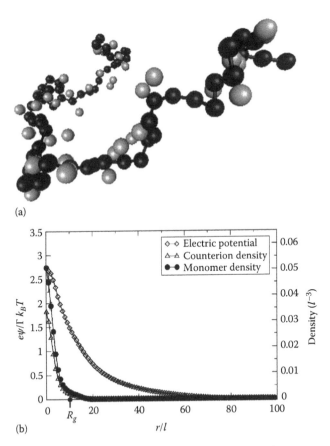

(a)

(b)

Figure 4.2 (a) A snapshot of a flexible polyelectrolyte with its counterions for $N = 60$ at $\Gamma = 2.8$ and the average segment concentration $5.63 \times 10^{-6}\ell^{-3}$. (b) Profiles of the electric potential, monomer density, and the counterion density against the radial distance from the center of mass of the chain.

around the chain backbone are constantly changing as the polymer conformations change with time. The electric potential (ψ in units of $k_B T \Gamma / e$) at a radial distance r from the center of mass of the chain is given in Figure 4.2b, along with the average density profiles of the monomers and the counterions. As is clearly evident, the electric potential is strongly correlated with the monomer distribution and is significant even at distances comparable to four times the radius of gyration of the chain. It must be noted that there are no mathematical divergences as long as we do not consider the unphysical situation of segments with zero radius.

4.1.4 Counterion Worm

The number of counterions adsorbed on the chain backbone is a difficult quantity to measure experimentally. However, computer simulations have helped to gain an understanding of the counterion cloud around the chain. Since the strongest attractive electric potential for the counterions is generally near the contour of the chain, the ion cloud would dress the chain along its contour. As a result, the chain with its counterions would look like a worm.

In order to obtain a quantitative measure of the number of counterions inside this worm, we make the following construction. Consider the skeletal chain model made of the united atoms as in Figure 2.2b. We construct a tube around the chain backbone from the position coordinates of the united atoms. The tube is the nonoverlapping superposition of spheres of fixed radius r_c centered at each united atom, as shown in Figure 4.3. All ions other than the monomers inside this tube constitute the ion cloud. Most of these ions are the counterions, and so we shall call the tube as the counterion worm for a particular polymer conformation. Knowing the number and charges of all ions inside this worm and adding the charges of all monomers of the chain, the net charge is obtained. Averaging over many conformations in equilibrium gives the average polymer

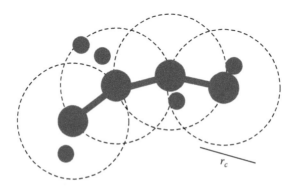

Figure 4.3 Counterion worm, constructed from the nonoverlapping superposition of spheres of radius r_c centered at each united atom of a skeletal chain. Small filled circles represent counterions.

charge Q_{eff} and the average degree of ionization of the polymer chain,

$$\alpha = \frac{Q_{eff}}{eN_0} = \frac{Q_{eff}}{ez_pN},$$

(4.7)

where N_0 is the number of ionizable repeat units and N is the number of Kuhn segments per chain.

The value of Q_{eff} depends on the choice of the cutoff radius r_c. One of the convenient choices of r_c is the value r_0 at which the electrostatic energy of a pair of monovalent ions ($\ell_B k_B T / r_0$) is comparable to the kinetic energy ($3k_B T/2$) of an ion. We shall use this choice for r_c in discussing the simulation results for the effective polymer charge below.

4.1.5 Dielectric Mismatch

One of the factors that significantly affect the extent of counterion adsorption on the chain is the dielectric constant of the solution. However, the dielectric constant of the solution is not uniform, due to the backbone structures of large polymer chains being oil-like and the solvent being polar. The attractive electrostatic energy between a pair of oppositely charged monovalent ions, separated by a distance r, in a solvent of uniform dielectric constant ϵ is from, Equation 3.5,

$$u(r) = -\frac{e^2}{4\pi \epsilon_0 \epsilon r}.$$

(4.8)

On the other hand, the dielectric constant in the region of counterion binding to the pendant-charged groups of the polymer can be quite different from that in the solvent. A typical local conformation of a few monomers of polystyrene sulfonate to which the sodium counterion binds is sketched in Figure 4.4a. In this region, the local dielectric constant can be substantially smaller, due to the fact that most of the materials made out of the polymer backbone without charges have dielectric constants in the range of 2–3. In fact, for biological

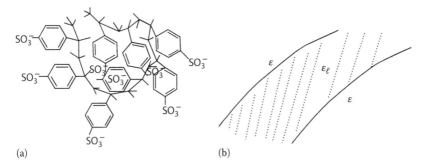

(a)   (b)

Figure 4.4 (a) Sketch of counterion binding to pendant groups at oily backbone. (b) Local dielectric constant ϵ_ℓ is lower than the bulk value ϵ.

macromolecules, it has been recognized (Mehler and Eichele 1984, Lamm and Pack 1997, Gong et al. 2008) that the effective dielectric constant varies rapidly from a low value near the chain backbone to the high bulk value. As a result, we imagine that the dielectric constant inside the counterion worm is ϵ_ℓ, which is substantially lower than the bulk value. This is sketched in Figure 4.4b, where the dielectric constant is ϵ_ℓ inside the worm and is ϵ outside the worm. Furthermore, the distance d between the bulky-charged group of the monomer and the counterion, when an ion-pair is formed in the counterion adsorption step, is usually different from the optimum distance between two small ions in the solvent. Therefore, the adsorption energy for one monomer–counterion pair is

$$u_{local}(r) = -\frac{e^2}{4\pi \epsilon_0 \epsilon_\ell d}. \tag{4.9}$$

It must be cautioned that even the notion of the dielectric constant at such nanoscopic length scales is not accurate and it is necessary to compute the polarization forces. However, such calculations are yet to be performed for the ill-structured heterogeneous suspensions of polyelectrolyte molecules. Therefore, ϵ_ℓ is taken to be different and smaller than the bulk value, in recognizing the existence of dielectric heterogeneity in these solutions. Since we do not know the value of ϵ_ℓ and the ion-pair distance d inside the worm, we combine these two quantities and define a parameter called the dielectric mismatch parameter δ as

$$\delta \equiv \frac{\epsilon}{\epsilon_\ell} \frac{\ell}{d}. \tag{4.10}$$

This parameter is the enhancement factor for the formation of ion pairs near the polymer backbone, as the ion-pair energy follows from Equation 4.9 as the product of the Bjerrum length and δ,

$$\frac{u_{local}(r)}{k_B T} = -\frac{\ell_B}{\ell}\delta. \tag{4.11}$$

In view of the discussion in Section 3.2.6, the value of ϵ_ℓ in the region of counterion binding to the pendant-charged groups of the polymer in aqueous solutions is somewhat in the range of 30 (Lamm and Pack 1997). Furthermore, the distance d between the charged monomer and the counterion in an ion pair can be slightly larger than the distance between two consecutive ionizable monomers on the chain backbone. As a result, the value of the dielectric mismatch parameter δ is expected to be larger than unity, but of order unity. In view of the lack of adequate understanding of the polarization forces at short distances for polyelectrolyte solutions, we take δ only as a parameter. In addition to parametrizing the dielectric heterogeneity in the solutions, δ reflects the specificity of the counterions, because the ion-pair distance d depends on the hydrated ionic radii of the counterions involved in the ion-pair formation. Also, δ is directly related to the logarithm of the ionization equilibrium constant of the monomer.

4.1.6 Electrostatic Excluded Volume

As described in Section 2.3, the polymer segments undergo excluded volume interactions when two nonbonded segments are in close proximity to each other. When the segments carry similar charges, the intersegment interaction is repulsive and can be long-ranged. As a result, the chain is expected to expand if the intrachain electrostatic interaction is not fully screened out by the salt ions and counterions. Analogous to the excluded volume interaction treated in Sections 2.3 and 2.5 for uncharged polymers, we now consider the additional excluded volume interaction due to electrostatic repulsion between the segments. According to the Debye–Hückel theory, the electrostatic interaction energy between two segments i and j, each of charge ez_p, separated by the distance r_{ij}, follows from Equation 3.20 as

$$\frac{u_{ij}(r_{ij})}{k_B T} = z_p^2 \ell_B \frac{e^{-\kappa r_{ij}}}{r_{ij}}. \tag{4.12}$$

Here, the inverse Debye length κ arises only from the counterions and salt ions and not from the charged monomers of the polymer, as given by Equation 4.5.

It can be shown that for sufficiently large values of κ, such that $\kappa R_g > 1$ (R_g being the radius of gyration of the polymer), the screened Coulomb interaction energy becomes short-ranged. In this limit, the result of Equation 4.12 can be written exactly (Muthukumar 1987) as

$$\frac{u_{ij}(r_{ij})}{k_B T} = \frac{4\pi z_p^2 \ell_B}{\kappa^2} \delta(r_{ij}), \tag{4.13}$$

where $\delta(r_{ij})$ is the Dirac delta function. On the other hand, for $\kappa = 0$, we have the Coulomb result,

$$\frac{u_{ij}(r_{ij})}{k_B T} = z_p^2 \ell_B \frac{1}{r_{ij}}. \tag{4.14}$$

The situation of $\kappa = 0$ is unphysical, because there are always counterions present in the solution of finite volume to meet the electroneutrality condition. Nevertheless, the behavior of Equation 4.14 is approachable in the asymptotic limit of the Debye length being larger than the radius of gyration of the polymer, $\kappa R_g \ll 1$. Combining the above two limits,

$$\frac{u_{ij}(r_{ij})}{k_B T} = \begin{cases} \frac{4\pi z_p^2 \ell_B}{\kappa^2} \delta(r_{ij}), & \kappa R_g \gg 1 \\ z_p^2 \ell_B \frac{1}{r_{ij}}, & \kappa R_g \ll 1 \end{cases} \tag{4.15}$$

We shall use these two limits as guides in interpreting the experimental data. For values of κR_g intermediate between the limits, the full form of Equation 4.12 is needed. A convenient result of the above argument is that the intersegment electrostatic interaction energy becomes short-ranged as for the uncharged polymers, provided the solutions are at high enough salt concentrations. For such conditions, the electrostatic part can be simply added to the neutral part of the excluded volume effect (Section 2.5). Therefore, by combining Equations 2.66 and 4.13,

$$\frac{u_{ij}(r_{ij})}{k_B T} = \left(v + \frac{4\pi z_p^2 \ell_B}{\kappa^2} \right) \delta(r_{ij}). \tag{4.16}$$

The first term is due to the uncharged part and the second term is due to the electrostatic part. We call the term $4\pi z_p^2 \ell_B/\kappa^2$ as the electrostatic excluded volume parameter, in the limit of high salt concentrations.

In addition to the electrostatic repulsion among the charged monomers (that are free from counterion adsorption), there are electrostatic interactions between the ion pairs formed by monomer–counterion binding. Since these ion pairs are constantly changing their locations along the chain contour and the chain itself adopts many random conformations, the interaction energy between a pair of ion pairs can be assumed to be that of a freely rotating pair of dipoles (Muthukumar 2004). The interaction energy between two freely rotating dipoles is short-ranged (Israelachvili 1992) and attractive. The contribution from the dipolar attraction among the ion pairs becomes significant when the polymer chain collects sufficient number of counterions and consequently contracts in size. When the polymer coil shrinks, there is more counterion binding due to the lower local dielectric constant. Thus, there is a cascade mechanism by which counterion adsorption escalates as the polymer collapses due to the attraction arising from ion-pair interactions (Khokhlov and Kramarenko 1994, Brilliantov et al. 1998, Kundagrami and Muthukumar 2010). The situation becomes quite complicated when multivalent counterions are involved in the formation of ion pairs (de la Cruz et al. 1995, Nguyen et al. 2000, Kundagrami and Muthukumar 2008). For the monovalent counterions, the effect from the ion-pair interactions can be treated (Muthukumar 2004) as the short-ranged excluded volume effect by adding a negative term to the excluded volume parameter v in Equation 4.16 that is dependent on ℓ_B, δ, and d. Since our primary focus is going to be in the translocation of non-globular polymer conformations, we shall not dwell too much on the consequences of ion-pair interactions in this book.

4.1.7 Electrostatic Persistence Length

Many polyelectrolyte molecules such as dsDNA possess intrinsic stiffness along the molecule arising from hydrogen bonding between the strands constituting the molecule and rotational barriers between the nearest monomers. When they carry charges, the chain stiffness can be enhanced due to the repulsion between the charges. Consider a rod-like chain of length L, which is less than its intrinsic persistence length ℓ_p (Section 2.4.4). Also, let the charge separation distance along the chain be smaller than the Debye length, which in turn is taken to be smaller than the persistence length (Figure 4.5a). Let this chain bend slightly with a constant curvature radius so that the directions of the ends subtend an angle θ ($\ll 1$). If the chain is uncharged, the bending energy is given by Equation 2.60. Since $R_c = L/\theta$ (from an elementary consideration of geometry), the bending energy for the whole length L follows from Equations 2.60, 2.61, and 2.63 as

$$\frac{U_b}{k_B T} = \frac{\ell_p}{2} \frac{\theta^2}{L}. \tag{4.17}$$

If the chain is uniformly charged, consideration (Odijk 1977, Skolnick and Fixman 1977) of electrostatic repulsion within the Debye–Hückel approximation leads to an additional term for $U_b/k_B T$, which is proportional to $\theta^2/2L$. Therefore, the coefficient of the additional contribution can be identified as a persistence length,

$$\frac{U_b}{k_B T} = \frac{\ell_p}{2} \frac{\theta^2}{L} + \frac{\ell_{pe}}{2} \frac{\theta^2}{L}, \tag{4.18}$$

where ℓ_{pe} is called the electrostatic persistence length. It is given by the Odijk–Skolnick–Fixman theory (Odijk 1977, Skolnick and Fixman 1977) as

$$\ell_{pe} = \frac{\ell_B}{4} \frac{\xi_D^2}{\ell^2}, \tag{4.19}$$

where ℓ_B, ξ_D, and ℓ are the Bjerrum length, Debye length, and the charge separation length, respectively, along the rod. The total persistence length $\ell_{p,eff}$ is given by

$$\ell_{p,eff} = \ell_p + \ell_{pe}. \tag{4.20}$$

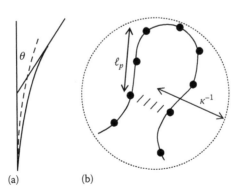

(a)   (b)

Figure 4.5 (a) Electrostatic stiffening leads to an additional contribution to the persistence length. (b) Electrostatic swelling leads to more swollen semiflexible coil.

It is difficult to experimentally observe the proportionality between the electrostatic persistence length and the inverse salt concentration, due to the result of Equation 4.19 being only a perturbation theory in the limit $\ell_p > L$. For semiflexible chains with L larger than ℓ_p (Figure 4.5b), the role of intrachain electrostatic interaction can be conveniently treated as electrostatic swelling in the vein of Section 4.1.6.

4.1.8 Hydrophobic Interaction

As discussed in Sections 2.3 and 2.5, there is an effective potential interaction between any two segments of the polymer mediated by the solvent molecules, even in the absence of charges. We call all of these short-range excluded volume interactions as the hydrophobic interaction. As described in Section 2.5, this is parametrized as the excluded volume parameter v (or equivalently w), or the Flory–Huggins χ parameter through Equation 2.65. The chemical specificities of the uncharged part of the monomer in the polyelectrolyte and the solvent are presumed to be adequately captured by the single parameter $w = (1 - 2\chi)/2$. For ideal conditions corresponding to the Flory temperature, the excluded volume parameter w is zero ($\chi = 1/2$), due to a cancelation between the net attractive and repulsive interactions among the solvent molecules and the polymer segments. If $w < 0$ ($\chi > 1/2$), polymer segments attract themselves more than the attraction between the polymer and solvent. As a result, the polymer contracts into globular conformations, as seen in Section 2.6. Since most of the polyelectrolyte backbones are immiscible with the polar solvents typically used in polyelectrolyte solutions, the chemical mismatch parameter χ is about 0.5 or higher ($w < 0$).

All of the above concepts must be addressed in interpreting the experimental results on the various measures of the equilibrium structures of polyelectrolyte chains in dilute solutions.

4.2 EXPERIMENTAL RESULTS

We first review the experimental data on the size of a polyelectrolyte molecule in dilute solutions in terms of its dependence on chain length and salt concentration. The main methodology for the determination of polymer size in solutions is the use of light scattering. Most of the experiments dealing with polyelectrolyte solutions are at polymer concentrations not low enough to avoid the long-ranged interference from other chains (Rice and Nagasawa 1961, Dautzenberg et al. 1994). Also, interpretation of data from light-scattering measurements on polyelectrolyte solutions at very low salt concentrations is quite difficult (Drifford and Dalbiez 1985, Forster and Schmidt 1995, Volk et al. 2004). Therefore, there has been a scarcity of reliable experimental data on polyelectrolyte size in infinitely dilute solutions. It is necessary to choose polymer concentrations low enough and the salt concentration not to be too

TABLE 4.1

Dependence of R_g (nm) of Et – PVP – Br on Contour Length and Salt
Concentration of NaBr

L (nm)	1 M	0.3 M	0.1 M	0.03 M	0.01 M	0.003 M	0.001 M
400	15.2 ± 0.6	17.6 ± 0.5	22.1 ± 0.5	28.8 ± 0.5	34.6 ± 0.8	41.6 ± 0.7	50.7 ± 0.8
1000	27.5 ± 0.5	29.3 ± 0.5	36.2 ± 0.5	46.6 ± 0.6	58.4 ± 1	75 ± 0.8	95 ± 2
1800	39.6 ± 0.5	43.8 ± 0.5	54.6 ± 0.7	69.9 ± 0.7	87.3 ± 1	116 ± 1.5	145 ± 2

Source: Beer, M. et al., *Macromolecules*, 30, 8375, 1997. With permission.

TABLE 4.2

Dependence of R_g (nm) of Et – PVP – Br on Specificity of Counterion from
Added Salt. Polymer Contour Length $L = 1000$ nm

C_s (mol/L)	NaF	NaCl	NaBr	NaI
0.1	47.2 ± 0.5	41.6 ± 0.8	36.2 ± 0.5	28.3 ± 0.8
0.03	53.7 ± 1	50.6 ± 0.5	46.6 ± 0.5	41.2 ± 0.4
0.01	66 ± 1	61.5 ± 0.5	60 ± 1	54.9 ± 0.7
0.003	86.5 ± 1	78.5 ± 0.7	75 ± 0.7	73.1 ± 0.6
0.001	107 ± 1	97 ± 1	95 ± 1.5	93.5 ± 1.5

Source: Beer, M. et al., *Macromolecules*, 30, 8375, 1997. With permission.

low to ensure that interchain correlations are absent. As an example, precise light-scattering results (Beer et al. 1997) on the radius of gyration R_g of poly(vinylpyridinium) cations in very dilute solutions are given in Tables 4.1 and 4.2 in terms of the polymer contour length L and concentration c_s of the added salt. The same data are plotted in Figure 4.6.

As expected, R_g increases as the chain length is increased at a fixed salt concentration. Also, for a fixed chain length, R_g increases as the salt concentration is reduced. This is totally consistent with our expectation based on electrostatic expansion. At lower salt concentrations, there is lower extent of screening of the intrachain electrostatic repulsion, and as a result the chain swells more. One could intuitively expect a flexible polyelectrolyte molecule to adopt a rod-like conformation due to repulsive intrachain electrostatic interaction in salt-free solutions, because the Debye length for such conditions can be larger than the radius of gyration. This limit seems never to be approached in experiments measuring R_g in dilute solutions, due to the omnipresence of counterions and counterion adsorption to the polymer. It is seen from the experimental data of Table 4.1 and Figure 4.6 that the radius of gyration increases only by a factor of about three when the salt concentration is reduced by three orders of magnitude, for each of the chain lengths.

The experimental data in Table 4.2 show that the specificity of the counterions plays a significant role in determining the polymer size. The

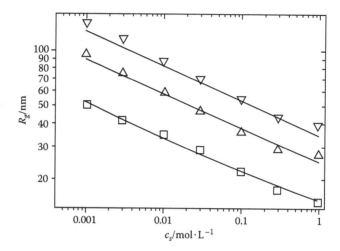

Figure 4.6   Dependence of radius of gyration of Et − PVP − Br on concentration of NaBr. $L = 400\,\text{nm}$ (□), $1000\,\text{nm}$ (△), and $1800\,\text{nm}$ (▽). The solid lines are from Equation 4.32 with $\alpha = 0.17$ and $w_0 = -0.03$. (From Beer, M. et al., *Macromolecules*, 30, 8375, 1997. With permission.)

ability of the counterions to compactify the chain increases in the order, $F^- < Cℓ^- < Br^- < I^-$, although all of these ions are monovalent. Thus, it is clear that the effective charge density of the polymer decreases with increasing polarizability of the counterions. A proper treatment of the polarizability of counterions is beyond the scope of the current theories of polyelectrolytes. Furthermore, the effect of multivalent counterions on the polymer size is much stronger than that from monovalent counterions. This is due to a combination of the fact that a multivalent counterion is a stronger adsorbing ion to the polymer than the monovalent ion, and the ability of multivalent ions to bridge multiple monomers to make transient crosslinks and as a result collapsing the polymer dramatically (Huber 1993, Ikeda et al. 1998, Prabhu et al. 2004). If the binding of multivalent counterions to the polymer is strong, even the net charge of the polymer can be reversed (Besteman et al. 2007). It is also experimentally known that the effective charge density of the polymer decreases with increasing hydrophobicity of the polymer backbone (Essafi et al. 1995).

4.3 SIMULATION RESULTS

As mentioned above, experimental protocols are challenging in order to directly probe isolated polyelectrolyte chains, such as their sizes, counterion distributions, and electric potential variations inside and outside the coils. These quantities are sometimes deduced from measurements of other quantities, such as the electrophoretic mobility. The interpretation of data in these indirect measurements also depends heavily on reliable theories. The theoretical formulation of the internal correlations within a polyelectrolyte molecule is also difficult, as discussed in Section 4.1. Faced with such challenges, computer simulations have played a significant role in helping to understand the basic features of charge correlations in polyelectrolytes.

There have been several simulation techniques implemented in modeling polyelectrolyte chains, including molecular dynamics, Brownian dynamics, Langevin dynamics, and Monte Carlo (Severin 1993, Stevens and Kremer 1995, Winkler et al. 1998, Liu and Muthukumar 2002, Holm et al. 2004a). The fundamental concepts derived from these simulations are naturally uniform, independent of the particular technique used. Since we are presently interested in the key concepts about the polyelectrolyte molecule and not in simulation techniques, we shall only illustrate the results from a generic model (Liu and Muthukumar 2002, Liu et al. 2003, Ou and Muthukumar 2005) of a polyelectrolyte chain and its properties as obtained from the Langevin dynamics simulations. The simulation system is made of n freely jointed chains each with N spherical beads of point unit electric charge $-e$, nN/z_c counterions (z_c being the valency of the counterion), n_+ cations of added salt with valency z_+ and n_+z_+/z_- anions of added salt with valency z_-, all placed in a cubic medium of permittivity $\epsilon_0\epsilon$ and volume V. The bond length between any two successive beads along a chain is allowed to fluctuate about the equilibrium value ℓ. The beads are allowed to interact among themselves with excluded volume and electrostatic interactions. The excluded volume contribution is the nonelectrostatic part of the potential interaction between nonbonded beads of the chain and is taken as a purely repulsive Lennard-Jones (LJ) potential,

$$u_{LJ} = \begin{cases} \epsilon_{LJ}\left[\left(\frac{\sigma}{r}\right)^{12} - 2\left(\frac{\sigma}{r}\right)^6 + 1\right], & r \leq \sigma \\ 0, & r > \sigma \end{cases} \tag{4.21}$$

where ϵ_{LJ} is the strength, σ is the hardcore distance, and r is the distance between two nonbonded beads.

With the repulsive LJ potential, polymer collapse due to hydrophobic effect is not addressed. The same form of Equation 4.21 is also used to capture the nonelectrostatic excluded volume interactions among the polymer beads and counterions, the difference appearing in the choice of the hardcore distance σ. The electrostatic interaction among the charged beads and ions is taken to be the Coulomb energy,

$$u_c\left(r_{ij}\right) = \frac{z_i z_j e^2}{4\pi\epsilon_0\epsilon r_{ij}} = \frac{z_i z_j \ell_B k_B T}{r_{ij}}, \tag{4.22}$$

where r_{ij} is the distance between the ions i and j, z_k is the valency of the kth ion, and ℓ_B is the Bjerrum length. The key control parameter in these simulations is the Coulomb strength parameter Γ defined in Equation 4.3,

$$\Gamma = |z_p z_c|\frac{\ell_B}{\ell} = |z_p z_c|\frac{e^2}{4\pi\epsilon_0\epsilon k_B T\ell} = |z_c|\frac{\ell_B}{\ell}, \tag{4.23}$$

where z_p is taken as -1. As noted already, the experimentally relevant range of Γ is $3.2 > \Gamma > 2.4$ for aqueous solutions ($0 < T < 100°C$) of polyelectrolytes with chemical charge separation along chain backbone of about 0.25 nm and $z_c = 1 = z_p$. Since the vast majority of experiments on polyelectrolytes uses water as the solvent, $\Gamma \approx 3$ is of experimental relevance. For multivalent counterions, this range of Γ is expanded by the multiple of z_c. Naturally, the values of Γ outside the above range represent solvents different from water and charge separation along the chain much different from ≈ 0.25 nm. Due to the temperature dependence of the permittivity of the polyelectrolyte solution and since the product $T\epsilon$ appears in the definition of ℓ_B, we must consider $T\epsilon$ as the temperature variable instead of T alone.

4.3.1 Radius of Gyration

The radius of gyration of a freely jointed chain of $N = 100$ beads in the absence of any added salt is given in Figure 4.7 as a function of the Coulomb strength parameter Γ. The case of monovalent counterions is presented in Figure 4.7a. Data for R_g when the chain is uncharged and with only repulsive LJ interaction are included in this figure, to illustrate the role of electrostatics on R_g. For very high values of $T\epsilon$ ($\Gamma \to 0$), only weak electrostatic repulsion is present. Consequently, R_g is slightly higher than that for the LJ chain. As the value of $T\epsilon$ is lowered, the electrostatic repulsion between beads becomes stronger and consequently R_g begins to increase with Γ. As $T\epsilon$ is decreased even further (i.e., Γ is close to almost 0.5), the intrachain electrostatic repulsion begins to be mitigated by electrostatic attraction between beads and counterions. The rate of chain swelling with a decrease in $T\epsilon$ begins to decrease.

As the value of $T\epsilon$ is lowered even more (i.e., Γ goes above roughly 1), there are a significant number of counterions close to the chain backbone, creating many dipoles. The interaction between these dipoles leads to intrachain attraction, working against the intrachain swelling arising from the uncompensated charges on the chain backbone. The net result is that R_g decreases as Γ increases (i.e., ϵT decreases). Yet, until $\Gamma \approx 5.0$, R_g of the polyelectrolyte chain is bigger than the volume expected for an uncharged chain in good solvents. The chain begins to collapse as Γ is increased beyond 5. Typical configurations of the chain are given in Figure 4.8 at different values of Γ. The shape factor, as given by R_g/R_h (where the hydrodynamic radius R_h is defined in Section 2.1), is plotted versus Γ in Figure 4.9. It is obvious from Figures 4.8 and 4.9 that the chain is highly anisotropic for Γ values around unity. In Figures 4.7a, 4.8, and 4.9, $N = 100$, $z_c = 1$, and the monomer density is $8 \times 10^{-4}\ell^{-3}$. By monitoring the N-dependence of R_g, the effective size exponent ν is reported to be 0.98, 0.85, 0.59, and 0.33 for $\Gamma = 1.0$, 3.0, 5.0, and 20.0, respectively (Liu and Muthukumar 2002). In spite of the exponents approaching the rod-like and globule-like limits, the actual size of the chain is less stretched than a rod (at $\Gamma = 1.0$) and less compact than a compact sphere (at $\Gamma = 20.0$). If the

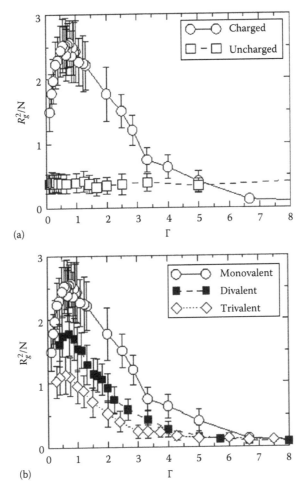

Figure 4.7 R_g for $N = 100$. (a) Freely jointed polyelectrolyte chain with monovalent counterions in comparison with an athermal uncharged chain. (b) Comparison of electrostatic swelling for monovalent, divalent, and trivalent counterions. (From Liu, S. and Muthukumar, M., *J. Chem. Phys.*, 116, 9975, 2002. With permission.)

counterions are multivalent, the chain expansion is substantially weaker, as illustrated in Figure 4.7b.

4.3.2 Counterion Adsorption and Effective Charge

When the Coulomb interaction parameter is large enough such that $\Gamma \geq 0.5$, attraction between the oppositely charged beads and some counterions begins to contribute significantly. Close examination of the position of counterions near

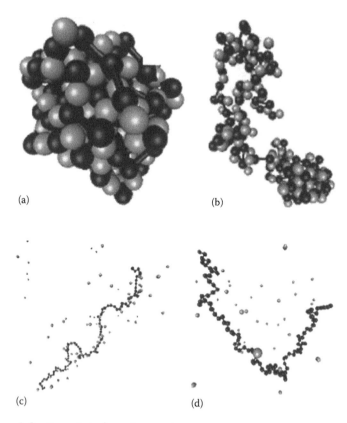

(a)

(b)

(c)

(d)

Figure 4.8 Snapshots from the simulations: (a) $\Gamma = 20.0$, (b) $\Gamma = 7.0$, (c) $\Gamma = 1.0$, and (d) $\Gamma = 0.13$. (Pictures are not in the same scale.) (From Liu, S. and Muthukumar, M., *J. Chem. Phys.*, 116, 9975, 2002. With permission.)

the chain backbone reveals that these counterions are not frozen at some fixed positions. Instead, they are found to undergo dynamics by which they move along the chain backbone and eventually exchange with other counterions that are not originally in the proximity of the chain backbone. A dynamic equilibrium of counterion distribution around the polyelectrolyte is maintained with a higher average density of counterions around the polyelectrolyte backbone than in the bulk.

By following the procedure described in Section 4.1 to construct the counterion worm, the average degree of ionization α as defined by Equation 4.7 is computed. The dependence of α on the electrostatic interaction parameter Γ is given in Figure 4.10 for several values of N, by fixing the monomer density at $\rho = 8 \times 10^{-4} \ell^{-3}$ and $z_c = 1$. As is seen clearly, α decreases smoothly without any discontinuity as Γ is increased. For the aqueous solutions of flexible polyelectrolytes at room temperatures, Γ is about 3, and α is around 0.2 due to the mechanism of counterion adsorption on the polymer

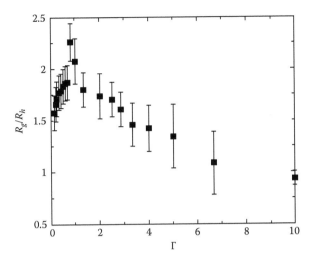

Figure 4.9   The shape factor R_g/R_h for $N = 100$, $z_c = 1$, and $\rho = 8\times10^{-4}\ell^{-3}$. The chain is anisotropic at $\Gamma \approx 1$ with an effective size exponent $v = 0.98$. The shape is self-avoiding-walk-like for $\Gamma \approx 3$ and globule-like for $\Gamma \approx 10$. (From Liu, S. and Muthukumar, M., *J. Chem. Phys.*, 116, 9975, 2002. With permission.)

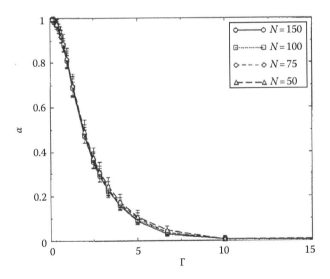

Figure 4.10   Charge fraction of a single flexible polyelectrolyte chain due to adsorption of monovalent counterions in salt-free solutions. (From Liu, S. and Muthukumar, M., *J. Chem. Phys.*, 116, 9975, 2002. With permission.)

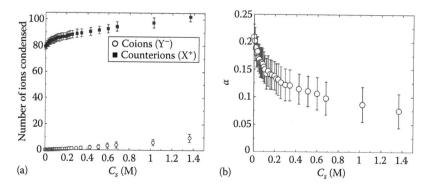

Figure 4.11   Decrease in the effective charge of the polymer with increasing salt concentration for $N = 100$ at $\Gamma = 3$. (a) Number of counterions and coions inside the counterion worm and (b) net degree of ionization. (From Liu, S. et al., *J. Chem. Phys.*, 119, 1813, 2003. With permission.)

backbone. In spite of copious reference in the literature to the Manning model (Section 3.2.5) of counterion condensation on an infinitely long and infinitely thin rod-like line charge, to be presumably valid even for flexible coil-like chains, the discontinuity of Figure 3.16b is not seen.

When a fixed quantity of salt is added to an equilibrated polyelectrolyte chain with a monovalent counterion, there is an additional counterion adsorption and α decreases further. This result is illustrated in Figure 4.11, where the monovalent counterion is labeled as X^+ and the monovalent added salt is XY (Y^- being the coion to the polymer). In these figures, $\Gamma = 3, N = 100$, and the polymer concentration $c_p = 8 \times 10^{-4}\ell^{-3}$. The numbers of X and Y ions inside the counterion worm are plotted in Figure 4.11a, as the salt concentration c_s of the XY salt is increased from 0 to 1.36 M. As already noted in Figure 4.10 for salt-free solutions, about 80% counterions (X^{+1}) are adsorbed inside the counterion worm. As the salt concentration increases, the number of adsorbed counterions around the polymer backbone increases. At higher salt concentrations, as the counterion concentration inside the counterion worm increases, an increasing number of coions (Y^-) are also brought inside the worm in an effort to maintain electroneutrality even at local scales. By adding the numbers of X^+ and Y^- ions and their charges and combining with the bare polymer charge, the effective degree of ionization of the polymer is obtained as given in Figure 4.11b. The degree of ionization decreases monotomically with the salt concentration.

When the counterion is multivalent, the nature of counterion adsorption and the consequent chain contraction are more drastic. The results for salt-free solutions are given in Figure 4.12 for the average charge fraction and in Figure 4.7b for the radius of gyration. As expected, the degree of ionization of the chain decreases with the valency of the counterions at all values of Γ. This decrease is in addition to the effect of z_c contained in the definition of $\Gamma = |z_p z_c|\ell_B/\ell_0$.

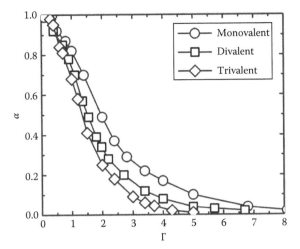

Figure 4.12 Dependence of α on the valency of the counterion in salt-free solutions for $N = 100$. (From Liu, S. and Muthukumar, M., *J. Chem. Phys.*, 116, 9975, 2002. With permission.)

When a salt of type AY_2 (A^{2+} is the counterion and Y^- is the coion) is added to an equilibrated flexible polyelectrolyte chain with a monovalent counterion (X^+), the divalent counterion competitively adsorbs on the polymer backbone by displacing the already adsorbed monovalent counterions. Typical results of this competition and the net effective charge of the polymer are given in Figure 4.13 as functions of the salt concentration c_s, for $N = 100$ and $c_p \ell^3 = 8 \times 10^{-4}$. As soon as even small amount of divalent counterions are

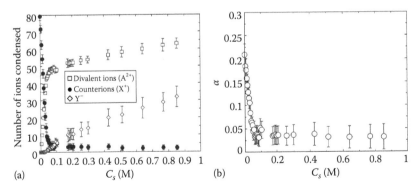

Figure 4.13 Competitive adsorption by divalent counterions against monovalent counterions. (a) Partitioning of monovalent counterions, divalent counterions, and monovalent coions inside the counterion worm as a function of the concentration of the AY_2-type salt. (b) The net charge fraction of the polymer. (From Liu, S. et al., *J. Chem. Phys.*, 119, 1813, 2003. With permission.)

present, these effectively replace the adsorbed monovalent counterions. As a result, α is reduced sharply by the divalent ions of the salt in comparison with the case of monovalent ions of the salt. In fact, depending on the value of Γ, ion size, and multivalent salt concentration, the counterion adsorption can lead to overcharging with a charge reversal of the net charge of the polymer.

4.4 ELECTROSTATIC SWELLING WITH FIXED POLYMER CHARGE

We shall now consider mean field arguments, resulting in analytical formulas, to capture the various experimental and simulation results seen above. One set of arguments assumes that the polymer charge is a fixed quantity. The other set of arguments allows the polymer charge to self-regulate as the experimental conditions change, which will be presented in Section 4.5.

In view of the counterion condensation that is always present in a polyelectrolyte solution with finite total volume, and because of the continuous exchange of counterions within the counterion worm, we assume that the adsorbed counterions are uniformly smeared along the polymer contour. Thus, the average degree of ionization of the polymer α is presumed to be true throughout the chain contour. Equivalently, we assume that the charge of a segment is $e\alpha z_p$, instead of the bare chemical charge $e z_p$. In other words, the effective charge density along the polymer is now given by

$$q_{\ell,eff} = \frac{e\alpha}{\ell_0} = \frac{e\alpha z_p}{\ell}, \tag{4.24}$$

instead of Equation 4.1. Therefore, the electrostatic interaction energy between two segments follow from Equation 4.12 as

$$\frac{u_{ij}(r_{ij})}{k_B T} = \alpha^2 z_p^2 \ell_B \frac{e^{-\kappa r_{ij}}}{r_{ij}}. \tag{4.25}$$

In view of the arguments leading to the two limits of Equation 4.12, we shall consider the two limits of (a) high salt and (b) low salt as follows.

4.4.1 High Salt Limit

As discussed in Section 4.1.6, the intrachain electrostatic interaction becomes short ranged for high salt concentrations and its strength can be simply added to the excluded volume parameter without charges. It follows from Equations 4.12 and 4.16 that an effective excluded volume parameter v_{eff} can be defined as

$$v_{eff} = v + \frac{4\pi \alpha^2 z_p^2 \ell_B}{\kappa^2}. \tag{4.26}$$

By defining v as $w\ell^3$ (Equation 2.65), the effective excluded volume parameter $w_{eff} \equiv v_{eff}/\ell^3$ is given by

$$w_{eff} = w + \frac{4\pi\alpha^2 z_p^2 \ell_B}{\kappa^2 \ell^3}. \qquad (4.27)$$

Therefore, it is only the copying of the results derived in Section 2.5, with the substitution of w_{eff} for w that needs to be done to derive the dependence of polymer size on Bjerrum length, Debye length, and chain length. By adding the free energy contributions from the chain connectivity and the sum of excluded volume effects from hydrophobicity and electrostatic repulsion, we rewrite Equation 2.71 as

$$R_g \sim \ell \left[w + \frac{4\pi\alpha^2 z_p^2 \ell_B}{\kappa^2 \ell^3} \right]^{1/5} N^{3/5}, \qquad (4.28)$$

where R_g is the radius of gyration, which is proportional to the Flory radius for the present situation. If we consider only the electrostatic contribution, R_g depends on the salt concentration c_s as

$$R_g \sim \frac{N^{3/5}}{c_s^{1/5}}, \qquad (4.29)$$

where Equation 3.26 has been used. Thus, the radius of gyration of a polyelectrolyte chain in dilute salty solutions depends on the chain length with the Flory exponent of $3/5$, as for an uncharged polymer in good solvents, and is inversely proportional to $c_s^{1/5}$.

Analogous to the derivation of Equations 2.73 and 2.74 for an uncharged polymer, a crossover formula for the radius of gyration of a flexible polyelectrolyte in a solution with high enough salt is obtained as follows. Substituting Equation 2.45 for the free energy of chain connectivity and Equation 2.68 (with w replaced by w_{eff}) for the excluded volume effect in Equation 2.67, we get

$$\frac{F(R)}{k_B T} = \frac{3}{2}\left(\frac{R^2}{N\ell^2} - 1 - \ln\frac{R^2}{N\ell^2} \right) + \frac{4}{3}\left(\frac{3}{2\pi} \right)^{3/2}\left[w + \frac{4\pi\alpha^2 z_p^2 \ell_B}{\kappa^2 \ell^3} \right]\frac{N^2\ell^2}{R^3}, \qquad (4.30)$$

where $F(R)$ is the free energy of a chain with its end-to-end distance at R. Minimization of $F(R)$ with respect to R yields (Muthukumar 1987)

$$\left(\frac{R^2}{N\ell^2} \right)^{5/2} - \left(\frac{R^2}{N\ell^2} \right)^{3/2} = \frac{4}{3}\left(\frac{3}{2\pi} \right)^{3/2}\left[w + \frac{4\pi\alpha^2 z_p^2 \ell_B}{\kappa^2 \ell^3} \right]\sqrt{N}, \qquad (4.31)$$

for the high salt limit. For the radius of gyration R_g, the above equation can be rewritten as (Beer et al. 1997)

$$\alpha_S^5 - \alpha_S^3 = \left[w_0 + \frac{134}{35} \sqrt{\frac{6}{\pi}} \frac{\ell_B}{\ell} \frac{\alpha^2 z_p^2}{\kappa^2 \ell^2} \right] \sqrt{N}, \qquad (4.32)$$

where w_0 is proportional to w and α_S is the expansion factor by which the R_g of the chain swells over its value at the Flory temperature $(R_{g0} = (N\ell^2/6)^{1/2})$,

$$\alpha_S \equiv \frac{R_g}{R_{g0}}. \qquad (4.33)$$

The simple theory given above, based on the assumption of uniform radial expansion of the coil due to fixed polymer charge, appears to capture the experimental data of Figure 4.6. By taking w_0 and α as fitting parameters, the curves in Figure 4.6 are given by Equation 4.32. One set of values, $\alpha = 0.17$ and $w_0 = -0.03$, seems to fit the data for various chain lengths and salt concentrations over three orders of magnitude. The slightly negative value for w_0 is reasonable as the polymer backbone is immiscible with water and therefore the excluded volume parameter must be negative. The degree of ionization being around 0.2 is also consistent with simulation results. Nevertheless, the degree of ionization α is taken here only as a fitting parameter, and we shall return to this issue in Section 4.5.

Similar approximate formulas have been derived (Ghosh et al. 2001) for semiflexible chains by treating the electrostatic interaction as electrostatic excluded volume. The validity and relevance of such formulas for experimental systems remain to be tested.

4.4.2 Low Salt Limit

For solutions of polyelectrolyte chains without added salt, the electrostatic interaction is expected to be relatively long-ranged, although the Debye length is finite due to the omnipresence of counterions in finite volumes. In order to ascertain the maximum limit for chain swelling due to intrachain electrostatic repulsion, let us assume that $\kappa = 0$, corresponding to the scenario where all counterions have left the polymer to explore their translational degrees of freedom in an infinite volume. In this limit, the electrostatic contribution to the free energy of a chain with its end-to-end distance at R follows from Equation 4.15 as

$$\frac{F_{el}}{k_B T} = \frac{\alpha^2 z_p^2 \ell_B}{2} \sum_i \sum_{j \neq i} \frac{1}{r_{ij}}, \qquad (4.34)$$

where r_{ij} is the distance between the segments i and j, and α is the uniform degree of ionization, due to the intrinsic nature of ionizable repeat units of the polymer.

On dimensional grounds, the double sum is proportional to N^2/R, where R is a typical length characterizing the polymer radius. Combining this result with Equation 2.69 yields the Flory scaling form,

$$\frac{F(R)}{k_B T} = \frac{3}{2\ell^2}\frac{R^2}{N} + \frac{w\ell^3}{2}\frac{N^2}{R^3} + \frac{\alpha^2 z_p^2 \ell_B}{2}\frac{N^2}{R}. \tag{4.35}$$

The first term, due to chain entropy, favors a lower value of R. The electrostatic repulsion favors larger values of R, even more strongly than the short-ranged excluded volume effect. The optimum value of R, $R^\star$, is obtained by minimizing $F(R)$ with respect to R,

$$\left.\frac{dF(R)/k_B T}{dR}\right|_{R^\star} = \frac{3R^\star}{N\ell^2} - \frac{\alpha^2 z_p^2 \ell_B N^2}{2\,(R^\star)^2} = 0 \tag{4.36}$$

so that

$$\frac{R^\star}{\ell} = \left(\frac{\alpha^2 z_p^2 \ell_B}{6\ell}\right)^{1/3} N \tag{4.37}$$

In the above results, we have taken $w = 0$, to illustrate the electrostatic effect.

By including the correct prefactors (consistent with perturbation theory) and the correct expression for chain entropy Equation 2.45, $F(R)$ becomes (Muthukumar 1987)

$$\frac{F(R)}{k_B T} = \frac{3}{2}\left[\frac{R^2}{N\ell^2} - 1 - \ln\left(\frac{R^2}{N\ell^2}\right)\right] + \frac{4}{3}\left(\frac{3}{2\pi}\right)^{3/2}\frac{w\ell^3 N^2}{R^3}$$

$$+ \frac{8}{5\sqrt{3}}\left(\frac{1}{2\pi}\right)^{1/2}\frac{f^2 Z_p^2 \ell_B N^2}{R}, \tag{4.38}$$

where R is the end-to-end distance. The optimum value for the end-to-end distance is

$$\frac{R^\star}{\ell} = \left(\frac{8}{15\sqrt{6\pi}}\frac{\alpha^2 z_p^2 \ell_B}{\ell}\right)^{1/3} N. \tag{4.39}$$

The N-dependence of $R^\star$ is rod-like for the asymptotic limit of $\kappa = 0$. However, the prefactor $\left(\alpha^2 z_p^2 \ell_B/\ell\right)^{1/3}$ is smaller than unity. In the regime of Equation 4.39 being valid, the contribution from the hydrophobic interaction w becomes negligible as the second term of Equation 4.38 is of order $1/N$, whereas the other terms are proportional to N. Similar to the dimensionless excluded volume parameter (Fixman parameter) identified in Section 2.5, the strength of the intrachain electrostatic interaction in salt-free solutions is given by the dimensionless electrostatic excluded volume parameter $z_{el,<}$,

$$z_{el,<} \equiv \frac{\alpha^2 z_p^2 \ell_B N^{3/2}}{\ell}. \tag{4.40}$$

The subscript $<$ denotes the low salt condition. It is the above combination of α, z_p, ℓ_B, and N that dictates the strength of the interaction instead of merely individually. When the value of this parameter is large, the third term in Equation 4.38 dominates so that $R \sim N$ and otherwise $R \sim \sqrt{N}$. Therefore, the electrostatic interaction is weak if $z_{el,<}$ is less than unity,

$$\frac{\alpha^2 z_p^2 \ell_B N^{3/2}}{\ell} < 1. \tag{4.41}$$

In other words, there exists a threshold value of charge fraction $\alpha^\star$

$$\alpha^\star \equiv \frac{1}{z_p N^{3/4} \left(\frac{\ell_B}{\ell}\right)^{1/2}} \tag{4.42}$$

to delineate regimes of weak and strong electrostatic interactions. If $\alpha < \alpha^\star$, the intrachain electrostatic interactions are weak. Instead of defining the threshold in terms of α, it can be equivalently defined in terms of N. The electrostatic interaction is weak if the number of Kuhn segments in the chain is smaller than the threshold value $N^\star$.

$$N^\star = \frac{1}{\left(\alpha^2 z_p^2 \frac{\ell_B}{\ell}\right)^{2/3}}. \tag{4.43}$$

In other words, if $N < N^\star$, the chain does not have significant contribution from the electrostatic swelling.

4.4.3 Electrostatic Blob

It is sometimes convenient to couch the results of Section 4.4.2 in terms of a graphical representation. Let us consider a chain of N segments. As seen from Equation 4.43, the electrostatic interaction among segments within a small section along the chain is expected to be weak. Therefore, we can imagine the chain to be made up by several contiguous sections within which the electrostatic interaction is weak and beyond which the electrostatic interaction is significant. Let us call these sections as electrostatic blobs (Figure 4.14). Let the number of segments in each of the blobs be g and the typical linear size of the blob be ξ_e. We choose g and ξ_e by stipulating that the electrostatic energy of one blob is comparable to $k_B T$. It follows from the third term on the right-hand side of Equation 4.38 that the electrostatic energy of g monomers in a blob of size ξ_e in units of $k_B T$ is $\alpha^2 z_p^2 \ell_B g^2 / 2\xi_e$. Therefore, the condition of blob energy being comparable to $k_B T$ is given by

$$\left(\alpha z_p g\right)^2 \frac{\ell_B}{\xi_e} \simeq 1. \tag{4.44}$$

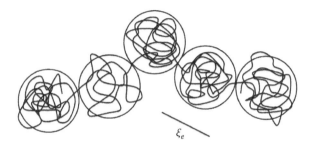

Figure 4.14 A chain is imagined to be made of electrostatic blobs of linear size ξ_e. Within the blob, electrostatic interactions are weak and the chain statistics appropriate for the uncharged backbone is applicable. Outside the blobs, electrostatic repulsion dominates and the chain is extended into a rod-like conformation.

Within the blob, the Gaussian statistics is applicable for the case of $w = 0$, so that

$$\xi_e^2 \sim g\ell^2. \tag{4.45}$$

Combining Equations 4.44 and 4.45 yields

$$\frac{\xi_e}{\ell} \sim \left(\frac{\alpha^2 z_p^2 \ell_B}{\ell}\right)^{-1/3} \tag{4.46}$$

and

$$g \sim \left(\frac{\alpha^2 z_p^2 \ell_B}{\ell}\right)^{-2/3}. \tag{4.47}$$

Since the electrostatic interaction between the blobs is assumed to be strong, we expect the chain to be stretched into a rod-like conformation with N/g blobs so that the end-to-end distance of the chain is

$$R \sim \left(\frac{N}{g}\right)\xi_e. \tag{4.48}$$

Substitution of Equations 4.46 and 4.47 gives

$$\frac{R}{\ell} \sim \left(\frac{\alpha^2 z_p^2 \ell_B}{\ell}\right)^{1/3} N, \tag{4.49}$$

which is the same result as in Equation 4.39, within the numerical prefactor.

The simplicity of the geometrical derivation of the scaling form (Equation 4.49), is appealing. Although the particular scaling law derived here is not useful due to the experimental difficulty in realizing the limit of $\kappa \to 0$, we have taken this opportunity to introduce the notion of thermal blobs. This illustrates an alternate method to derive scaling laws (de Gennes 1979), instead of minimizing a derived free energy expression. We shall take advantage of this method in the following chapter in deriving scaling laws for chains under confinement.

4.4.4 Crossover Formula

The limits of high salt and low salt have been addressed in Sections 4.1.1 and 4.1.2, respectively. In experiments, the salt concentrations are not necessarily in such extreme limits. Similarly, the dimensionless excluded volume parameter $w\sqrt{N}$ and the electrostatic excluded volume parameter defined in Equation 4.40 can assume intermediate values, instead of being either zero or very large. The crossover formula for the free energy that recovers the limits of Equations 4.30 and 4.38 is (Muthukumar 1987)

$$\frac{F(R)}{k_B T} = \frac{3}{2}\left(\tilde{\ell}_1 - 1 - \ln\tilde{\ell}_1\right) + \frac{4}{3}\left(\frac{3}{2\pi}\right)^{3/2}\frac{w\sqrt{N}}{\tilde{\ell}_1^{3/2}} + 2\sqrt{\frac{6}{\pi}}\frac{\alpha^2 z_p^2 \tilde{\ell}_B N^{3/2}}{\sqrt{\tilde{\ell}_1}}\Theta_0(\tilde{a})\,,$$

(4.50)

where

$$\Theta_0(\tilde{a}) = \frac{\sqrt{\pi}}{2}\left(\frac{2}{\tilde{a}^{5/2}} - \frac{1}{\tilde{a}^{3/2}}\right)e^{\tilde{a}}\text{erfc}\left(\sqrt{\tilde{a}}\right) + \frac{1}{3\tilde{a}} + \frac{2}{\tilde{a}^2} - \frac{\sqrt{\pi}}{\tilde{a}^{5/2}} - \frac{\sqrt{\pi}}{2\tilde{a}^{3/2}}.$$

(4.51)

Here, $\tilde{\ell}_1$ is defined as the square of the expansion factor $R^2/N\ell^2$,

$$\tilde{\ell}_1 \equiv \frac{\ell_1}{\ell} \equiv \frac{R^2}{N\ell^2},$$

(4.52)

and

$$\tilde{\ell}_B = \frac{\ell_B}{\ell}, \quad \tilde{a} = \frac{\kappa^2\ell^2 N\tilde{\ell}_1}{6},$$

(4.53)

and *erfc* is the complimentary error function (Abramowitz and Stegun 1965). The root mean square end-to-end distance R is

$$R = \ell\sqrt{N\tilde{\ell}_1},$$

(4.54)

where $\tilde{\ell}_1$ is given by the crossover formula (Muthukumar 1987)

$$\tilde{\ell}_1^{5/2} - \tilde{\ell}_1^{3/2} = \frac{4}{3}\left(\frac{3}{2\pi}\right)^{3/2}w\sqrt{N} + \frac{4}{3}\sqrt{\frac{6}{\pi}}\alpha^2 z_p^2 \tilde{\ell}_B \tilde{\ell}_1 N^{3/2}\Theta(\tilde{a})\,,$$

(4.55)

with Θ being

$$\Theta\left(a\right) = \frac{\sqrt{\pi}}{2a^{5/2}}\left[\left(a^2 - 4a + 6\right)e^a erfc\left(\sqrt{a}\right) - 6 - 2a + \frac{12}{\sqrt{\pi}}\sqrt{a}\right]. \quad (4.56)$$

With the assumption of uniform expansion used in deriving the above crossover formula, the radius of gyration R_g is $\ell\sqrt{N\tilde{\ell}_1/6}$. Therefore, Equation 4.55 can be used to calculate the radius of gyration of a flexible polyelectrolyte at all values of the Bjerrum length, Debye length, chain length, and the excluded volume parameter for a fixed polymer charge.

4.5 SELF-REGULARIZATION OF POLYMER CHARGE

As seen in Section 4.3.2, computer simulations clearly demonstrate, for parameters representing dilute aqueous solutions of flexible polyelectrolytes, that there is a continuous exchange of counterions between the neighborhood of the polymer and the background and that any counterion adsorbed on the polymer can be mobile along the polymer backbone. The counterion condensation around a polymer is reminiscent of adsorption of a gas on a lattice, except that now the lattice is a topologically correlated polymer chain. The optimization between the translational entropy of the counterions (resulting in less adsorption) and the electrostatic attraction of the counterions by the polymer (resulting in more adsorption) is influenced by the conformational fluctuations of the polymer. As a result, the polymer charge self-regulates as experimental conditions change, and its value is dictated self-consistently by the compatibility between polymer conformations and counterion adsorption.

The net polymer charge is unique to a polymer conformation (Muthukumar 2004). For example, a polymer chain cannot be assumed to bear the same net charge as it undergoes coil–globule transition, a common assumption prevalent in the polyelectrolyte literature. As discussed in Sections 4.2 and 4.3, the extent of conformation-dependent counterion condensation is affected by many factors including the nature of the counterion (size and valency), concentration of added salt, polymer concentration, solvent dielectric constant, and temperature. This effect has been addressed theoretically in the contexts of isolated chains (Khokhlov 1980, Muthukumar 2004), polyelectrolyte brushes (Pincus 1991), and gels (Khokhlov and Kramarenko 1994). We give below a brief discussion of self-regularization of polymer charge by addressing all of the concepts introduced in Section 4.1.

Let us consider a single chain of N monomers in volume V. Each monomer is monovalently charged, and ℓ is the distance between two successive monomers along the polymer. Due to the electroneutrality condition, there are N monovalent counterions. Let M be the number of counterions adsorbed on the polyelectrolyte so that M/N is the degree of counterion adsorption and $\alpha = 1 - (M/N)$ is the degree of ionization of the polyelectrolyte. In addition, let c_s be the number concentration of an added salt, which is fully dissociated into n_+ counterions and n_- coions ($c_s = n_+/V = n_-/V$). The dissolved ions are assumed to be monovalent, and the counterion from the salt is chemically identical to that of the polymer. The coupling between α and the radius of gyration R_g is calculated self-consistently as follows (Muthukumar 2004).

The free energy F has six contributions F_1, F_2, F_3, F_4, F_5, and F_6, related, respectively, to (1) entropy of adsorbed counterions on the polymer backbone, (2) translational entropy of unadsorbed counterions and all other ions (except the polymer) distributed in volume V, (3) fluctuations arising from interactions among all dissociated ions except the polymer, (4) gain in energy due to the formation of ion pairs accompanying counterion adsorption, (5) free energy of the polyelectrolyte with $N - M$ charges and M dipoles on its backbone interacting with the neutralizing background composed of $N - M$ counterions and salt ions in a solution of monomer density $\rho = N/V$, and (6) correlations among the ion pairs on the polymer.

1. *Entropy of adsorbed ions.* Since $(1-\alpha)N$ counterions are adsorbed to each chain on an average and there are $N!/[((1-\alpha)N)!][(\alpha N)!]$ ways of formation of ion pairs, the free energy associated with the entropy of adsorbed counterions, with the Stirling approximation, is

$$\frac{F_1}{Nk_BT} = [\alpha \log \alpha + (1 - \alpha) \log(1 - \alpha)]. \tag{4.57}$$

All nonlinear effects resulting from possible cooperative features associated with placement of counterions along the polymer contour are ignored.

2. *Entropy of unadsorbed ions.* The free energy due to the translational entropy associated with $(N - M + n_+)$ unadsorbed counterions and n_- coions in volume V is the familiar entropy of mixing term

$$\frac{F_2}{Vk_BT} = (\alpha\tilde{\rho} + \tilde{c}_s) \ln(\alpha\tilde{\rho} + \tilde{c}_s) + \tilde{c}_s \ln \tilde{c}_s - (\alpha\tilde{\rho} + 2\tilde{c}_s), \tag{4.58}$$

where $\tilde{c}_s$ and $\tilde{\rho}$ are made dimensionless

$$\tilde{c}_s \equiv c_s\ell^3, \quad \tilde{\rho} \equiv \rho\ell^3. \tag{4.59}$$

3. *Correlations among dissociated ions.* Using the Debye–Hückel result given by Equation 3.44, we get

$$\frac{F_3}{k_BT} = -V\frac{\kappa^3}{12\pi} \tag{4.60}$$

where

$$\kappa^2 = 4\pi\ell_B(\alpha\rho + 2c_s). \tag{4.61}$$

This contribution is due to the Coulomb interactions among the ions in the counterion cloud around the whole polymer.

4. *Energy of adsorbed ions.* The gain in energy due to an adsorbed ion at the chain backbone depends on the microscopic details such as the ionic radii and the local dielectric constant, as discussed in Section 4.1.5. Using the phenomenological dielectric mismatch parameter defined in Equation 4.10, the free energy associated with the formation of ion pairs follows from Equation 4.11 as

$$\frac{F_4}{Nk_BT} = -(1-\alpha)\frac{\ell_B}{\ell}\delta. \tag{4.62}$$

5. *Chain free energy.* The free energy of a flexible chain with degree of ionization α and radius of gyration $R_g/\ell = \sqrt{N\tilde{\ell}_1/6}$ is given by Equation 4.50.

6. *Interaction between ion pairs.* As discussed in Section 4.1.6, the ion pairs resulting from the adsorbed counterions may be assumed to be randomly distributed along the chain backbone with random orientations. The interaction among these ion pairs leads to a short-ranged attractive contribution and can be absorbed into the two-body excluded volume interaction parameter w as long as the polymer has not substantially collapsed into a globule (Muthukumar 2004).

The total free energy F of the chain is

$$F = F_1 + F_2 + F_3 + F_4 + F_5, \tag{4.63}$$

where F_1, F_2, F_3, and F_4 are given by Equations 4.57 through 4.61, and F_5 is given by Equation 4.50. By minimizing F with respect to the degree of ionization α and the expansion factor $\tilde{\ell}_1$,

$$\frac{\partial F}{\partial f} = 0 = \frac{\partial F}{\partial \tilde{\ell}_1} \tag{4.64}$$

the optimum effective charge of the polymer $N\alpha$ and the optimum radius of gyration of the polymer $R_g = \ell\sqrt{N\tilde{\ell}_1/6}$ are obtained.

The results from this self-consistent procedure are given in Figure 4.15, showing the dependencies of the chain expansion factor $\tilde{\ell}_1$ and the degree of ionization α on the Bjerrum length. The effect of monovalent salt concentration $\tilde{c}_s \equiv c_s\ell^3$ on $\tilde{\ell}_1$ and α is given in Figure 4.16. These results capture all trends observed in computer simulations discussed in Section 4.3 for the polymer size and the effective polymer charge.

A simplification arises in the self-consistent calculation, if we were to consider only highly swollen flexible polyelectrolyte molecules. Inspection of numerical values of the various terms in Equation 4.63 shows that the contributions from chain entropy (F_5) and ion correlations (F_3) are weak in comparison with the other three terms, implying that counterion adsorption and polymer conformations are only weakly coupled for swollen coils (Kundagrami and Muthukumar 2010). By taking only F_1, F_2, and F_4 in Equation 4.63, the degree of ionization is given by the formula

$$\alpha = \frac{1}{2\tilde{\rho}}\left\{-\left(\tilde{c}_s + e^{-\delta\tilde{\ell}_B}\right) + \sqrt{\left(\tilde{c}_s + e^{-\delta\tilde{\ell}_B}\right)^2 + 4\tilde{\rho}e^{-\delta\tilde{\ell}_B}}\right\}. \tag{4.65}$$

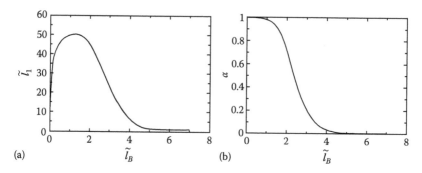

Figure 4.15 Dependence of (a) expansion factor $\tilde{\ell}_1$ and (b) the effective degree of ionization on the Coulomb strength $\tilde{\ell}_B$ for $N = 1000, c_s = 0, w = 0$, and $\delta = 3.5$.

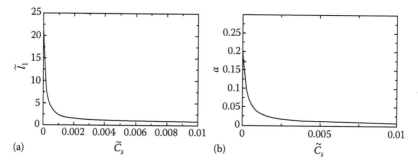

Figure 4.16 Dependence of (a) expansion factor $\tilde{\ell}_1$ and (b) the effective degree of ionization on monovalent salt concentration for $N = 1000$, $\tilde{\ell}_B = 3$, $w = 0$, and $\delta = 3.5$. (From Muthukumar, M., J. Chem. Phys., 120, 9343, 2004. With permission.)

This closed-form result for α readily gives its dependence on $\tilde{\ell}_B$ (i.e., $T\epsilon\,(T)$), the salt concentration c_s, the monomer concentration ρ, and the dielectric mismatch parameter δ. The value of α given by Equation 4.65 is substituted into Equation 4.55 to calculate the expansion factor $\tilde{\ell}_1$. A comparison of α and $\tilde{\ell}_1$ thus calculated with the separation approximation and with the numerical computation with full coupling reveals that the separation approximation is very good as long as the chain is not collapsed below the Gaussian chain size.

The full self-consistent calculation is necessary if the chain changes its conformation significantly as the experimental conditions are varied. As an example, the polymer chain collects most of the counterions when it undergoes coil-to-globule transition (Kundagrami and Muthukumar 2010) as seen in experiments (Loh et al. 2008). The polyelectrolyte chain is essentially uncharged in its collapsed state. The role of multivalent counterions and salt ions can be addressed in an analogous manner, and the behavior of polyelectrolyte chains becomes very rich (de la Cruz et al. 1995, Wittmer et al. 1995, Schiessel 1999, Nguyen et al. 2000, Kundagrami and Muthukumar 2008).

4.6 CONCENTRATION EFFECTS

As the polymer concentration increases, even in salt-free solutions where similarly charged polymers should repel each other upon close approach, the chains would eventually interpenetrate into each other with a substantial gain in entropy, by overcompensating the electrostatic interchain repulsion. Therefore, even for similarly charged flexible polyelectrolyte chains in salt-free solutions, we can define an overlap concentration $c^\star$ for the polymer, above which the chains interpenetrate. This is analogous to the uncharged polymer solutions discussed in Section 2.7. When the solution contains enough salt, then the situation is anyway akin to uncharged polymers in good solutions (Section 4.1.6).

For polyelectrolyte concentrations above the overlap concentration, which in turn depends on the chain length and salt concentration, the intersegment potential interactions are significantly modified by the topological correlations of the monomers due to chain connectivity. We have already seen in Section 2.7 that the excluded volume effect among the segments of a labeled chain in solutions of uncharged polymer molecules is screened by the interpenetration by other chains. As a result, the size exponent becomes $1/2$ in semidilute and concentrated solutions although the solvent in these solutions is a good solvent. The same screening effect occurs for polyelectrolyte solutions as well (de Gennes et al. 1976).

There are basically two kinds of long-ranged correlations in polyelectrolyte solutions as far as equilibrium properties are concerned. First is the electrostatic correlation among the ions, which results in the Debye screening. Second is the excluded volume screening due to the monomer correlations arising from chain connectivity. Both screenings occur simultaneously in polyelectrolyte solutions. This double screening (Muthukumar 1996a) must be addressed self-consistently. As a result, the effective pairwise interaction between any two segments of a labeled chain has a new correlation length, which is a functional of the Debye length. In fact, the pairwise interaction energy between two similarly charged segments can be attractive at intermediate length scales comparable to the coil radius, under certain conditions, although the second virial coefficient is positive. Due to the double screening, the size exponent for the radius of gyration of a labeled chain attains the value ($\nu = 1/2$) corresponding to the Gaussian chain value (Prabhu et al. 2003), although the polymer carries a net charge proportional to its length,

$$R_g \sim \sqrt{N}, \quad c > c^\star. \tag{4.66}$$

The key results of the double screening theory of solutions of flexible polyelectrolyte chains are summarized in Table 4.3, for $w = 0$. As for uncharged

TABLE 4.3

R_g and ξ in Dilute, Semidilute, and Concentrated Solutions of Flexible Polyelectrolytes, for $w = 0$

Quantity	c_s	Dilute	Semidilute	Concentrated
R_g	Low	$\ell_B^{1/3} N$	$\ell_B^{1/12} c^{-1/4} \sqrt{N}$	$\sqrt{N}$
R_g	High	$\left(\dfrac{\ell_B}{\kappa^2}\right)^{1/5} N^{3/5}$	$\left(\dfrac{\ell_B}{\kappa^2}\right)^{1/8} c^{-1/8} \sqrt{N}$	$\sqrt{N}$
ξ	Low	R_g	$\ell_B^{-1/6} c^{-1/2}$	$(\ell_B c)^{-1/4}$
ξ	High	R_g	$\left(\dfrac{\ell_B}{\kappa^2}\right)^{-1/4} c^{-3/4}$	$\left(\dfrac{\ell_B}{\kappa^2} c\right)^{-1/2}$

Source: Muthukumar, M., *J. Chem. Phys.*, 105, 5183, 1996a.

polymers, three regimes of polymer concentration may be identified. The case of dilute solutions has been our major focus, as described above. For concentrations larger than the overlap concentration, there are two distinct regimes, namely semidilute and concentrated. The concentration dependencies of the radius of gyration of a labeled chain and the correlation length of monomer density fluctuations are different in these regimes. These dependencies vary additionally with the salt concentration. Approximate crossover formulas connecting the various entries in Table 4.3 are known (Muthukumar 1996a). These results have also been established experimentally (Nierlich et al. 1979, Kaji et al. 1988, Nishida et al. 2001, Prabhu et al. 2003).

We shall use some of the results given in Table 4.3 in our later discussions dealing with polyelectrolyte dynamics. Also, analogous to Equation 2.97 for uncharged polymers, the free energy of a crowded polyelectrolyte solution can be derived (Muthukumar 2002a, Lee and Muthukumar 2009, Muthukumar et al. 2010), by combining the entropy of mixing, pairwise interaction energies among the various species, the free energy due to fluctuations in monomer density and ion densities, and adsorption energy inside the counterion worms. The free energy of mixing ΔF of a polyelectrolyte solution can be written as

$$\Delta F = \Delta F_0 + \Delta F_{ad} + \Delta F_{fl,ions} + \Delta F_{fl,p}. \tag{4.67}$$

where ΔF_0 is the mean field free energy of mixing of all species in the solution, analogous to Equation 2.94, ΔF_{ad} is the contribution from the adsorption of counterions in the counterion worm around the chains, $\Delta F_{fl,ions}$ is due to the correlations of the dissociated ions as given by Equation 3.43, and $\Delta F_{fl,p}$ is due to monomer density correlations and is given by Equation 2.96, except that ξ is now dictated by the double screening as given in Table 4.3. The osmotic pressure of a polyelectrolyte solution in a confining region can be derived from the net free energy.

When the polyelectrolyte molecules are rod-like, they can undergo a first-order phase transition from an isotropic state to a nematic state, depending on the polymer concentration, salt concentration, and the thickness of the chain backbone (Onsager 1949, Odijk 1986, Carri and Muthukumar 1999). We will ignore the interchain orientational correlations in the context of polymer translocation.

4.7 SUMMARY

In aqueous solutions of polyelectrolyte molecules, a significant amount of counterions adsorb on the macromolecules. The counterion worm along the contour of the polymer is dynamic in nature, with continuous exchanges in the identities of the counterions inside the worm and their locations along the chain backbone. The number of adsorbed counterions on a chain depends self-consistently on the chain conformation. As the experimental conditions change, the average polymer conformation changes, accompanied by a change in the average number of adsorbed counterions too. As a result, the average net polymer charge self-regulates with changes in experimental conditions and is substantially lower than the nominal chemical charge.

The effective degree of ionization of a chain depends on all experimental variables, including the flexibility of the polymer backbone, polymer concentration, size and valency of the counterion, identity and amount of dissolved simple electrolytes, dielectric constant of the solvent, local dielectric constant in the vicinity of the polymer backbone, and temperature. Approximate closed-form expressions, based on mean field theories, are provided to address these contributing factors. Due to the collective effects arising from the above contributing factors, the net polymer charge in aqueous solutions is only about one-fifth of the chemical charge that might be surmised from the titration curves for the polyelectrolyte.

In spite of the complicated nature of self-regulation of polymer charge, the net effect of electrostatic interactions between segments can be treated as a short-ranged electrostatic excluded volume effect for aqueous solutions containing high enough monovalent salt in the range of 0.1–1 M. For these high salt conditions, the radius of gyration of a flexible or a semiflexible polyelectrolyte chain, with chain length much larger than the persistence length, can be approximately described by

$$R_g \sim \ell \left[w + \frac{4\pi\alpha^2 z_p^2 \ell_B}{\kappa^2 \ell^3} \right]^{1/5} N^{3/5}, \qquad (4.68)$$

where w is the hydrophobic excluded volume parameter, α is the effective degree of ionization of the polymer, z_p is the chemical charge per Kuhn length, ℓ_B is the Bjerrum length, κ is the inverse Debye length, ℓ is the Kuhn length, and N is the number of Kuhn lengths per chain. Only the counterions and small electrolyte ions, and not the charged macromolecules, contribute to the definition of the Debye length for polyelectrolyte solutions.

The omnipresence of the counterion worm around a polymer chain controls both the equilibrium and dynamical behavior of the polymer. When a charged polymer is forced to move in a direction due to an externally imposed electric field, the counterion worm must move generally in the opposite direction and at the same time rejuvenating a new counterion worm around the newer polymer conformation. Such cooperativity is responsible for the electrophoretic mobility of polyelectrolytes being independent of their chain length even in salty solutions where they are coil-like. We shall return to this theme when we discuss the dynamics of polyelectrolyte chains undergoing the translocation process.

5

CONFINEMENT, ENTROPIC BARRIER, AND FREE ENERGY LANDSCAPE

When a polymer molecule is spatially confined in a volume less than its molecular volume in free solution, the number of polymer conformations is reduced, and at the same time, the intersegment interactions become more pronounced. As a result, the free energy of the molecule generally increases, due to a decrease in chain conformational entropy and an increase in energy arising from the intrachain excluded volume effect. Therefore, as a chain is squeezed from one region in a free solution into another region through a spatially restricting pore or channel, the chain must go through a free energy barrier. The free energy barrier is additionally contributed by the enthalpic interactions between the polymer and the inside walls of the pore that may bear specific chemical decorations. In this chapter, we shall estimate the free energy barriers relevant to the geometries sketched in Figure 5.1 and derive the free energy landscape as a function of the extent of translocation. The translocation process can occur either as a single-file conformation or as a multiply folded conformation depending on the size of the pore. We shall first discuss single-file translocation through a tiny pore embedded in a thin planar membrane (Figure 5.1a) and through a pore connecting two spherical chambers (Figure 5.1b). One of the major features of the single-file translocation process is the free energy penalty associated with the obligatory search by one chain end to find the pore entrance. The examples of Figure 5.1a and b allow us to capture the essential elements of the free energy landscape associated with the translocation process. In experiments involving solid-state nanopores and channels, the spatial restrictions can be big enough to allow multiply folded conformations. We shall also discuss this scenario by considering wider pores (Figure 5.1c) and channels (Figure 5.1d). Here we will explore the effect arising from chain stiffness as well. For a given pore geometry, different stages of translocation by a macromolecule can be written as a composite of several components, such as one part of the chain hanging in the donor compartment, another part hanging in the receiver compartment, and the rest residing inside the pore. We shall summarize key formulas for the free energies of polymer conformations corresponding to these different components, which can be used to construct the free energy landscape for any particular mode of translocation. Our construction of the free energy landscape is based on equilibrium considerations.

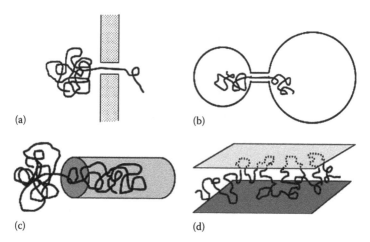

Figure 5.1 Sketches of geometries for polymer translocation: (a) a tiny hole in a thin planar membrane, (b) two spherical cavities connected by a bridge, (c) a cylindrical pore, and (d) an infinitely wide channel.

5.1 HOLE IN A WALL

Most of the basic features of the free energy landscape for polymer translocation through a pore embedded in a planar membrane can be illustrated by considering an ideal model shown in Figure 5.2a. We imagine that the pore diameter and the membrane thickness are so small in comparison with the radius of gyration and length of the polymer molecule that the pore can be mimicked by a small hole in an infinite wall (Sung and Park 1996, DiMarzio and Mandell 1997, Muthukumar 1999). The wall separates the donor (region *I*) and receiver (region *II*) compartments and is taken to be purely repulsive to the polymer. The polymer can pass from the donor compartment to the receiver compartment only through the hole. We take both compartments to

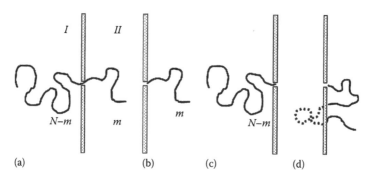

Figure 5.2 (a) Polymer translocation in transition. (b, c) Polymer tails in semi-infinite spaces. (d) Disallowed polymer conformation due to the impenetrability of the surface to the chain.

be filled with an aqueous electrolyte solution and the polymer to be a flexible polyelectrolyte chain of N Kuhn segments, each with segment length ℓ. Also, we assume that the hole is so tiny that it can accommodate only one monomer at a time. Therefore, the chain can go through the hole only as a single file.

We ask the question of what the free energy of the chain is as the chain undergoes translocation through the hole from region I to region II. As an example of an intermediate state during translocation, let there be m segments in the receiver compartment and the remaining $N - m$ segments be in the donor compartment (Figure 5.2a). Since the wall is impenetrable to the polymer, the strand of m segments in region II is in semi-infinite space, being unable to cross the wall. Such a strand with one end anchored on a wall is called a "tail." Similarly, there is another tail of $N - m$ segments in region I. Therefore, the chain conformation in Figure 5.2a is a composite of two tail-like conformations given in Figure 5.2b and c. The partition sum $\mathcal{Z}$ associated with the total number of allowed conformations of the chain with m segments in region II, and $N - m$ segments in region I is the product of the partition sums for the two tails,

$$\mathcal{Z} = \mathcal{Z}_I(N - m)\mathcal{Z}_{II}(m). \qquad (5.1)$$

Analogous to Equation 2.12 for the total number of conformations in free space, the partition sum for a tail of large enough N segments in the semi-infinite (half) space is given by (Eisenriegler 1993)

$$\mathcal{Z}_{half}(N) = N^{\gamma'-1}e^{-\mu N/k_BT}, \qquad (5.2)$$

where μ is the chemical potential per segment in the half space and γ' is another critical exponent, and its value depends on the solution conditions in the half space. The effective values of γ' for the three major experimental limits are (Eisenriegler 1993)

$$\gamma' = \begin{cases} 1/2, & \text{ideal } \theta \text{ solutions} \\ \simeq 0.69, & \text{good solutions (high salt)} \\ 1, & \text{low salt} \end{cases} \qquad (5.3)$$

The prime on the exponent γ is to emphasize that this surface exponent is quite different from the full space exponent. For example, in good solutions, $\gamma \simeq 1.16$ (Equation 2.78), whereas $\gamma' \simeq 0.69$. The number of allowed conformations for a tail is substantially reduced due to the prohibition of some polymer conformations as shown in Figure 5.2d.

The free energy F_{half} of a tail of N segments is

$$F_{half}(N) = -k_BT \ln \mathcal{Z}_{half}(N) \qquad (5.4)$$

so that it follows from Equation 5.2 as

$$\frac{F_{half}(N)}{k_B T} = (1 - \gamma') \ln N + \frac{N\mu}{k_B T}. \tag{5.5}$$

The total free energy of the chain $F(m)$, with m segments translocated into region II, follows from Equations 5.1 and 5.4 as the sum of free energies of the two tails,

$$\frac{F(m)}{k_B T} = (1 - \gamma_2') \ln m + (1 - \gamma_1') \ln (N - m) - m\frac{\Delta\mu}{k_B T}, \tag{5.6}$$

where Equation 5.5 has been used and the unnecessary constant terms are ignored. Here, γ_1' and γ_2' are the values of γ' in regions I and II, respectively. $\Delta\mu \equiv \mu_1 - \mu_2$, where μ_1 and μ_2 are the chemical potentials of the polymer per segment in regions I and II, respectively. In the translocation experiments with an applied electric voltage difference between the two regions, $\Delta\mu$ is the electrochemical potential difference between the donor and receiver compartments per polymer segment. In general, $\Delta\mu$ can arise from various sources, such as polymer concentration asymmetry, gradients in salt concentration and pH, differential adsorption of the polymer on the membrane surfaces, complexation with other macromolecules, etc. (DiMarzio and Mandell 1997, Park and Sung 1998b, Muthukumar 1999, 2001). The first two terms on the right-hand side of Equation 5.6 originate from the chain entropy of the two tails, and the third term is the gain in free energy if chain connectivity were to be absent.

A plot of $F(m)$ against m gives the free energy curve for polymer translocation as a function of the extent of translocation. According to Equation 5.6, the curve depends on the electrochemical potential gradient $\Delta\mu$ driving the process, chain length N, and the exponents γ_1' and γ_2', which in turn depend on the ionic strength and solvent quality in the two compartments. A typical result from Equation 5.6 is given in Figure 5.3a for $N = 100$, $\Delta\mu/k_B T = 0.01$, $\gamma_1' = 0.69$, and $\gamma_2' = 0.69$. There is clearly a free energy barrier, with its maximum value $F^\star$ occurring at a critical value $m^\star$ of the number of segments translocated into region II. This barrier arises due to the presence of the first two terms on the right-hand side of Equation 5.6, which are entropic in origin. The value of $m^\star$ is obtained as the solution of the condition, $\partial F(m)/\partial m = 0$,

$$\frac{m^\star}{N} = \frac{(\tilde{\mu} + 2 - \gamma_1' - \gamma_2') - \left[(\tilde{\mu} + 2 - \gamma_1' - \gamma_2')^2 - 4\tilde{\mu}(1 - \gamma_2')\right]^{1/2}}{2\tilde{\mu}}, \tag{5.7}$$

where $\tilde{\mu} \equiv N\Delta\mu/k_B T$. Substitution of this value of $m^\star$ in Equation 5.6 gives $F^\star$. For the parameters used in constructing Figure 5.3a, the free energy maximum has a value of $2.09 k_B T$ at $m^\star = 22$. The occurrence of a free energy maximum at some critical value of m implies that the translocation process

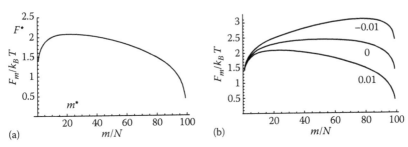

Figure 5.3 Plot of F_m/k_BT against m/N, according to Equation 5.6. (a) Free energy barrier at the critical number m^* of segments to be nucleated for successful translocation. (b) Entropic barriers for downhill ($\Delta\mu/k_BT = 0.01$) and uphill ($\Delta\mu/k_BT = -0.01$) processes. The barrier is symmetric in the absence of the driving force.

is like a nucleation phenomenon (Muthukumar 2001). If the number of segments translocated into the receiver compartment is less than the critical value m^*, then the subsequent transfer of segments from the donor is an unfavorable process. There is a greater tendency for the segments to go back to the donor compartment. On the other hand, if the number of monomers transferred in the initial stochastic process is larger than m^*, then further transfer is more favorable as the free energy landscape now is down-hill. Therefore, there exists a nucleation barrier for polymer translocation. A critical number of segments m^* must be nucleated in the receiver compartment for the eventual translocation process to become successful.

The values of m^* and the free energy maximum F^* depend on all parameters appearing in Equation 5.6. The effect of the sign of the driving force is illustrated in Figure 5.3b. When the driving force is absent ($\Delta\mu = 0$), the free energy barrier is symmetric. If the driving force is in the opposite direction to the translocation process, then the barrier is much bigger. Also, even if the barrier is crossed, the chain in region *II* is only metastable and the chain will eventually return to the donor compartment. The dependencies of m^* and F^* on $N\Delta\mu/k_BT$ are given in Figure 5.4a and b, respectively. For large values of $N\Delta\mu/k_BT$, the value of m^* becomes progressively small. As seen in Figure 5.4b, the free energy maximum also decreases with $N\Delta\mu/k_BT$. The effect of chain length is included in Figure 5.4b. As the chain length increases, the barrier becomes higher.

The role of solution conditions in the two compartments on m^* and F^* can be readily calculated from Equations 5.6 and 5.7. By fixing γ_2' at 0.69 corresponding to a high salt concentration in the receiver compartment, the effect due to γ_1' on m^* and F^* is given in Figure 5.5 for $N\Delta\mu/k_BT = 2$. The effect of γ_2' for a fixed value of γ_1' is given in Figure 5.6. In both cases, the value of the free energy maximum decreases as the equilibrium polymer conformation is changed from Gaussian-like to rod-like. On the other hand, the value of m^*

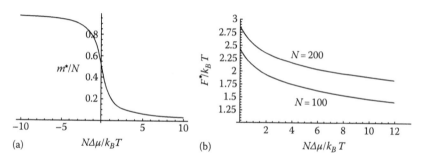

Figure 5.4 Dependence of (a) critical number m^* for nucleation and (b) free energy F^* at the barrier height on $N\Delta\mu/k_BT$ for $\gamma_1' = 0.69 = \gamma_2'$.

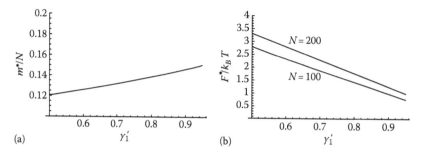

Figure 5.5 Dependence of (a) m^* and (b) F^* on γ_1' for $N\Delta\mu/k_BT = 2$ and $\gamma_2' = 0.69$.

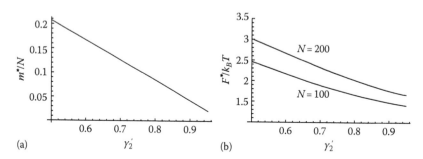

Figure 5.6 Dependence of (a) m^* and (b) F^* on γ_2' for $N\Delta\mu/k_BT = 2$ and $\gamma_1' = 0.69$.

increases with γ_1' for a fixed γ_2', and decreases with γ_2' for a fixed γ_1', although the variations are only slight.

The above results are based on equilibrium conformations of polymer chains in transit through the hole, after one segment has already reached the receiver compartment. While the presence of entropic barrier is evident, it is only weak in the order of a few k_BT units. In fact, this barrier can become insignificant if the driving force is enormous as in most of the single-molecule electrophysiology experiments. For such strong driving forces, the free energy landscape can be approximated as a ramp (Lubensky and Nelson 1999). Nevertheless, for translocation processes relevant to biological conditions with relatively weak driving forces, the control of free energy barrier by regulating polymer conformations can be significant.

In the above construction of the free energy landscape, the polymer conformations correspond to intermediate stages of a translocation event, where at least one segment is present in either of the two regions. As we shall see in Section 5.2, the search by one chain end to place itself at the pore entrance suffers from a significant free energy barrier. This barrier, due to the loss of translational degrees of freedom of the chain end, can overwhelm the subsequent barrier due to conformational entropy discussed above.

5.2 SPHERICAL CAVITIES

Consider the translocation of a polymer chain of N segments from a spherical cavity of radius R_1 to another spherical cavity of radius R_2, through a narrow pore of length $M\ell$ (Figure 5.7a). Let the pore be shorter than the polymer length ($M < N$) and narrow enough to allow only single-file translocation. When the segments are inside the pore, let the average interaction energy between the pore and a segment be ϵ_p. The driving force for translocation is the electrochemical potential difference $\Delta\mu$ between the donor and receiver compartments.

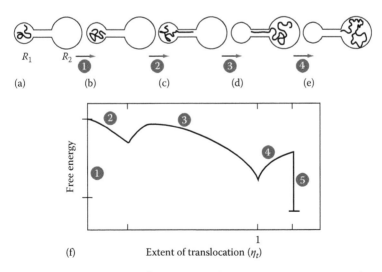

Figure 5.7 Key steps (a–e) of translocation between two compartments through an interactive pore and the accompanying free energy landscape (f).

The translocation process and the associated free energy landscape have five essential stages (Figure 5.7). The first stage, which is entropically most unfavorable, is the placement of one of the chain ends at the pore entrance. The second stage corresponds to filling the pore with the polymer. This step depends on the energy of interaction between the polymer segments and the pore wall, and any electric potential gradient present across the pore. If the net charge of the pore wall is opposite to that of the polymer, then the second stage is energetically favorable, even in the absence of electric fields. In the third stage, the rest of the monomers left behind in the donor compartment are transferred to the receiver compartment by the conformational entropic barrier mechanism. The fourth stage corresponds to the depletion of the polymer from the pore. If the pore is attractive to the polymer, then the fourth stage is energetically unfavorable, in the absence of driving forces. In the fifth stage, the polymer slips into the acceptor compartment.

If the polymer chain obeys Gaussian chain statistics (Section 2.4.3), exact formulas for the free energy landscape can be derived (Park and Sung 1998a, Muthukumar 2003). In spite of the fact that the ideal Gaussian chain behavior is not observed except at the Flory condition (Section 2.3), exact results allow us to gain insight into the key features of the translocation process. In view of this, we shall discuss translocation of a Gaussian chain in the next section. The effects of excluded volume for uncharged polymers and electrostatic interactions for polyelectrolytes and counterion worms will be discussed in Sections 5.2.2 and 5.2.3, respectively.

5.2.1 Gaussian Chain

The statistical properties of a Gaussian chain in confined geometries can be derived by drawing an analogy with standard exercises in the theory of heat conduction (Carslaw and Jaeger 1986). Basically, the probability $P(\mathbf{r}, \mathbf{r}_0; N)$ that the ends of a chain of length N Kuhn segments are at $\mathbf{r}$ and $\mathbf{r}_0$ is given by

$$\left(\frac{\partial}{\partial N} - \frac{\ell^2}{6} \nabla_{\mathbf{r}}^2 \right) P(\mathbf{r}, \mathbf{r}_0; N) = 0, \tag{5.8}$$

with the boundary condition that P is zero if $\mathbf{r}$ or $\mathbf{r}_0$ is on the surface of the confining cavity for all values of N. Using the standard mathematical techniques (Ozisik 1993) to solve equations of the above type, exact answers can be written down for various confining boundaries.

The total probability of a Gaussian chain, with the condition that all of its N segments can be anywhere inside a confining sphere of radius R (Figure 5.8a) without hitting the wall, is given by (e.g., Muthukumar 2003)

$$P_0(N, R) = \frac{8R^3}{\pi \ell^3} \sum_{m=1}^{\infty} \frac{1}{m^2} \exp\left[-m^2 \pi^2 \left(\frac{R_g}{R} \right)^2 \right]. \tag{5.9}$$

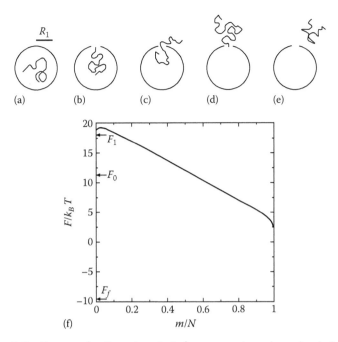

Figure 5.8 Escape of a Gaussian chain from one sphere through a hole. (a–e) Represent different stages of translocation. (f) Plot of the entropic contribution to $F/k_B T$ against m/N for $N = 260$ and $R = 5\ell$. The initial and final free energies are $F_0 = 11.34 k_B T$ and $F_f = -9.67 k_B T$. The largest contributor to the barrier is the first stage with $(7.37 + \epsilon_p) k_B T$. (From Muthukumar, M., *J. Chem. Phys.*, 118, 5174, 2003. With permission.)

where R_g is the radius of gyration $\ell \sqrt{N/6}$. The entropy S_0 of confinement of a Gaussian chain is given by $S_0 = k_B \ln P_0$. Therefore, the free energy of the chain under confinement $F_0 (= -TS_0)$ follows from Equations 1.1 and 5.9 as

$$\frac{F_0}{k_B T} = -\ln\left(\frac{4\pi R^3}{3\ell^3}\right) + F_c\left(\frac{R_g}{R}\right), \tag{5.10}$$

where F_c is defined as

$$F_c\left(\frac{R_g}{R}\right) = -\ln\left[\frac{6}{\pi^2}\sum_{m=1}^{\infty}\frac{1}{m^2}\exp\left[-m^2\pi^2\left(\frac{R_g}{R}\right)^2\right]\right]. \tag{5.11}$$

The first term on the right-hand side of Equation 5.10 is due to the fact that one end of the chain can be placed anywhere inside the volume of the cavity. The second term, called the confinement free energy, arises from all allowed conformations due to chain connectivity and confinement effects, provided that one end is already placed somewhere inside the cavity. For large values of N such that $R_g > R$, the leading term in Equation 5.11 dominates the sum. The

approximation of keeping only the leading term, called the ground-state dominance (GSD) approximation, is an excellent approximation for $R_g \geq 0.3R$. With
the GSD approximation, F_c becomes

$$F_c\left(\frac{R_g}{R}\right) \simeq \ln\frac{\pi^2}{6} + \pi^2 \left(\frac{R_g}{R}\right)^2. \tag{5.12}$$

Next, we consider the situation of Figure 5.8b, where one end is anchored
at a distance a, comparable to the segment length ℓ, in front of the hole in the
cavity. The probability $P_1(N, R)$ of realizing all conformations corresponding
to the situation of Figure 5.8b is (e.g., Muthukumar 2003)

$$P_1(N, R) = \frac{2a}{R} \sum_{m=1}^{\infty} \exp\left[-m^2\pi^2\left(\frac{R_g}{R}\right)^2\right]. \tag{5.13}$$

The entropy S_1 of the state in Figure 5.8b is $k_B \ln P_1$ and the free energy F_1
follows from Equation 1.1 as

$$\frac{F_1}{k_B T} = \frac{\epsilon_p}{k_B T} - \ln P_1(N, R). \tag{5.14}$$

Here, ϵ_p is the free energy of interaction of the chain end with the pore. This
term also includes a term $k_B T \ln \bar{z}$ associated with the loss of one segment in
the extensive part of Equation 2.12.

In going from the state of Figure 5.8a to the state of Figure 5.8b, there is a
decrease in entropy and the consequent free energy increase is

$$\frac{F^\dagger}{k_B T} = \frac{(F_1 - F_0)}{k_B T} = \frac{\epsilon_p}{k_B T} - \ln\left(\frac{P_1}{P_0}\right), \tag{5.15}$$

where P_0 and P_1 are given by Equations 5.9 and 5.13. Within the GSD approximation for Gaussian chains, $F^\dagger$ is independent of N and it depends on only the
size of the cavity according to

$$\frac{F^\dagger}{k_B T} \simeq \frac{\epsilon_p}{k_B T} + \ln\left(\frac{4}{\pi}\frac{R^4}{a\ell^3}\right). \tag{5.16}$$

Therefore, the entropic barrier associated with bringing one end of a Gaussian
chain to the pore mouth from a spherical cavity of radius R increases logarithmically with R. As we shall see below, this entropic barrier constitutes most of
the free energy barrier for polymer translocation (Figure 5.8f). This barrier can
of course be mitigated by dialing ϵ_p with specific binding of the chain end at
the pore entrance.

For a tail-like conformation hanging outside at a small distance a from the
pore in a spherical cavity of radius R, as in Figure 5.8c and d, the probability

$P_2(N, R)$ for a Gaussian tail of N segments is (Hiergeist and Lipowsky 1996, Park and Sung 1998a)

$$P_2(N, R) = \frac{a}{R}\left(1 + \frac{R}{\sqrt{\pi}R_g}\right). \qquad (5.17)$$

For the state of Figure 5.8c, where m segments are outside the sphere and $N - m$ segments are inside the sphere, the probability is given by

$$P_m(N, R) = P_1(N - m, R)P_2(m, R). \qquad (5.18)$$

Since only one segment is interacting with the hole, the corresponding free energy is

$$\frac{F_m}{k_B T} = \frac{\epsilon_p}{k_B T} - \ln P_m(N, R), \qquad (5.19)$$

where Equations 5.13 and 5.17 have to be used. The cases of $m = 1$ and $m = N - 1$ correspond to the situations of Figure 5.8b and d, respectively. Similar to the result in Section 5.1, there is a conformational entropic barrier as a function of the number of segments m translocated out of the spherical cavity. The free energy F_f of the final state (Figure 5.8e) is given by Equation 5.9, by taking the radius of the sphere to be a large value to mimic the volume outside the spherical cavity.

   The above results for the sequence of events in Figure 5.8 constitute the free energy landscape for the translocation of a Gaussian chain from a confining spherical cavity through a hole, as given in Figure 5.8f (Muthukumar 2003). The values of the parameters in constructing the free energy curve are $N = 260$ and $R = 5\ell$. The localization distance a for the chain end is taken as $\ell/2$, and the segment density in the outside chamber is chosen as $0.01664\ell^{-3}$, for illustrative purposes. The free energy values of $F_0/k_B T$, $(F_1 - \epsilon_p)/k_B T$, and $F_f/k_B T$ are respectively 11.34, 18.71, and -9.67. Therefore, it is obvious that most of the free energy barrier arises from having to place the first segment at the pore entrance. For the example given in Figure 5.8f, the entropic part of this barrier is $7.37k_B T$. The actual barrier is of course this part plus the pore-segment interaction energy ϵ_p. The additional contribution to the net free energy barrier, from the conformational entropic change during the translocation process (third stage in Figure 5.7), is mild in comparison with the barrier associated with the search of the chain end for the pore (first stage in Figure 5.7).

   The above calculation can be readily implemented for the process depicted in Figure 5.9, where the chain undergoes translocation from one spherical cavity of radius R_1 to another spherical cavity of radius R_2 through a hole. For this situation, F_0 and F_f are given by Equation 5.10 with $R = R_1$ and $R = R_2$, respectively. F_1 is again given by Equation 5.14 with $R = R_1$. The

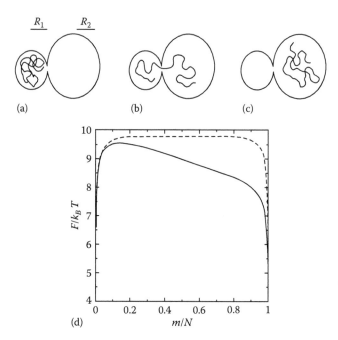

Figure 5.9 (a–d) Entropic barrier for translocation of a Gaussian chain from one sphere of radius R_1 to another sphere of radius R_2 through a hole. $N = 100$. $R_1 = 5\ell = R_2$ for the dashed curve; $R_1 = 5\ell$ and $R_2 = 6\ell$ for the solid curve. (From Muthukumar, M., J. Chem. Phys., 118, 5174, 2003. With permission.)

free energy F_m for the intermediate conformations as in Figure 5.9 follows from Equations 5.13 and 5.14 as

$$\frac{F_m}{k_B T} = \frac{\epsilon_p}{k_B T} - \ln[P_1(N - m, R_1) P_1(m, R_2)]. \qquad (5.20)$$

The free energy curve as a function of the extent of translocation depends on the radii of the donor and receiver compartments, as illustrated in Figure 5.9d. A smaller cavity creates an entropic driving force to expel the chain into a larger cavity, and the driving force follows from the above-described GSD approximation as

$$\frac{F_f - F_0}{k_B T} = -3 \ln\left(\frac{R_2}{R_1}\right) - \frac{\pi^2 R_g^2}{R_1^2}\left(1 - \frac{R_1^2}{R_2^2}\right). \qquad (5.21)$$

Let us now consider the effect of the pore of length $M\ell$ connecting two spheres of radii R_1 and R_2, as in Figure 5.7a. The free energies of the states of Figure 5.7a, b, and e are exactly the same as considered above. In an intermediate state between Figure 5.7b and c, let s segments be inside the pore and $N - s$ segments be in the donor sphere. The free energy of this state is

$$\frac{F_s}{k_B T} = s\frac{\epsilon_p}{k_B T} - \ln P_1(N - s, R_1), \qquad (5.22)$$

where P_1 is given by Equation 5.13. Same expression is valid if s segments are inside the pore and $N - s$ segments are in the recipient compartment, by replacing R_1 by R_2. Similarly, for the case of M segments inside the pore, m segments in the receiver sphere of radius R_2, and $N - M - m$ segments in the donor sphere of radius R_1, the free energy is

$$\frac{F_m}{k_B T} = M\frac{\epsilon_p}{k_B T} - \ln[P_1(N - M - m, R_1)P_1(m, R_2)]. \qquad (5.23)$$

The free energy curve for the process of Figure 5.7b through e, given by Equations 5.22 and 5.23, is plotted in Figure 5.10 against the fraction η_t of segments depleted from the donor sphere for $N = 100, M = 10, \epsilon_p/k_B T = -0.1, R_1 = 5\ell$, and $R_2 = 6\ell$. The translocation order parameter η_t is $0, 0.1$, and 1.0 for the states of Figure 5.7b through d, respectively. During the last stage ($1.0 < \eta_t < 1.1$), the chain is expelled from the pore, and the fraction of segments in the receiver sphere increases from 0.9 to 1.0.

In illustrating the effect of pore–polymer interaction, we have assumed that it is attractive by choosing a negative value for ϵ_p. The presence of an interacting pore of finite length significantly modifies the free energy landscape for polymer translocation, as evident from a comparison of Figures 5.10 and 5.9d. The

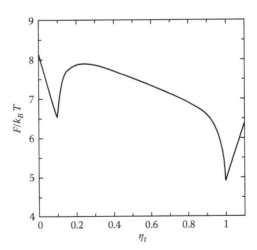

Figure 5.10   Effect of an interactive pore. Free energy curve for the transition from Figure 5.7b through e, as a function of the translocation order parameter. $N = 100, M = 10, \epsilon_p = -0.1, R_1 = 5\ell$, and $R_2 = 6\ell$. (From Muthukumar, M., J. Chem. Phys., 118, 5174, 2003. With permission.)

single barrier of Figure 5.9d (for $M = 0$) is replaced by three steps. An adsorbing polymer reduces the free energy of the chain as the latter gets sucked into the pore. However, this attraction results in another barrier during chain expulsion from the pore at later stages of translocation. These two processes of suction and expulsion of the chain are separated by the usual entropic barrier associated with the transfer of $N - M$ segments.

In addition to the above features arising from pore geometries and pore–polymer interactions, there are the usual electrochemical potential gradients that drive the translocation from the donor reservoir to the recipient reservoir. This effect will be present as an additional $\Delta\mu/k_BT$ term per segment in the recipient reservoir, as discussed in Section 5.1. Also, if there is a linear electric potential drop $\Delta\psi$ across the pore in Figure 5.7, then there are additional electrostatic energy terms in the above equations (Ambjornsson et al. 2002, Muthukumar 2003). As an example, consider an intermediate state in the second stage of Figure 5.7b. At a distance $s'\ell$ from the pore entrance, the potential energy of a segment with an effective charge $e \mid z_p \mid \alpha$ (as discussed in Chapter 4) is $-ez_p\alpha s'\Delta\psi/M$. Therefore, when s segments are inside the pore and the rest are in the donor sphere, the free energy is given by Equation 5.22 by replacing ϵ_p by (Muthukumar 2003)

$$\epsilon_p \rightarrow \epsilon_p - \frac{e \mid z_p \mid \alpha s\Delta\psi}{2M}. \tag{5.24}$$

We shall return to these results when we discuss the kinetics of polymer translocation under an electric field in later chapters.

The above discussion offers a convenient methodology to compose an approximate free energy landscape for polymer translocation and strategies to modify the landscape with experimental handles such as pore length, strength of pore–polymer interaction, volumes of donor and recipient compartments, and electrochemical potential gradients, for a given polymer length.

5.2.2 Uncharged Polymer with Excluded Volume

While the general features described above based on Gaussian chain statistics are valid, the intrachain excluded volume interactions significantly affect the value of the entropic barrier associated with the first stage of search for the pore entrance by a chain end (Kong and Muthukumar 2004). The tendency of the chain to expand due to excluded volume forces, but being confined in a small volume, results in a lower barrier for the placement of the chain end at the pore entrance. For all experimental conditions, except the special Flory condition, excluded volume effect is present among all segments (Section 2.3). The role of intrachain excluded volume effect on a confined chain is addressed by rewriting Equation 5.8 as

$$\left(\frac{\partial}{\partial N} - \frac{\ell^2}{6}\nabla_{\mathbf{r}}^2 + \mathcal{V}[P]\right) P(\mathbf{r}, \mathbf{r}_0; N) = 0, \tag{5.25}$$

with the same boundary conditions as in Equation 5.8. The third term on the left-hand side of Equation 5.25 is the extra potential generated by intrachain excluded volume effect, and we shall call Equation 5.25 as the Edwards equation (Edwards 1966). This term depends on the probability P itself, whose solution is sought by solving the above equation. Therefore, P needs to be solved self-consistently. For a chain confined inside a spherical cavity, the self-consistent solution requires numerical work (Kong and Muthukumar 2004).

First, we consider the confinement free energy of a chain with excluded volume effect inside a spherical cavity of radius R. This situation is equivalent to a polymer solution, with the boundary of the cavity playing the role of the container of the solution (Grosberg and Khokhlov 1994). The free energy of the system is therefore given by the same expression as in Equation 2.97. Within the mean field theory of Flory–Huggins, where the fourth term in the right-hand side of Equation 2.97 arising from monomer density fluctuations is ignored, the confinement free energy F_c is

$$\frac{F_c}{k_B T} \sim V\phi^2 \sim R^3 \left(\frac{N}{R^3}\right)^2 \sim \frac{N^2}{R^3}, \tag{5.26}$$

where the two-body excluded volume interaction is assumed to dominate over the entropy of mixing. For the present case, the volume of the solution is the volume of the cavity, and the volume fraction ϕ of the polymer is proportional to N/R^3, since there is only one chain inside the cavity. In fact, we have used the same result as above in deriving the Flory size exponent in Section 2.5. On the other hand, if the chain length is such that the segment concentration inside the cavity corresponds to semidilute conditions, then the term of $1/(24\pi\xi^3)$ in Equation 2.97 contributes significantly to the confinement free energy. Here ξ depends on the monomer concentration c, according to $\xi \sim c^{-\nu/(3\nu-1)}$, as discussed in Section 2.7. If the contribution from fluctuations in monomer densities dominates over the rest of the terms in Equation 2.97, the confinement free energy becomes

$$\frac{F_c}{k_B T} \sim \frac{V}{\xi^3} \sim R^3 \left(\frac{N}{R^3}\right)^{3\nu/(3\nu-1)} \sim \left(\frac{R_g}{R}\right)^{3/(3\nu-1)}, \tag{5.27}$$

where the relation defining the size exponent, $R_g \sim N^\nu$, is used. For a good solution, where the excluded volume parameter $w\sqrt{N}$ is large and positive (Section 2.5), $\nu \simeq 3/5$, so that the confinement free energy is

$$\frac{F_c}{k_B T} \sim \left(\frac{R_g}{R}\right)^{15/4} \sim \frac{N^{9/4}}{R^{15/4}}. \tag{5.28}$$

Careful numerical simulations (Cacciuto and Luijten 2006) show that Equation 5.28 is more accurate than the Flory result (Equation 5.26), for monomer concentrations in the semidilute regime. However, as the chain length increases so that the monomer concentration is in the concentrated regime (Table 2.2), the contribution from $\xi^{-3} \sim c^{3/2}$ is weak in comparison with the $\sim c^2$ term in Equation 2.97. Under these circumstances, the scaling law of Equation 5.26 is expected to be valid. In general, the free energy of confinement is a crossover function given by Equation 2.97.

Let us now consider the entropic barrier associated with the placement of one end of a chain with excluded volume in front of the pore. Here, analytical formulas are not available and approximations and numerical computations need to be made. For the localization of chain end at the pore entrance, corresponding to going from Figure 5.8a to b, the free energy barrier $F^{\dagger}$ is given by Equation 5.15 for a Gaussian chain. This result is plotted in Figure 5.11 as a function of chain length N for a fixed radius of the sphere ($R = 5\ell$). For small values of N, $F^{\dagger}$ increases with the chain length, as the monomer correlation due to chain connectivity makes it harder for the chain end to find the pore entrance. On the other hand, after reaching the space-filling value of N, such that $R_g \geq R$, the barrier depends only on R independent of N as given by Equation 5.16.

This result for the Gaussian chain is changed significantly by excluded volume interactions among monomers. With the same definition of the excluded volume parameter w as in Section 2.5, the dependence of $F^{\dagger}$ on w is given in Figure 5.11. Unlike the case of Gaussian chains, the free energy barrier depends nonmonotonically on the chain length. $F^{\dagger}$ first increases, reaches a maximum, and then decreases. For small values of N, the trend is the same as for the Gaussian chain, because the confining effect from the cavity is not significant. For larger values of N, the pressure due to excluded volume builds up inside the sphere, thus reducing the barrier for placing one chain end at the pore on the surface. As a result, $F^{\dagger}$ decreases with an increase in N. Also, for a

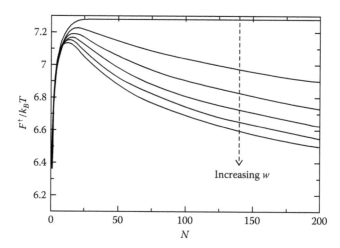

Figure 5.11   Entropic barrier for placing one chain end at the pore decreases with chain length due to intrachain excluded volume interaction. (From Kong, C.Y. and Muthukumar, M., *J. Chem. Phys.*, 120, 3460, 2004. With permission.)

given value of chain length large enough, the barrier decreases with a stronger repulsive intrachain excluded volume interaction. Therefore, as the polymer length is increased in a confining region, the search of a chain end for the pore entrance becomes easier in the presence of repulsive excluded volume interactions among the monomers. This result is also valid for a polyelectrolyte chain captured in a small volume of an electrolyte solution, because the electrostatic interaction for salty solutions becomes short-ranged excluded volume interaction (Section 4.1.6).

The intrachain excluded volume effect plays a significant role in the subsequent stages of translocation as well. As an example, let us revisit the process given in Figure 5.9. The free energy barriers in Figure 5.9d can develop a local free energy minimum. When the excluded volume effect is absent, $w = 0$, the free energy barrier is a maximum at the midpoint of translocation ($m = N/2$) for the symmetric radii of the spherical cavities ($R_1 = R_2$) as seen in Figure 5.9d. On the other hand, as w increases, a local free energy minimum develops at $m = N/2$, as shown in Figure 5.12a. Therefore, a metastable intermediate conformation of the polymer, where the monomers are equally partitioned between the donor and recipient spheres, arises due to excluded volume interactions. Such metastable states have been seen in experiments (Nykypanchuk et al. 2002). The degree of partitioning of segments in the metastable state depends on the driving force for translocation. When the driving force arises from the asymmetry of the radii of the donor and recipient spheres, the number of segments m^* in the recipient compartment, corresponding to the metastable state, is obtained with the use of dominance of two-body interactions as

$$\frac{F_m}{k_B T} \simeq w \left[\frac{(N-m)^2}{R_1^3} + \frac{m^2}{R_2^3} \right], \tag{5.29}$$

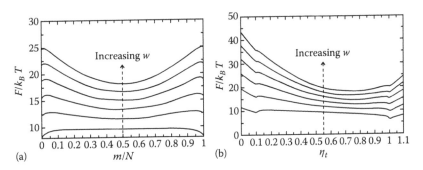

(a)   (b)

Figure 5.12 Excluded volume effect creates a metastable state. (a) Same as Figure 5.9 with $w = 0.0, 0.1, 0.2, 0.3, 0.4$, and 0.5; $R_1 = R_2 = 5\ell$, and $N = 100$. (b) Free energy barrier for going from Figure 5.7b through d with $w = 0.0, 0.1, 0.2, 0.3, 0.4$, and 0.5; $N = 100$, $M = 10$, $\epsilon = -0.1$, $R_1 = 4\ell$, and $R_2 = 5\ell$. (From Kong, C.Y. and Muthukumar, M., J. Chem. Phys., 120, 3460, 2004. With permission.)

and then minimizing this expression with respect to m. The result is

$$\frac{m^\star}{N} = \frac{R_2^3}{R_1^3 + R_2^3}. \tag{5.30}$$

The same result of Equation 5.30 is obtained if we use Equation 5.28 instead of Equation 5.26 in the derivation. The number of segments in the recipient compartment for the metastable conformation is simply the ratio of the volume of the recipient compartment to the total volume of the donor and recipient compartments. This formula turns out to be a good approximation for partitioning of segments when a chain straddles between two cavities (Kong and Muthukumar 2004).

The excluded-volume effect can generate a local free-energy minimum also for the situation of Figure 5.7 and the corresponding free-energy curve in Figure 5.10. The result is illustrated in Figure 5.12b for $N = 100, M = 10, \epsilon_p/k_B T = -0.1, R_1 = 4\ell$, and $R_2 = 5\ell$. Here, the free energy is plotted against the translocation order parameter η_t, as in Figure 5.10. The lowest curve is for a Gaussian chain, and the development of local free-energy minimum inside a barrier becomes more prominent as the excluded-volume parameter increases.

5.2.3 Polyelectrolyte Chain

When a polyelectrolyte chain is confined inside a cavity, the entropy of chain conformations, electrostatic and hydrophobic interactions among the segments, counterion adsorption, local dielectric mismatch, entropy of small ions and solvent, and correlations in monomer density and counterion worms contribute to the free energy (Chapter 4). Since the situation of a chain inside a spherical cavity is equivalent to a solution as mentioned above, the free energy of the solution given by Equation 4.67 is the confinement free energy. In addition to this formulation based on the double screening theory (Section 4.6), numerical calculations have been carried out by solving Equation 5.25 self-consistently. Now, there is an additional contribution to the potential term $V[P]$. This contribution arises from the various charges in the system and is given by the Poisson–Boltzmann formalism (Section 3.1.2). The self-consistent treatment of the small ions with the Poisson–Boltzmann equation (Equation 3.13) and the polyelectrolyte by the Edwards equation (Equation 5.25) demands heavy numerical work (Kumar and Muthukumar 2008, Kumar et al. 2009). Based on these calculations, it turns out that for a solution containing one polyelectrolyte molecule inside a sphere, the free energy of confinement is dominated by the translational entropy of small ions and the solvent molecules over all other contributing factors for polymer volume fractions of up to 0.7. Only for higher volume fractions than 0.7, chain conformational entropy and hydrophobic interaction energy become important (Kumar and Muthukumar 2008). Also, the electrostatic interaction energy plays only a minor role at all polymer concentrations inside the cavity.

Basically, the confinement free energy of a polyelectrolyte chain inside a cavity is due to the translational entropy of counterions, electrolyte ions, and solvent molecules. Simple scaling formulas based on the radius of gyration of the polyelectrolyte, analogous to Equations 5.28 and 5.26, are not applicable for the confinement free energy of a polyelectrolyte in spherical cavities, unlike the case of uncharged polymers.

Although the conformational entropy of a confined polyelectrolyte is only a weak contributor to the confinement free energy, the situation is quite different for the free energy barrier to put one chain end at the pore. By solving the coupled Poisson–Boltzmann–Edwards equations for the polyelectrolyte, counterions, salt ions, and the incompressible solvent, the free energies of the states in Figure 5.8a and b have been computed (Kumar and Muthukumar 2009). The difference in these free energies is the free energy barrier $F^\dagger$ for the first stage of Figure 5.7. For a confined polyelectrolyte chain, $F^\dagger$ consists of five contributions: electrostatic energy, ΔE_e; excluded volume interaction energy, ΔE_w; entropy of counterions and electrolyte ions, $-T\Delta S_{ions}$; solvent entropy, $-T\Delta S_{solvent}$; and the polyelectrolyte conformational entropy, $-T\Delta S_{poly}$. $F^\dagger$ and its various contributions are plotted in Figure 5.13a as functions of the volume fraction of the polymer in the cavity. It is evident that the free energy barrier is entropic and is essentially due to the conformational changes in the polyelectrolyte molecule. All other contributions, being very close in their values for the two states of Figure 5.8a and b, cancel out when the difference in free energies is taken. By exploring a wide range of values for the degree of ionization, Bjerrum length, Flory–Huggins χ parameter, and salt concentration, the entropic barrier for the first stage of translocation is calculated to be about $7 \pm 1 k_B T$.

As we have seen in Sections 4.1.6 and 4.2, the chain statistics of a polyelectrolyte molecule in an electrolyte solution is qualitatively similar to that of an uncharged polymer with repulsive excluded volume interaction. The numerical work (Kumar and Muthukumar 2009) with Poisson–Boltzmann–Edwards equations for a confined polyelectrolyte chain shows that the free energy barrier $F^\dagger$, for moderate and high salt concentrations, is quantitatively equivalent to that of an uncharged flexible polymer in good solutions. This is illustrated in Figure 5.13b, where the free energy barriers for an uncharged polymer and a flexible polyelectrolyte are plotted against the polymer volume fraction for different radii of the spherical cavity. Experimentally relevant values of the various parameters (Bjerrum length $\ell_B = 3\ell$, degree of ionization $\alpha = 0.1$, monovalent salt concentration $c_s = 0.1$ M, and the Flory–Huggins chemical mismatch parameter $\chi = 0.45$) were used in these calculations.

Analysis of the numerical data on the free energy barrier associated with the end of a chain, confined in volume $\sim R^3$, searching for the pore entrance shows that (Kong and Muthukumar 2004, Kumar and Muthukumar 2009)

$$\frac{F^\dagger}{k_B T} \sim \frac{1}{N^\beta}, \quad \beta \simeq 0.2 \pm 0.1, \tag{5.31}$$

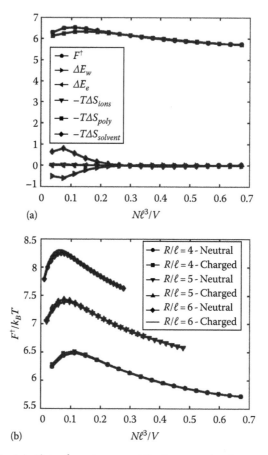

Figure 5.13 (a) Plot of various contributions to the free energy barrier against the segment volume fraction $N\ell^3/V$ for $R = 4\ell$. (b) Comparison between uncharged and charged polymer chains. In both (a) and (b), $\ell_B = 3\ell$, $\alpha = 0.1$, $c_s = 0.1$ M, and $\chi = 0.45$. (From Kumar, R. and Muthukumar, M., *J. Chem. Phys.*, 131, 194903, 2009. With permission.)

as long as the segment density inside the confining volume is large enough (polymer volume fraction ≥ 0.1). The uncertainty in the apparent exponent β is due to the deliberate variations in the values of the parameters (ℓ_B, α, c_s, and χ) used in constructing the above equation. This result becomes useful in interpreting experimental data on the molecular-weight dependence of translocation processes.

5.3 CYLINDRICAL PORES

If the pore is wide enough to allow multiple monomers at a cross-section of the pore, the chain can fold back and forth, and it can assume deformed conformations as sketched in Figure 5.1c. When a chain is confined in a cylindrical pore, the chain conformation becomes anisotropic, and the free energy must depend on the pore diameter. If the chain were to obey Gaussian statistics, analytical formulas can be derived by solving Equation 5.8 for a cylindrical cavity, and free energy of confinement in terms of the chain length and pore diameter (Casassa 1967, 1972, Casassa and Tagami 1969, Wong and Muthukumar 2008a). For chains with excluded-volume effects, Equation 5.25 needs to be solved by a self-consistent procedure mentioned in Section 5.2.2, which requires numerical work. However, scaling arguments (de Gennes 1979) provide simple and conceptually transparent results for confined chain behavior. We now derive the scaling results for flexible and semiflexible chains inside a uniform cylindrical pore.

5.3.1 Flexible Chain

Consider a flexible chain of N segments confined in an infinitely long cylindrical pore of uniform diameter D_p. As sketched in Figure 5.14a, the chain is deformed and average polymer dimension along the pore axis is $R_\parallel$. The scaling law for $R_\parallel$ in terms of the chain length, pore diameter, and the solvent quality can be readily written based on two basic results: (1) if there is no confinement,

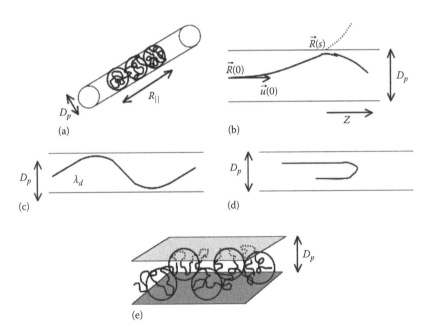

(a)   (b)   (c)   (d)   (e)

Figure 5.14 (a) A confined chain inside a pore is a set of blobs of diameter D_p. (b) A semiflexible chain is deflected by the pore at a contour distance of s, called the deflection length λ_d. (c) A long semiflexible chain is a series of deflection lengths. (d) Hairpin formation by semiflexible chains. (e) A chain confined in a channel is a two-dimensional self-avoiding-walk of blobs of diameter D_p.

$R_{\parallel}$ is proportional to R_g in dilute solutions; and (2) when the chain is strongly confined inside the pore, the chain conformation would have to adopt rod-like conformation so that $R_{\parallel}$ is proportional to N. In view of the first result, we write $R_{\parallel}$ as

$$R_{\parallel} \sim R_g f_c(D_p), \tag{5.32}$$

where the scaling function $f_c(D_p)$ represents the confinement effect. It is a constant in the limit of $D_p \to \infty$, corresponding to the bulk solution. Since the typical length representing the polymer size is R_g, the argument of the function $f_c(D_p)$ is rewritten as the dimensionless variable D_p/R_g, so that the above equation becomes

$$R_{\parallel} \sim R_g f\left(\frac{D_p}{R_g}\right), \tag{5.33}$$

where $f(D_p/R_g)$ is now the scaling function. Next, we use the second result $R_{\parallel} \sim N$ for strong confinements, $D_p \ll R_g$, in identifying the form of $f(D_p/R_g)$ in this limit. Since $R_g \sim N^{\nu}$, the only way $R_{\parallel}$ can be proportional to N in the right-hand side of Equation 5.33 is when the above function f is an exponent of its argument, say x. Therefore,

$$R_{\parallel} \sim R_g \left(\frac{D_p}{R_g}\right)^x. \tag{5.34}$$

Using the dilute solution behavior $R_g \sim N^{\nu}$, and the above expectation $R_{\parallel} \sim N$, we get

$$N \sim N^{\nu} \left(\frac{D_p}{N^{\nu}}\right)^x. \tag{5.35}$$

By matching the exponent of N on both sides,

$$1 = \nu - x\nu, \tag{5.36}$$

and hence

$$x = -\frac{1-\nu}{\nu}. \tag{5.37}$$

Substituting this value of x in Equation 5.34 yields the scaling law in the strong confinement limit

$$R_{\parallel} \sim \frac{N}{D_p^{1/\nu-1}}, \quad D_p \ll R_g. \tag{5.38}$$

Therefore, the scaling function $f(D_p/R_g)$ is a crossover function with the limits of $(D_p/R_g)^{-(1-\nu)/\nu}$ and a constant, for very small and very large values of the argument (D_p/R_g), respectively. Hence, Equation 5.33 gives the scaling law for $R_\parallel$ with the asymptotic results,

$$R_\parallel = \begin{cases} R_g, & R_g \ll D_p, \\ \dfrac{N}{D_p^{1/\nu-1}}, & R_g \gg D_p. \end{cases} \tag{5.39}$$

Specifically, for a good solution $(\nu \simeq 3/5)$, $R_\parallel$ in the strong confinement limit is

$$R_\parallel \sim \frac{N}{D_p^{2/3}}. \tag{5.40}$$

It is to be noted that, as the size exponent increases from 1/2 to 3/5 due to polymer–solvent excluded volume interactions, the chain extends more inside the pore for the same values of D_p and N, while maintaining the one-dimensional nature of $R_\parallel \sim N$ for strong confinements.

The free energy of confinement F_c of the chain is derived similarly (de Gennes 1979). Now F_c/k_BT is dimensionless and its dependence on the pore diameter can be written as a function $f_1(D_p/R_g)$ in terms of the only dimensionless variable for length that depends on the pore diameter and the polymer size,

$$\frac{F_c}{k_BT} \equiv f_1\left(\frac{D_p}{R_g}\right). \tag{5.41}$$

For strong confinements, we expect that F_c is directly proportional to N. Therefore the scaling function f_1 must be a power of its argument, say x', so that

$$\frac{F_c}{k_BT} \sim N \sim \left(\frac{D_p}{R_g}\right)^{x'} \tag{5.42}$$

Since $R_g \sim N^\nu$,

$$N \sim \left(\frac{D_p}{N^\nu}\right)^{x'}, \tag{5.43}$$

Matching the exponents of N on both sides gives $x' = -1/\nu$, and hence

$$\frac{F_c}{k_BT} \sim \frac{N}{D_p^{1/\nu}}, \quad D_p \ll R_g. \tag{5.44}$$

Specifically, for good solutions, where $\nu \simeq 3/5$,

$$\frac{F_c}{k_BT} \sim \frac{N}{D_p^{5/3}}, \quad D_p \ll R_g. \tag{5.45}$$

For Gaussian chains, where $\nu = 1/2$, the confinement free energy F_c is proportional to D_p^{-2}, in agreement with the exact results (Casassa 1967) based on solving Equation 5.8. Since the size exponent for a flexible polyelectrolyte chain in an electrolyte solution with some salt is about the same as for an uncharged polymer in a good solvent (Section 4.4), Equations 5.40 and 5.45 are sufficient in dealing with translocation of flexible polyelectrolytes through pores that allow non-single-file mobility.

The above scaling results can also be derived by devising the geometrical blob picture introduced in Section 4.4.3. Let us imagine that the confined polymer conformation in Figure 5.14a is a series of blobs, each of diameter D_p. The monomers inside the blob interact among themselves as if they are in the bulk without any confinement. The effect of confinement is manifest only at distances comparable to the pore diameter. Let each blob have g monomers, so that there are N/g blobs per chain. Since the chain extension $R_\parallel$ is now a one-dimensional excluded volume walk along the pore axis, with N/g steps, each of length D_p, we get

$$R_\parallel \sim \frac{N}{g} D_p. \qquad (5.46)$$

As pointed out above, the statistics inside each blob is the same as in bulk, so that the linear size of the blob D_p scales with the number of monomers g inside it as $\sim g^\nu$,

$$g \sim D_p^{1/\nu}. \qquad (5.47)$$

Combining Equations 5.46 and 5.47, we get

$$R_\parallel \sim \frac{N}{g} D_p \sim \frac{N}{D_p^{1/\nu}} D_p \sim \frac{N}{D_p^{(1/\nu)-1}}, \qquad (5.48)$$

which is the same result as Equation 5.38. Also, the number of blobs is

$$\frac{N}{g} \sim \frac{N}{D_p^{1/\nu}}. \qquad (5.49)$$

Comparing this result with Equation 5.44, we conclude that the confinement free energy is $k_B T$ times the number of blobs in the confined chain,

$$F_c \sim (\text{number of blobs}) \times k_B T. \qquad (5.50)$$

This is a general result for all situations amenable to devising the above blob argument.

5.3.2 Semiflexible Chain

Let us consider a wormlike chain discussed in Section 2.4 to be trapped inside a cylindrical pore of diameter D_p. The bending energy of a wormlike chain is given by Equation 2.61 and the mean square end-to-end distance of a chain with total contour length L is given by the Kratky–Porod formula (Equation 2.56), in terms of the persistence length ℓ_p (Equations 2.58 and 2.63). For narrow pores, the bending of the chain would result in its collision with the pore wall. First, we estimate the average distance along the chain at which the chain is likely to collide with the wall, as a function of the pore diameter and the persistence length. Let us consider a sector of length s along the polymer contour confined along the z-direction inside the pore (Figure 5.14b). For this sector, the mean square end-to-end distance follows from Equation 2.56, in the limit of $\ell_p > s$, as

$$\langle [\mathbf{R}(s)]^2 \rangle = s^2 \left(1 - \frac{s}{3\ell_p} + \cdots \right), \quad s < \ell_p. \tag{5.51}$$

Similarly, the mean square projection of the end-to-end distance vector on the direction $\mathbf{u}(0)$ of the first bond is obtained from Equation 2.55 (Yamakawa 1997) as

$$\langle [\mathbf{R}(s) \cdot \mathbf{u}_0]^2 \rangle = s^2 \left(1 - \frac{s}{\ell_p} + \cdots \right), \quad s < \ell_p. \tag{5.52}$$

With z-coordinate along the pore axis, and x- and y-axes being perpendicular to the pore axis, the mean square end-to-end distance is given by

$$\langle [\mathbf{R}(s)]^2 \rangle = \langle x^2(s) \rangle + \langle y^2(s) \rangle + \langle [\mathbf{R}(s) \cdot \mathbf{u}_0]^2 \rangle, \tag{5.53}$$

where the first two terms on the right-hand side are the x- and y-components. When the sector of length s collides with the wall, both $\langle x^2(s) \rangle$ and $\langle y^2(s) \rangle$ are proportional to the pore diameter,

$$\langle x^2(s) \rangle = \langle y^2(s) \rangle \sim D_p^2. \tag{5.54}$$

Combining Equations 5.51 through 5.54 yields

$$D_p^2 \sim \frac{s^3}{\ell_p}, \quad s < \ell_p. \tag{5.55}$$

We define the arc length s, at which a semiflexible chain-contour collides with the pore wall on an average, as the deflection length λ_d (Odijk 1983), satisfying the relation of Equation 5.55

$$\lambda_d^3 \sim D_p^2 \ell_p, \quad \lambda_d \ll \ell_p. \tag{5.56}$$

The direction of the chain orientation gets deflected after a contour length of λ_d (Figure 5.14b).

Let us now consider a semiflexible chain of length L larger than its persistence length ℓ_p, which in turn is larger than the deflection length λ_d

(Figure 5.14c). The chain conformation can be imagined to be a series of sectors of rod-like conformation of length λ_d. If a rod of length λ_d is in free solution, the number of allowed orientations is proportional to the surface area of a sphere carved by a radius of $\lambda_d/2$. When this rod is confined in a pore, the allowed orientations are only within a cap of the sphere with an area proportional to the diameter of the pore. Therefore, the fraction of allowed orientations inside the pore with reference to the free solution is proportional to $(D_p/\lambda_d)^2$. The negative logarithm of this ratio gives the free energy of confinement (in units of k_BT) for this sector, due to the reduction in orientation entropy. Since there are L/λ_d sectors in the confined chain, the entropic part of the free energy of confinement is

$$\frac{F_{c,entropy}}{k_BT} \simeq \frac{L}{\lambda_d} \ln \left(\frac{\lambda_d}{D_p} \right)^2, \qquad (5.57)$$

where the orientational fluctuations of the different sectors are assumed to be uncorrelated. Substituting Equation 5.56, the above equation can be equivalently written as

$$\frac{F_{c,entropy}}{k_BT} \simeq \frac{L}{\lambda_d} \ln \left(\frac{\ell_p}{\lambda_d} \right). \qquad (5.58)$$

In terms of the pore diameter and the persistence length of the polymer, the result is

$$\frac{F_{c,entropy}}{k_BT} \simeq \frac{L}{D_p^{2/3} \ell_p^{1/3}} \ln \left(\frac{\ell_p}{D_p} \right)^{2/3}. \qquad (5.59)$$

In addition to the entropic part to the free energy of confinement, there are energetic penalties associated with the bending of the chain at the deflection points. At one deflection point, the bending energy can be estimated by using Equations 2.60 through 2.63. Since the local radius of curvature for the chain at the deflection point is bounded by the radius of the pore $D_p/2$, and the arc length that bends here is a maximum of $\pi D_p/2$, the bending energy at one deflection point is

$$\frac{U_b}{k_BT} \Big|_{bend} \sim \ell_p \int_0^{\pi D_p/2} ds \left(\frac{1}{D_p/2} \right)^2 \sim \frac{\ell_p}{D_p}. \qquad (5.60)$$

Furthermore, if the chain length is too long, there are $L/\lambda_d - 1$ bends, and the total bending energy is

$$\frac{U_b}{k_BT} \sim \frac{L}{\lambda_d} \frac{\ell_p}{D_p} \sim \frac{L\ell_p^{2/3}}{D_p^{5/3}}, \qquad (5.61)$$

where Equation 5.56 is used. Therefore, the energy contribution from the deflections has a different dependence on D_p from the entropic contribution given by Equation 5.59. It is noteworthy that for a long semiflexible chain deflecting multiple times at the pore wall, the energy contribution is proportional to the confinement free energy for a flexible chain in good solutions (Equation 5.45). Adding the entropic and bending energy contributions, the total confinement free energy of a long semiflexible chain ($L > \ell_p > \lambda_d$) follows as

$$\frac{F_c}{k_B T} \simeq \frac{L}{D_p^{2/3} \ell_p^{1/3}} \ln\left(\frac{\ell_p}{D_p}\right)^{2/3} + \frac{L \ell_p^{2/3}}{D_p^{5/3}}. \tag{5.62}$$

Clearly, these terms are strictly valid only in some limits, and the experimental situations might correspond to crossover behavior among these limits. Furthermore, when a stiff polyelectrolyte such as a long piece of dsDNA is forced to undergo translocation through a narrow pore, the chain can easily make hairpin conformations as sketched in Figure 5.14d. Here, the bending energy penalty associated with the hairpins is easily countered by the gain in electrostatic energy of the folded conformations inside a favorable electric potential gradient across the pore. Once the chain enters the pore in such folded conformations, it is unlikely to unfold and explore other possible conformations.

5.4 INFINITELY WIDE CHANNELS

We follow the same scaling analysis as above for a flexible chain confined between two parallel plates separated by a distance D_p (Figure 5.14e). This construct is also called a slab, or a slit, or an infinitely wide channel of finite thickness. Again, we imagine that the confined chain conformation is a collection of blobs of linear size D_p, each containing g monomers. Inside each blob, the chain statistics is the same as in the unconfined free solution, $D_p \sim g^\nu$. There are N/g blobs in the chain. As these connected blobs are confined as a two-dimensional arrangement inside the channel, a chain conformation is equivalent to a self-avoiding-walk of N/g steps, with step length D_p, in two dimensions. Therefore, the radial extent of the confined chain $R_\parallel$ follows from Equations 2.81 through 2.83 as

$$R_\parallel \sim \left(\frac{N}{g}\right)^{3/4} D_p \sim \frac{N^{3/4}}{D_p^{(3/4\nu)-1}}, \tag{5.63}$$

where $g \sim D_p^{1/\nu}$ has been used. Also, the number of blobs is again given by Equation 5.49, so that the free energy of confinement follows from Equation 5.50 as

$$\frac{F_c}{k_B T} \sim \frac{N}{D_p^{1/\nu}}, \quad D_p \ll R_g, \tag{5.64}$$

which is exactly the same result as for cylindrical pores. This result can of course be alternatively derived from the fact that the confinement free energy is extensive (that is, $F_c \sim N$) in the strongly confined regime as done with Equations 5.41 through 5.44.

For an uncharged polymer in a good solution or a flexible polyelectrolyte chain in an aqueous solution with moderate amount of salt, the size exponent ν is $\simeq 3/5$. For such cases, the radial extent of the confined chain in a channel geometry and the confinement free energy are given by

$$R_\parallel \sim \frac{N^{3/4}}{D_p^{1/4}}, \quad D_p \ll R_g, \tag{5.65}$$

and

$$\frac{F_c}{k_B T} \sim \frac{N}{D_p^{5/3}}, \quad D_p \ll R_g. \tag{5.66}$$

Whereas the free energy of confinement follows the same scaling law for both the cylindrical and channel geometries, the size extent in the direction orthogonal to confinement depends on D_p differently reflecting on the particular lower spatial dimension arising from the confinement. The above results are generally valid even for branched polymers (de Gennes 1999, Sakaue and Raphael 2006).

5.5 SUMMARY

We have mapped the translocation process to a nucleation and growth phenomenon, characterized by a free energy barrier. The principal source of this free energy barrier is conformational entropy of the polymer. Additional contributions to the overall free energy landscape arise from solvent conditions, polymer–pore interactions, geometry of the pore, and the electrochemical potential gradient driving the whole process.

We have explored these factors analytically by considering a number of idealized situations such as a structureless hole, a narrow pore, and a wide channel.

The initial step of placing one chain end at the pore entrance is the primary source of the entropic barrier for polymer translocation. The typical value of this entropic barrier is about $7k_B T$, which is within the range of energy associated with the hydrolysis of one ATP molecule. The free energy barrier for polymer entrance at the pore is further modulated by the specific interaction between the chain end and the pore.

When the pore or channel is wide, the polymer undergoes non-single-file translocation. For such geometries, simple scaling formulas are derived for the confinement free energy and the spatial extent of the polymer, as functions of chain length, pore diameter or channel thickness, and the size exponent for the polymer reflecting the solvent quality. The differences between a flexible and an inflexible chain under confinement are also addressed.

It is important to note that all of the results presented in this chapter are based on equilibrium arguments. These arguments appear to be justified by the experimental data to date. The free energy landscapes we have derived within this premise will be applied to the kinetics of polymer translocation in later chapters.

6

RANDOM WALKS, BROWNIAN MOTION, AND DRIFT

Consider a large polymer molecule being dragged into a solvent made of relatively small molecules. The instantaneous movement of the polymer is difficult to predict due to collisions by a very large number of solvent molecules in random directions. As a result, the polymer undergoes random motion. The response of the polymer molecule to the random forces acting on it is additionally modulated by the presence of any externally imposed fields such as electric fields, thermal gradients, and gravitation. This phenomenon is ubiquitous in all situations dealing with polymers and colloidal particles dispersed in solutions. In this chapter, we shall summarize the key conclusions in general terms and implement these results in later chapters.

The subject of the movement of large molecules and particles in a solvent has a glorious history. Experimentally, the random motion was first observed in 1828 by Robert Brown in suspensions of pollen grains in water (Brown 1828). This phenomenon became known as the Brownian motion. On the theoretical side, Karl Pearson formulated the following problem in 1905 (Pearson 1905a): "A man starts from a point O and walks l yards in a straight line; he then turns through any angle whatever and walks another l yards in a second straight line. He repeats this process n times. I require the probability that after n of these stretches he is at a distance between r and $r + \delta r$ from his starting point O." The solution to this problem was already known in Rayleigh's earlier work (Rayleigh 1880, 1899). After learning about Rayleigh's results, Karl Pearson concluded (Pearson 1905b) that "The lesson of Lord Rayleigh's solution is that in open country the most probable place of finding a drunken man who is at all capable of keeping on his feet is somewhere near his starting point." This kind of walk was subsequently known as the Pearson walk. The Pearson walk and the Brownian motion are one and the same in terms of the global properties we are interested in. By idealizing Brownian motion as a random walk, Einstein, Smoluchowski, and Langevin derived relationships for the time dependence of the position of the particle. The main results of these theories are simply known as the Einsteinian relations for Brownian motion.

The major conclusion of the Einsteinian relation is that the mean square displacement of the particle is proportional to the time elapsed,

$$\langle (\text{distance})^2 \rangle \sim \text{time}, \quad (\text{Einsteinian}) \qquad (6.1)$$

while the average displacement is zero. The angular brackets indicate that the random motion of the particle is averaged over many realizations of the process. We call the behavior of suspended particles obeying the above law "diffusion." The Einsteinian result is qualitatively different from the result for a particle with uniform velocity, as given by Newton's law,

$$(\text{distance})^2 \sim (\text{time})^2, \quad (\text{Newtonian}) \tag{6.2}$$

with the distance traveled being velocity times the time elapsed. In the context of polymer dynamics and translocation, we call this behavior as the drift.

All experiments dealing with large-scale dynamics of suspended particles in solutions are bounded by the above two limits. The Einsteinian law is manifest when collisions by the solvent molecules due to their thermal motion dominate the dynamics of the particle. On the other hand, if the externally imposed forces on the particle dominate, then the Newtonian limit is approached. In general, the behavior is "diffusion–drift."

In order to describe the crossover behavior of a particular system (between the above two limits) under experimental investigation, powerful theoretical techniques have been developed. These include the formalisms of Langevin and Fokker–Planck. In addition to describing the interplay between the diffusion and drift for the average displacement of the particle, these formalisms enable us to address experimentally relevant issues such as the time needed by a particle, initially captured inside a pore region, to exit through a particular boundary and what the probability of this kind of exit is. Although a discussion of these formalisms might appear too technical, we give only a summary of the important formulas for the various experimentally measured quantities, in order to be useful for implementing predictions for particular experiments. We shall illustrate the results with the simple example of a structureless particle undergoing drift and diffusion in one dimension. For more details, there are excellent resources in the literature (Chandrasekhar 1943, Cox and Miller 1968, Berg 1983, Gardiner 1985, Risken 1989, Redner 2001, Mazo 2002, van Kampen 2007). We shall repeatedly refer to the formulas in this chapter in our later discussion of polyelectrolyte dynamics and translocation kinetics. If necessary, some of the technical sections can be skipped in the first reading and we can return to them in the following chapters.

6.1 BIASED RANDOM WALK

Consider a sequence of N independent steps in one dimension, with each step having a probability p to go forward (right) and probability $q(=1-p)$ to go backward (left). In one realization of the walk of N steps (Figure 6.1a), let n_1 be the number of steps that moved forward and the rest $(N - n_1 = n_2)$ are the number of steps that moved backward. The probability of realizing the combination of n_1 forward steps and n_2 backward steps among N independent steps is the binomial distribution

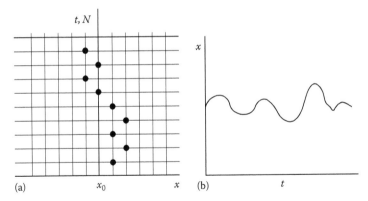

Figure 6.1 (a) A realization of a random walk in one dimension. (b) For very short step lengths and time intervals between steps, the trajectory of a random walk can be represented as a continuous curve.

$$P_N(n_1) = \frac{N!}{n_1! n_2!} p^{n_1} q^{n_2}. \tag{6.3}$$

Averaging over all possible realizations of N steps, the average of n_1 and n_1^2 is given by

$$\langle n_1 \rangle = Np, \tag{6.4}$$

and

$$\langle n_1^2 \rangle = (Np)^2 + Npq. \tag{6.5}$$

$\langle n_1^2 \rangle - \langle n_1 \rangle^2$ is the variance σ_N^2, where σ_N is the standard deviation. For the binomial distribution given in Equation 6.3,

$$\sigma_N^2 = Npq. \tag{6.6}$$

If the walk involves very large number of steps ($N \to \infty$) with finite values of p (so that $pN \to \infty$), $P_N(n_1)$ of Equation 6.3 is approximated by the Gaussian distribution,

$$P_N(n_1) \simeq \frac{1}{\sqrt{2\pi}\,\sigma_N} e^{-\frac{1}{2}\frac{(n_1 - \langle n_1 \rangle)^2}{\sigma_N^2}}. \tag{6.7}$$

This is derived by making the Stirling approximation for factorials and expanding $P_N(n_1)$ around its maximum value at $\langle n_1 \rangle$ and keeping only the quadratic term in the expansion.

If the step has unit length, then the net displacement of the walk is

$$x \equiv (n_1 - n_2) = 2n_1 - N. \tag{6.8}$$

Substituting this in Equation 6.3, we get

$$P(x,N) = \frac{N!}{\left(\frac{N+x}{2}\right)! \left(\frac{N-x}{2}\right)!} p^{\frac{N+x}{2}} q^{\frac{N-x}{2}}. \tag{6.9}$$

Master Equation

The above random walk process can be equivalently described as the probability that the walker is at x after $N + 1$ steps to be the sum of two terms: p times the probability of being at $x - 1$ after N steps, and q times the probability of being at $x + 1$ after N steps. This is stated as the "master equation,"

$$P(x, N + 1) = pP(x - 1, N) + qP(x + 1, N). \tag{6.10}$$

Analogous to Equation 6.7, for $N \to \infty$ and $pN \to \infty$, $P(x,N)$ is approximated by the Gaussian distribution,

$$P(x,N) \simeq \frac{1}{\sqrt{8\pi Npq}} e^{-\frac{1}{8}\frac{(x-\langle x \rangle)^2}{Npq}}, \tag{6.11}$$

where

$$\langle x \rangle = (p - q)N \tag{6.12}$$

and

$$\langle x^2 \rangle = (p - q)^2 N^2 + 4Npq. \tag{6.13}$$

6.1.1 Gaussian Chain Statistics

As discussed in Section 2.4, a trajectory of a random walk can be mapped into a polymer conformation by identifying the steps with the bonds. Presently, if the step length of the walk is ℓ, Equation 6.11 is written as

$$P(x,N) \simeq \frac{1}{\sqrt{8\pi Npq\ell^2}} e^{-\frac{1}{8}\frac{(x-\langle x \rangle)^2}{Npq\ell^2}}, \tag{6.14}$$

with

$$\langle x \rangle = (p - q)N\ell \tag{6.15}$$

and

$$\langle x^2 \rangle = (p - q)^2 N^2 \ell^2 + 4Npq\ell^2. \tag{6.16}$$

For $p = q$, we get,

$$P(x,N) \simeq \frac{1}{\sqrt{2\pi N\ell^2}} e^{-\frac{1}{2}\frac{x^2}{N\ell^2}}, \tag{6.17}$$

with the mean square displacement given by

$$\langle x^2 \rangle = N\ell^2, \quad v = \frac{1}{2}. \tag{6.18}$$

This is exactly the one-dimensional analog of the result for a Gaussian chain discussed in Section 2.4.3, and v is the size exponent, which is independent of the space dimension for a random walk.

6.1.2 Drift–Diffusion

If every step takes time δt, then

$$N \equiv \frac{t}{\delta t}, \tag{6.19}$$

where t is the elapsed time for the walk. Now, Equation 6.14 can be written as

$$P(x,t) = \frac{1}{\sqrt{4\pi Dt}} e^{-\frac{1}{4}\frac{(x-vt)^2}{Dt}}, \tag{6.20}$$

where

$$\langle x \rangle = vt, \tag{6.21}$$

and

$$\langle (x - \langle x \rangle)^2 \rangle = 2Dt, \tag{6.22}$$

with

$$v = \frac{(p-q)\ell}{\delta t}, \tag{6.23}$$

and

$$D = \frac{2\ell^2 pq}{\delta t}. \tag{6.24}$$

For ℓ and δt very small, the trajectory of the walk can be represented as a continuous curve as shown in Figure 6.1b. The probability distribution given by Equation 6.20 obeys the "drift–diffusion" equation,

$$\frac{\partial P(x,t)}{\partial t} = -v\frac{\partial P(x,t)}{\partial x} + D\frac{\partial^2 P(x,t)}{\partial x^2}. \tag{6.25}$$

Master Equation

The above equation can be obtained more directly from Equation 6.10,

$$P(x, t + \delta t) = pP(x - \delta x, t) + qP(x + \delta x, t), \tag{6.26}$$

where $\delta x = \ell$. The step length is taken to be infinitesimally small to emphasize the $N \to \infty$ limit. Expanding the P terms in this equation around $P(x, t) \equiv P$ as a Taylor series,

$$P + \frac{\partial P}{\partial t} \delta t = p \left(P - \frac{\partial P}{\partial x} \delta x + \frac{1}{2} \frac{\partial^2 P}{\partial x^2} (\delta x)^2 + \cdots \right)$$
$$+ q \left(P + \frac{\partial P}{\partial x} \delta x + \frac{1}{2} \frac{\partial^2 P}{\partial x^2} (\delta x)^2 + \cdots \right) \tag{6.27}$$

In view of $p + q = 1$, this gives the same equation as Equation 6.25,

$$\frac{\partial P}{\partial t} = -v \frac{\partial P}{\partial x} + D \frac{\partial^2 P}{\partial x^2}, \tag{6.28}$$

where

$$v = (p - q) \frac{\delta x}{\delta t}, \tag{6.29}$$

and

$$D = \frac{1}{2} \frac{(\delta x)^2}{\delta t}. \tag{6.30}$$

The first term on the right-hand side of Equation 6.25 is called the drift term, with v being the velocity of the random walker proportional to the bias in the probabilities to move forward and backward. As we shall see below, in the context of a particle or a polymer molecule undergoing biased random walk, the velocity is due to the presence of external forces. For example, $v = \mu E$, for a charged particle, where E is the electric field along a specified direction (say, pore axis) and the μ is the electrophoretic mobility. The second term on the right-hand side of Equation 6.25 is the diffusion term representing the omnipresent incessant collisions on the particle by the solvent molecules in the solution at finite temperatures. D is called the diffusion coefficient. If the drift is absent, the particle undergoes pure diffusion. Now, the particle does not go anywhere on average, and its mean square displacement follows the Einsteinian dynamics given by Equation 6.1. When drift is present, the average displacement follows the Newtonian dynamics as given by Equation 6.21, and the fluctuations in the position, as captured by Equation 6.22, superpose on top of the drift.

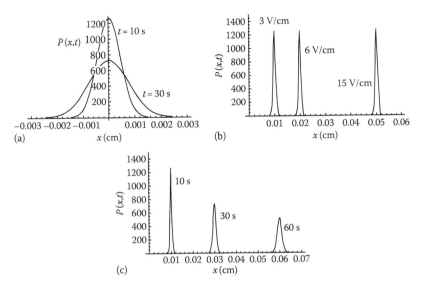

Figure 6.2 Probability distribution function of $\sim 10^4$ bp dsDNA in a buffer solution, with diffusion and electrophoretic drift. (a) Effect of diffusion in the absence of drift. A delta-function-like initial distribution spreads with time $t = 30$, and 60 s for $D = 10^{-8}$ cm^2/s and $v = 0$. (b) Effect of velocity ($v = 0.001, 0.002$, and 0.005 cm/s, corresponding, respectively, to $E = 3, 6$, and 15 V/cm. (c) Time evolution of the probability distribution for $v = 0.001$ cm/s and $D = 10^{-8}$ cm^2/s.

The relative roles of v and D are illustrated in Figure 6.2, by considering a polymer molecule (say, $\sim 10^4$ bp dsDNA in suitable buffer solutions at room temperatures) with $D = 10^{-8}$ cm^2/s and the electrophoretic mobility $\mu \simeq 3.33 \times 10^{-4}$ cm^2/V/s at electric fields of $E = 3, 6$, and 15 V/cm. Figure 6.2a gives the spreading of the probability of finding a particle, initially placed at the origin ($x = 0$), as a function of time in the absence of drift ($v = 0$). As time progresses, the mean square displacement (proportional to the width of the distribution) increases linearly with time, with the average value of x being zero at all times. The effect of drift on the particle's diffusion is presented in Figure 6.2b and c. For a fixed time $t = 10$ s, the combined effects of the velocity of the particle (corresponding to $E = 3, 6$, and 15 V/cm for the present example) and diffusion are given in Figure 6.2b. As the electric field strength increases, the location of the peak increases. The widths of the distribution are controlled by D. The time evolution of the distribution function for a fixed velocity ($v = 0.001$ cm/s, that is $E = 3$ V/cm) is given in Figure 6.2c. As time progresses, the width of the distribution increases due to diffusion, while the location of the peak moving linearly with time in accordance with the drift process.

6.2 BROWNIAN MOTION AND LANGEVIN EQUATION

We now present an alternate way of addressing the same phenomenon of biased random walk. Let us consider a rigid spherical particle of radius R suspended in a solvent (Figure 6.3a). If the radius of the particle is much greater than the size of a single solvent molecule, the solvent background may be taken as a hydrodynamic continuum with some shear viscosity, say η_0. Due to thermal energy, the solvent molecules in the vicinity of the surface of the particle collide perpetually against the particle from all directions, and in response, the particle tries to move in random directions. At any instant of time, the net force from the solvent molecules on the particle is random in both direction and magnitude. The response by the surface of the particle against the solvent molecules at the surface results in friction. There can be an external force $\mathbf{f}_{ext}$ such as the gravitational force acting on the particle as well. The equation of motion of the particle is given by the Newton's equation whereby the inertial force of the particle is balanced by all forces acting on the particle,

$$m\frac{d\mathbf{u}(t)}{dt} = -\zeta \mathbf{u}(t) + \mathbf{f}_{ext} + \boldsymbol{\Gamma}(t), \qquad (6.31)$$

where $\mathbf{u}(t)$ is the instantaneous velocity of the particle, and m and ζ are the mass and the friction coefficient of the particle, respectively. The first term on the right-hand side is the frictional force on the particle. $\boldsymbol{\Gamma}(t)$ is the random force from the perpetual collisions of the solvent molecules on the particle. Note that both $\boldsymbol{\Gamma}(t)$ and ζ arise from random collisions of the solvent molecules on the particle. As shown below, they are related by the fluctuation-dissipation theorem.

The above equation is called the Langevin equation, which is an extension of Newton's law to include random forces. The velocity of the particle, $\mathbf{u}(t)$, as the solution of the above equation is also random, due to the randomness of the force $\boldsymbol{\Gamma}(t)$. Since $\boldsymbol{\Gamma}(t)$ is expected to be highly fluctuating in the timescales of molecular collisions, the details of $\mathbf{u}(t)$, as formal solutions of Equation 6.31,

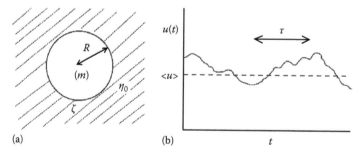

Figure 6.3 (a) A rigid spherical particle suspended in a liquid. (b) The instantaneous velocity of the particle fluctuates about its average with a correlation time of τ.

are also too fine to be precisely computed. However, we expect $\mathbf{u}(t)$ to fluctuate about some average value $\langle \mathbf{u} \rangle$, and the values of $\mathbf{u}(t)$ at two different times t_1 and t_2 to be correlated for $|t_1 - t_2|$ less than some characteristic time τ and uncorrelated for longer intervals (Figure 6.3b). We expect the timescales of self-correlations of the random force $\mathbf{\Gamma}(t)$ and the particle velocity to be separated by many orders of magnitude. We now address the dependence of the characteristic time and the time-correlation functions of $\mathbf{u}(t)$ on m and ζ.

In view of the random nature of $\mathbf{\Gamma}$ and hence $\mathbf{u}$, a solution of Equation 6.31 is implied only in some average sense. In fact, a formal solution of Equation 6.31 can be readily obtained but it is only useful for computing average properties of $\mathbf{u}$ based on the average properties of the random force. As an example of the simplest situation and at the same time of common occurrence in experiments, let us assume that the random force $\mathbf{\Gamma}(t)$ fluctuates independently of $\mathbf{u}(t)$ about zero average and is correlated essentially locally in time,

$$\langle \mathbf{\Gamma}(t) \rangle = 0, \tag{6.32}$$

$$\langle \mathbf{\Gamma}(t) \cdot \mathbf{\Gamma}(t') \rangle = \frac{6k_B T \zeta}{m^2} \delta(t - t'). \tag{6.33}$$

The angular brackets denote equilibrium averages over the distribution function of $\mathbf{\Gamma}(t)$. The locality in time for the correlation of $\mathbf{\Gamma}(t)$ is expressed as the Dirac delta function, $\delta(t - t')$. The random force $\mathbf{\Gamma}(t)$ satisfying the above two equations is generally known as the "white noise" or "Gaussian noise." The coefficient $6k_B T \zeta / m^2$ appearing in Equation 6.33 arises due to the requirement that the mean square velocity of the particle in equilibrium, calculated from Equations 6.31 through 6.33, must be equal to the result from the equipartition theorem,

$$\frac{1}{2} m \langle u^2(t) \rangle = \frac{3}{2} k_B T. \tag{6.34}$$

Since Equation 6.33 relates the correlation function of the fluctuating noise $\mathbf{\Gamma}(t)$ at temperature T with the friction coefficient ζ associated with the frictional dissipation of energy at the particle–solvent interface, it is referred to as the "fluctuation-dissipation theorem." Therefore, the random force and the friction coefficient are not arbitrary but must satisfy Equation 6.33 at a fixed temperature of the system.

6.2.1 Brownian Particle without External Force

The velocity correlation functions and averages related to the displacement of the particle can be derived by constructing these quantities from Equation 6.31, and then averaging the resulting expressions with the use of Equations 6.32 and 6.33. In the absence of any external forces ($\mathbf{f}_{ext} = 0$), the velocity correlation function of the particle is given by (Chandrasekhar 1943, Berne and Pecora 1976)

$$\langle \mathbf{u}(t) \cdot \mathbf{u}(t') \rangle = \frac{3k_BT}{m}e^{-\frac{\zeta}{m}|t-t'|}. \tag{6.35}$$

This is an exponentially decaying function, $\exp(-|t - t'|/\tau)$, with

$$\tau = \frac{m}{\zeta}. \tag{6.36}$$

Since τ is a measure of over how long $\mathbf{u}(t)$ and $\mathbf{u}(t')$ are correlated to each other, it is called the correlation time of the particle. The mean square displacement of the center of mass, $\mathbf{R}_{cm}(t)$, of the particle follows from Equation 6.35 as

$$\langle [\mathbf{R}_{cm}(t) - \mathbf{R}_{cm}(0)]^2 \rangle = \left\langle \left(\int_0^t dt' \mathbf{u}(t') \right)^2 \right\rangle \tag{6.37}$$

$$= \frac{6k_BT}{\zeta} \left[t - \frac{m}{\zeta} \left(1 - e^{-\frac{\zeta}{m}t} \right) \right]. \tag{6.38}$$

The two regimes of dynamics given by Equations 6.1 and 6.2, namely the Einsteinian and Newtonian dynamics, can be recognized from Equation 6.38. For short enough times in comparison with the particle's correlation time $\tau = m/\zeta$, Equation 6.38 reduces to

$$\langle [\mathbf{R}_{cm}(t) - \mathbf{R}_{cm}(0)]^2 \rangle = \frac{3k_BT}{m}t^2 = \langle u^2 \rangle t^2, \quad t \ll \tau, \tag{6.39}$$

where Equation 6.34 is used in writing the last equality. Therefore, the root-mean-square displacement increases linearly with time,

$$\langle [\mathbf{R}_{cm}(t) - \mathbf{R}_{cm}(0)]^2 \rangle^{1/2} = \langle u^2 \rangle^{1/2} t. \quad \text{(Newtonian)} \tag{6.40}$$

Evidently, the Newtonian dynamics emerges when inertial force dominates over the frictional force ($t < m/\zeta$).

On the other hand, for very long times compared to τ, Equation 6.38 becomes

$$\langle [\mathbf{R}_{cm}(t) - \mathbf{R}_{cm}(0)]^2 \rangle = \frac{6k_BT}{\zeta}t, \quad t \gg \tau. \tag{6.41}$$

In the limit of long enough times, the dynamics of the particle is Einsteinian where the mean square displacement (and not the root-mean-square displacement) is proportional to time,

$$\langle [\mathbf{R}_{cm}(t) - \mathbf{R}_{cm}(0)]^2 \rangle = 6Dt, \quad \text{(Einsteinian)} \tag{6.42}$$

where

$$D = \frac{k_B T}{\zeta}. \qquad (6.43)$$

Now, the particle is said to undergo diffusion, or Brownian dynamics, with the diffusion constant D defined by Equation 6.43. The Einsteinian dynamics emerges when the frictional force dominates over the inertial force ($t > m/\zeta$). In writing Equation 6.43, we have only defined D in terms of ζ. A derivation of this Einsteinian result is given in Section 6.4.

The friction coefficient ζ of the particle, which is a measure of the frictional resistance by the particle against the solvent, was derived by Stokes a long time ago (1851) to depend linearly on the particle radius and viscosity of the solvent (Landau and Lifshitz 1959),

$$\zeta = 6\pi \eta_0 R. \qquad (6.44)$$

The prefactor in this equation is for the nonslip boundary condition, where the velocity of the particle is the same as that of the solvent at its surface. The basic input in the derivation of the above Stokes law is how the velocity field in the fluid is modified by a surface element of the particle and how this modified field gets further modified by other surface elements, while the surface elements remain structurally correlated in size and shape (Figure 6.4). It may be equivalently stated that the Stokes law is a consequence of full hydrodynamic coupling among all surface elements of the sphere arising from velocity

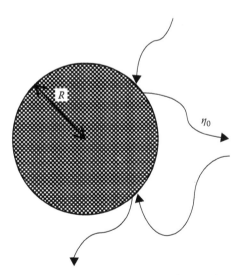

Figure 6.4 Origin of the Stokes law of friction lies in the correlated velocity fluctuations in the fluid arising from the surface of the particle with no-slip.

fluctuations in the solvent continuum. Combining Equations 6.43 and 6.44, we get the "Stokes–Einstein" law,

$$D = \frac{k_B T}{6\pi \eta_0 R}. \quad \text{(Stokes–Einstein)} \tag{6.45}$$

The key message is that the diffusion coefficient depends inversely on the linear size of the particle and not on its surface area or volume.

6.2.2 Brownian Particle under an External Force

Returning to the Langevin equation (Equation 6.31), for a Brownian particle under an external force, let us consider the one-dimensional version with the external force $\mathbf{f}_{ext} \equiv f(x)$ acting at the location x of the particle,

$$m\frac{du}{dt} = -\zeta u + f(x) + \sqrt{\zeta k_B T}\Gamma(t), \tag{6.46}$$

where u is now the x-component of $\mathbf{u}$. The x-component of the random force is defined as $\sqrt{\zeta k_B T}\Gamma(t)$, where $\Gamma(t)$ is random. Ignoring the inertia and noting that $dx/dt = u$, we get

$$\frac{dx}{dt} = \frac{f(x)}{\zeta} + \left(\frac{k_B T}{\zeta}\right)^{1/2}\Gamma(t). \tag{6.47}$$

It must be remarked that the coefficient of the noise term in Equation 6.46 was chosen so that the fluctuation-dissipation theorem for $\Gamma(t)$ satisfies the equations,

$$\langle \Gamma(t) \rangle = 0 \tag{6.48}$$

$$\langle \Gamma(t)\Gamma(t') \rangle = 2\delta(t - t'), \tag{6.49}$$

without any physical parameters appearing as prefactors.

We shall encounter Equation 6.47 and its solution given in the following sections many times in later chapters in the context of translocation. The external force $f(x)$ can arise generally from free energy barriers along the pore and the electric forces driving polymer translocation.

6.2.3 General Form of Langevin Equation

The above two examples of a Brownian particle are representative of a large class of stochastic processes. Let us consider a general stochastic process $\xi(t)$, without any memory, where the random variable ξ is time-dependent and the noise is white noise. The general Langevin equation describing the stochastic process $\xi(t)$ can be written as (Risken 1989)

$$\frac{d\xi}{dt} = h(\xi, t) + g(\xi, t)\Gamma(t), \tag{6.50}$$

where the Gaussian random noise $\Gamma(t)$ satisfies the following relations

$$\langle \Gamma(t) \rangle = 0 \tag{6.51}$$

$$\langle \Gamma(t)\Gamma(t') \rangle = 2\delta(t - t'). \tag{6.52}$$

The angular brackets indicate averaging over a sufficiently long period of time or equivalently an equilibrium average. $h(\xi, t)$ and $g(\xi, t)$ are functions corresponding to the specific stochastic process. With the definition of the correlation function given in Equation 6.52, the noise strength is absorbed into $g(\xi, t)$.

As an example, for the Brownian particle under an external force discussed in Section 6.2.2,

$$\xi = x \tag{6.53}$$

$$h = \frac{f(x)}{\zeta} \tag{6.54}$$

$$g = \sqrt{\frac{k_B T}{\zeta}} = \sqrt{D}, \tag{6.55}$$

where Equation 6.43 is used.

6.3 FOKKER–PLANCK–SMOLUCHOWSKI EQUATION

As seen above, a solution of the Langevin equation (Equation 6.50) (which is a nonlinear partial differential equation with random noise) consists of constructing the correlation functions of $\xi(t)$ from the equation and then averaging the expressions with the help of the properties of the noise $\Gamma(t)$. An alternative method of solution is to find the probability distribution function $P(x, t)$ for realizing a situation in which the random variable $\xi(t)$ has the particular value x at time t. $P(x, t)$ is an equivalent description of the stochastic process $\xi(t)$ and is given by the Fokker–Planck equation (Chandrasekhar 1943, Gardiner 1985, Risken 1989, Redner 2001, Mazo 2002)

$$\frac{\partial P(x, t)}{\partial t} = -\frac{\partial}{\partial x} J(x, t) \tag{6.56}$$

$$J(x, t) = A(x, t)P(x, t) - \frac{\partial}{\partial x}[B(x, t)P(x, t)]. \tag{6.57}$$

The mathematical form of this equation is exactly the same as Equation 6.25, with $J(x, t)$ being the flux of the probability distribution function, and the first and second terms on the right-hand side of Equation 6.57 being the drift and diffusion terms, respectively. For the general stochastic process given by

Equation 6.50, the drift and diffusion coefficients are (Risken 1989)

$$A(x,t) = h(x,t) + g(x,t)\frac{\partial g(x,t)}{\partial x} \qquad (6.58)$$

$$B(x,t) = g^2(x,t). \qquad (6.59)$$

As a specific example pertinent to our interests in polymer translocation, let us consider a Brownian particle undergoing diffusion in the presence of an external field arising from free energy barriers or/and an electric field, as described in Section 6.2.2. In view of Equations 6.53 through 6.55, we get

$$A(x,t) = \frac{f(x)}{\zeta}, \qquad (6.60)$$

$$B(x,t) = D, \qquad (6.61)$$

and the flux of the probability distribution is given by

$$J(x,t) = \frac{f(x)}{\zeta}P(x,t) - D\frac{\partial P(x,t)}{\partial x}. \qquad (6.62)$$

Using the Einsteinian relation (Equation 6.43), we get from Equations 6.56 and 6.62

$$\frac{\partial P(x,t)}{\partial t} = D\left\{ -\frac{\partial}{\partial x}\left[\frac{f(x)}{k_BT}P(x,t)\right] + \frac{\partial^2 P(x,t)}{\partial x^2} \right\}. \qquad (6.63)$$

This equation is also known as the Smoluchowski equation. If the force is a constant, then Df/k_BT is the uniform velocity of the particle. Now, the above equation is identical to Equation 6.25 derived in Section 6.1 for a biased random walk.

One of the main tools in describing the translocation process is embodied in Equation 6.63. When there is a free energy landscape $F(x)$ along the trajectory of the diffusing polymer molecules along the pore, the external force $f(x)$ is

$$f(x) = -\frac{\partial F(x)}{\partial x}. \qquad (6.64)$$

If there is an electrochemical potential gradient $\Delta\mu$ driving the mobility of the polymer across the pore, then $F(x)$ is directly related to $\Delta\mu$. In fact, all contributing factors to the interaction between the pore and the polymer and the electrochemical potential gradients across the pore can be combined into an effective free energy $F(x)$. Substitution of Equation 6.64 into Equation 6.63 yields

$$\frac{\partial P(x,t)}{\partial t} = D\left\{ \frac{\partial}{\partial x}\left[\frac{1}{k_BT}\frac{\partial F(x)}{\partial x}P(x,t)\right] + \frac{\partial^2 P(x,t)}{\partial x^2} \right\}. \qquad (6.65)$$

The input $F(x)$ in Equation 6.65 depends on the specifics of the transport process. For a given stochastic process of the general type given by Equation 6.50, or for a prescribed free energy landscape, Equation 6.65 is to be solved with appropriate boundary conditions. From such calculations, details about the probability distribution function and averages of the quantities associated with the translocation process can be obtained. We shall return to this calculational tool repeatedly for different experimental situations to be discussed in later chapters.

6.4 COLLECTION OF BROWNIAN PARTICLES

The Smoluchowski equation (Equation 6.63) for the probability distribution function of a diffusing particle under the influence of an external force is also applicable for the time evolution of a collection of particles. Let $c(\mathbf{r}, t)$ be the local concentration of the particles at the location $\mathbf{r}$ at time t. In view of the conservation of the total number of particles in the system, the rate of change of local particle concentration in a volume element is given by the continuity equation

$$\frac{\partial c(\mathbf{r}, t)}{\partial t} = -\nabla \cdot \mathbf{J}(\mathbf{r}, t), \tag{6.66}$$

where $\mathbf{J}(\mathbf{r}, t)$ is the flux of the particles into the volume element. The flux has two parts: diffusive and convective. When the particles behave independently of each other, the diffusive part is given by the Fick's law, $-D\nabla c(\mathbf{r}, t)$. The convective part is $c(\mathbf{r}, t)\mathbf{u}(\mathbf{r}, t)$, where $\mathbf{u}(\mathbf{r}, t)$ is the velocity of the particle. Therefore,

$$\mathbf{J}(\mathbf{r}, t) = -D\nabla c(\mathbf{r}, t) + c(\mathbf{r}, t)\mathbf{u}(\mathbf{r}, t). \tag{6.67}$$

In the presence of an external force $\mathbf{f}$, the velocity of the particle is given in the friction-dominated situations as

$$\mathbf{u} = \frac{1}{\zeta}\mathbf{f}, \tag{6.68}$$

where ζ is the friction coefficient of the particle. The force itself can be written as the negative gradient of the potential U, or as a generalization, that of the free energy $F(\mathbf{r})$,

$$\mathbf{f}(\mathbf{r}) = -\nabla U(\mathbf{r}) = -\nabla F(\mathbf{r}). \tag{6.69}$$

We have taken the potential and the free energy landscape acting on the particle to be time-independent. By substituting Equations 6.68 and 6.69 into Equation 6.67, we obtain

$$\mathbf{J}(\mathbf{r}, t) = -D\nabla c(\mathbf{r}, t) + \frac{1}{\zeta}c(\mathbf{r}, t)\mathbf{f}(\mathbf{r}, t), \tag{6.70}$$

and

$$\mathbf{J}(\mathbf{r}, t) = -D\nabla c(\mathbf{r}, t) - \frac{1}{\zeta} c(\mathbf{r}, t)\nabla F(\mathbf{r}). \tag{6.71}$$

In equilibrium, the flux must be zero, and the solution of Equation 6.71 must be the Boltzmann distribution function for the concentration of the particles at the location $\mathbf{r}$ where the effective potential is $F(\mathbf{r})$,

$$c(\mathbf{r}) \sim e^{-\frac{F(\mathbf{r})}{k_B T}}. \tag{6.72}$$

The stipulation that Equation 6.71 must be consistent with the equilibrium result of Equation 6.72 results in the Einsteinian result of Equation 6.43,

$$D = \frac{k_B T}{\zeta}. \tag{6.73}$$

The above description is a derivation of the Einsteinian result introduced in Equation 6.43. Using this result, the flux can be written as

$$\mathbf{J}(\mathbf{r}, t) = -D\left[\nabla c(\mathbf{r}, t) + c(\mathbf{r}, t)\nabla\left(\frac{F(\mathbf{r})}{k_B T}\right)\right]. \tag{6.74}$$

For a one-dimensional situation along the x-coordinate, the continuity equation is

$$\frac{\partial c(x, t)}{\partial t} = -\frac{\partial J(x, t)}{\partial x}, \tag{6.75}$$

and the flux is

$$J(x, t) = -D\left[\frac{\partial c(x, t)}{\partial x} + c(x, t)\frac{\partial}{\partial x}\left(\frac{F(x)}{k_B T}\right)\right], \tag{6.76}$$

and substitution of Equation 6.76 into Equation 6.75 gives

$$\frac{\partial c(x, t)}{\partial t} = D\left\{\frac{\partial}{\partial x}\left[\frac{1}{k_B T}\frac{\partial F(x)}{\partial x}c(x, t)\right] + \frac{\partial^2 c(x, t)}{\partial x^2}\right\}. \tag{6.77}$$

This is exactly the same Smoluchowski equation as Equation 6.65 for the probability distribution function for finding one particle at the location x and time t. We shall return to Equation 6.77 when we address ionic currents through nanopores.

6.5 EQUILIBRIUM VERSUS STEADY STATE

The spatial variation of concentration of the suspended particles is time-independent when the right-hand side of Equation 6.66 is zero. This can arise when either the flux $\mathbf{J}$ is zero or a constant. In equilibrium, $\mathbf{J}$ is zero, and there are no net flows of the particles into a volume element. When **J** is a constant, the system is in a steady state. Now, there are sources and sinks respectively, to generate and deplete the particles into the system in order to maintain a steady state. The consequences of equilibrium and steady-state conditions on the system can be quite different, as illustrated below for one-dimensional systems. In both circumstances, the concentration profile is independent of time. The same argument for the flux of the particles is also valid for the flux of the probability of a particle negotiating a free energy landscape.

6.5.1 Equilibrium

When a system is in equilibrium, the flux is zero. We get from Equation 6.76

$$\frac{1}{c(x)}\frac{\partial c(x)}{\partial x} = -\frac{\partial}{\partial x}\left(\frac{F(x)}{k_B T}\right). \tag{6.78}$$

Upon integration, we get

$$c(x) = c_0 e^{-\frac{F(x)}{k_B T}}, \tag{6.79}$$

where c_0 is the uniform particle concentration in the absence of the external potential $U(x)$ or the free energy $F(x)$ acting on the particle at x (Figure 6.5). This is the Boltzmann distribution valid for all equilibrium systems.

6.5.2 Steady State

By rearranging Equation 6.76, we get

$$J(x) = -De^{-\frac{F(x)}{k_B T}}\frac{\partial}{\partial x}\left[c(x)e^{\frac{F(x)}{k_B T}}\right]. \tag{6.80}$$

In the steady state, J is constant at all values of x. As a specific example, let us consider a domain of x between 0 and L, where $F(x)$ is a known function (Figure 6.5). In reality, this domain may correspond to the length of a pore and the $F(x)$ is the corresponding free energy profile for the mobility of the particle

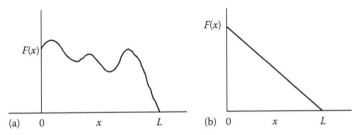

Figure 6.5 (a) An example of free energy profiles along the x-direction between $x = 0$ and $x = L$. (b) Free energy profile with a constant force.

along the pore. Integration of the above equation, with the assumption that D is a constant, yields

$$J = -D\frac{(c(L)e^{F(L)/k_BT} - c(0)e^{F(0)/k_BT})}{\int_0^L dx e^{F(x)/k_BT}}. \quad (6.81)$$

Here, the steady-state flux J is a constant, independent of x and t. It depends only on the free energy landscape, duration of the domain, and the diffusion constant of the particle.

The steady-state concentration profile $c(x)$ is obtained by integrating Equation 6.80,

$$-\int_0^x dx' \left(\frac{J}{D}\right) e^{F(x)/k_BT} = c(x)e^{F(x)/k_BT} - c(0)e^{F(0)/k_BT}. \quad (6.82)$$

Rearrangement of this equation with constant J and D yields

$$c(x) = e^{-F(x)/k_BT}\left[c(0)e^{F(0)/k_BT} - \left(\frac{J}{D}\right)\int_0^x dx' e^{F(x)/k_BT}\right]. \quad (6.83)$$

The particle concentration at x is given by the above equation with the substitution of Equation 6.81 for the constant flux J. It is readily seen from Equations 6.79 and 6.83 that the concentration profiles in the steady state and the equilibrium state are different.

6.6 FINITE BOUNDARIES AND FIRST PASSAGE TIME

Let us consider a particle undergoing a drift–diffusion stochastic process given by the Fokker–Planck equation (Equations 6.56 and 6.57)

$$\frac{\partial P(x,t)}{\partial t} = -\frac{\partial}{\partial x}[A(x)P(x,t)] + \frac{\partial^2}{\partial x^2}[B(x)P(x,t)]. \quad (6.84)$$

Let the initial position of the particle at $t = 0$ be x_0 (Figure 6.6a). When the particle reaches the boundaries at $x = a$ and $x = b$, the particle can either disappear from the system, or get reflected back into the system, or undergo a mixture of partial absorption and partial reflection. The basic question we ask is how long does it take for the particle to reach one of the boundaries without ever reaching the other boundary. This is called the first passage time (FPT). This itself is a stochastic quantity reflecting the stochastic nature of the drift–diffusion process of the particle. A couple of representative trajectories of the particle is sketched in Figure 6.6b. The corresponding FPTs are τ' and τ''. We seek the probability distribution for the FPT and the average FPT. Also, we are interested in the probability that the particle would exit through a specified boundary as a function of the initial location. The answers to all of these

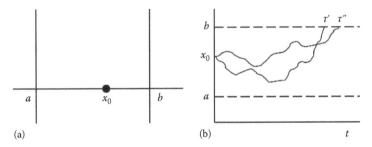

Figure 6.6 (a) A particle confined initially at x_0 between two barriers. (b) Two examples of trajectories of a drifting Brownian particle with the FPT being τ' and τ''.

questions depend on the drift $(A(x))$ and diffusion $(B(x))$ coefficients entering in the Fokker–Planck equation as well as the boundary conditions. After summarizing the general equations and results, we illustrate their utility for a simple example of a particle moving with a constant velocity and undergoing diffusion with a constant diffusion coefficient.

6.6.1 General Equations

Let the probability distribution function of the FPT (τ) be $g(x_0, \tau)$ and its moments are defined as

$$\tau_n(x_0) = \langle \tau^n \rangle = \int_0^\infty d\tau \, \tau^n g(x_0, \tau). \tag{6.85}$$

The probability distribution function can be shown to be related to the transition probability $p(x, t; x', t')$ as (Cox and Miller 1968, Gardiner 1985, Risken 1989, Redner 2001)

$$g(x_0, \tau) = -\frac{d}{d\tau} \int_a^b dx p(x, \tau; x_0, 0). \tag{6.86}$$

The transition probability $p(x, t; x', t')$ is the conditional probability of the process realizing the value x at time t if it has the value x' at another time t'. For stochastic processes without any memory (Markov processes), $p(x, t; x', t')$ satisfies the same equation as $P(x, t)$,

$$\frac{\partial p(x, t; x_0, 0)}{\partial t} = -\frac{\partial}{\partial x}[A(x)p(x, t; x_0, 0)] + \frac{\partial^2}{\partial x^2}[B(x)p(x, t; x_0, 0)], \tag{6.87}$$

or, alternatively (Cox and Miller 1968),

$$\frac{\partial p(x, t; x_0, 0)}{\partial t} = A(x_0)\frac{\partial}{\partial x_0}p(x, t; x_0, 0) + B(x_0)\frac{\partial^2}{\partial x_0^2}p(x, t; x_0, 0). \tag{6.88}$$

The above equations are known as the forward and backward Fokker–Planck equations. For a particular stochastic process (namely, prescribed $A(x)$ and $B(x)$), $p(x, t; x_0, 0)$ needs to be calculated with appropriate boundary conditions (Gardiner 1985). Substituting this result in Equation 6.86, $g(x_0, \tau)$ is then obtained.

The leading moments of the FPT distribution are the mean FPT,

$$\tau_1(x_0) = \langle \tau \rangle, \tag{6.89}$$

and

$$\tau_2(x_0) = \langle \tau^2 \rangle. \tag{6.90}$$

Both τ_1 and τ_2 are functions of the initial location x_0. For the general stochastic process of Equation 6.84, τ_1 and τ_2 obey (Cox and Miller 1968, Gardiner 1985)

$$\left\{ A(x_0) \frac{\partial}{\partial x_0} + B(x_0) \frac{\partial^2}{\partial x_0^2} \right\} \tau_1(x_0) = -1, \tag{6.91}$$

and

$$\left\{ A(x_0) \frac{\partial}{\partial x_0} + B(x_0) \frac{\partial^2}{\partial x_0^2} \right\} \tau_2(x_0) = -2\tau_1(x_0). \tag{6.92}$$

τ_1 and τ_2 are obtained by solving Equations 6.91 and 6.92 for any set of $A(x)$ and $B(x)$ with appropriate boundary conditions.

In the above equations, we have been considering the unconditional FPTs irrespective of which boundary at which the particle exits. We are also interested in the likelihood of exits through a particular boundary. For example, we can represent successful translocation events out of the pore as exits through an absorbing boundary at $x = b$. Let $\pi_b(x_0)$ be the probability and $\langle \tau_b(x_0) \rangle$ be the mean exit time for the particle to exit through $x = b$, starting from its initial location x_0. It can be shown that $\pi_b(x_0)$ and $\langle \tau_b(x_0) \rangle$ satisfy the equations (Gardiner 1985)

$$\left\{ A(x_0) \frac{\partial}{\partial x_0} + B(x_0) \frac{\partial^2}{\partial x_0^2} \right\} \pi_b(x_0) = 0, \tag{6.93}$$

and

$$\left\{ A(x_0) \frac{\partial}{\partial x_0} + B(x_0) \frac{\partial^2}{\partial x_0^2} \right\} [\pi_b(x_0) \langle \tau_b(x_0) \rangle] = -\pi_b(x_0), \tag{6.94}$$

with the boundary conditions

$$\pi_b(a) = 0, \quad \pi_b(b) = 1, \tag{6.95}$$

and

$$\pi_b(a) \langle \tau_b(a) \rangle = 0 = \pi_b(b) \langle \tau_b(b) \rangle. \tag{6.96}$$

The probability of exit through the other boundary, $x = a$, and the corresponding mean exit time $\langle \tau_a(x_0) \rangle$ are given by the same equations as above by interchanging the labels a and b. Naturally,

$$\pi_a(x_0) + \pi_b(x_0) = 1. \tag{6.97}$$

6.6.2 General Solutions

The mean FPT for a given stochastic process, specified by the drift and diffusion coefficients $A(x)$ and $B(x)$, is obtained by solving Equation 6.91, which requires two boundary conditions. Let us take the boundary at $x_0 = b$ to be absorbing so that $\tau_1(b) = 0$. For the other boundary condition at $x_0 = a$, we consider three possibilities. In the first case, $BC1$, we assume that the boundary is absorbing ($\tau_1(a) = 0$). In the second case, $BC2$, the boundary is reflecting, that is, $\tau_1'(a) = 0$ (where τ_1' is the first derivative of τ_1 with respect to x). In the third case, $BC3$, we assume a mixed boundary condition, usually called the radiation boundary condition,

$$-\tau_1'(a) + h\tau_1(a) = 0, \tag{6.98}$$

where h is a parameter representing the relative weight of absorption versus reflection. This boundary condition reduces, respectively, to the reflective and absorbing boundary conditions for $h = 0$ and $h \to \infty$. These three cases of boundary conditions at $x = a$ are summarized in Table 6.1.

The general results for the mean FPT are given below for the three boundary conditions. First, we define new functions $\psi(x)$ and $H(x)$ in terms of the drift and diffusion coefficients,

$$\psi(x) \equiv \exp\left(\int_a^x dx' \frac{A(x')}{B(x')} \right), \tag{6.99}$$

and

$$H(x) \equiv -\int_a^x dy \frac{\psi(y)}{B(y)}. \tag{6.100}$$

TABLE 6.1

Three Boundary Conditions for Mean FPT

BC1	$x = a$	Absorbing	$\tau_1(a) = 0$
	$x = b$	Absorbing	$\tau_1(b) = 0$
BC2	$x = a$	Reflecting	$\tau_1'(a) = 0$
	$x = b$	Absorbing	$\tau_1(b) = 0$
BC3	$x = a$	Radiating	$-\tau_1'(a) + h\tau_1(a) = 0$
	$x = b$	Absorbing	$\tau_1(b) = 0$

The results for the mean FPT are as follows. For $BC1$,

$$\tau_1(x_0) = \frac{\left[\left(\int_{x_0}^{b}\frac{dy}{\psi(y)}\right)\int_{a}^{x_0}\frac{dy'}{\psi(y')}H(y') - \left(\int_{a}^{x_0}\frac{dy}{\psi(y)}\right)\int_{x_0}^{b}\frac{dy'}{\psi(y')}H(y')\right]}{\int_{a}^{b}\frac{dy}{\psi(y)}}. \quad (6.101)$$

For $BC2$,

$$\tau_1(x_0) = \int_{b}^{x_0}\frac{dy}{\psi(y)}H(y). \quad (6.102)$$

For $BC3$,

$$\tau_1(x_0) =$$

$$\frac{-\int_{x_0}^{b}\frac{dy'}{\psi(y')}H(y') - h\left(\int_{a}^{x_0}\frac{dy}{\psi(y)}\right)\int_{x_0}^{b}\frac{dy'}{\psi(y')}H(y') + h\left(\int_{x_0}^{b}\frac{dy}{\psi(y)}\right)\int_{a}^{x_0}\frac{dy'}{\psi(y')}H(y')}{1 + h\int_{a}^{b}\frac{dy}{\psi(y)}}.$$

$$(6.103)$$

The results of $BC1$ and $BC2$ are recovered from Equation 6.103 in the limits of $h \to \infty$ and $h = 0$, respectively. Analogous equations can be derived for $\tau_2(x_0)$ (which is a measure of the width of the distribution function for the FPT) for the three boundary conditions.

The conditional probability that the particle obeying the stochastic process of Equation 6.84 exits through the boundary at $x = b$ is obtained by solving Equation 6.93 with the boundary conditions of Equation 6.95 as

$$\pi_b(x_0) = \frac{\Psi(a, x_0)}{\Psi(a, b)}, \quad (6.104)$$

where

$$\Psi(x, y) \equiv \int_{x}^{y}\frac{dz}{\psi(z)}. \quad (6.105)$$

Similarly, the conditional probability to exit through the other boundary at $x = a$ is $\pi_a(x_0) = 1 - \pi_b(x_0)$ as given by

$$\pi_a(x_0) = \frac{\Psi(x_0, b)}{\Psi(a, b)}. \quad (6.106)$$

The mean FPT $\tau_b(x_0)$ for the particle to reach the boundary at $x = b$, without ever reaching the other boundary at $x = a$, can be derived as

$$\tau_b(x_0) = \frac{\Psi(a, x)\Phi_+(x, b) - \Phi_+(a, x)\Psi(x, b)}{\Psi(a, x)\Psi(a, b)}, \quad (6.107)$$

where

$$\Phi_+(p,q) = \int_p^q \frac{dy}{\psi(y)} \int_a^y dz \frac{\psi(z)\Psi(a,z)}{B(z)}. \tag{6.108}$$

Similarly, the mean FPT $\tau_a(x_0)$ for the particle to reach the boundary at $x = a$, without ever reaching the other boundary at $x = b$, is

$$\tau_a(x_0) = \frac{\Psi(a,x)\Phi_-(x,b) - \Phi_-(a,x)\Psi(x,b)}{\Psi(x,b)\Psi(a,b)}, \tag{6.109}$$

with

$$\Phi_-(p,q) = \int_p^q \frac{dy}{\psi(y)} \int_a^y dz \frac{\psi(z)\Psi(z,b)}{B(z)}. \tag{6.110}$$

We now have a compendium of key equations for the probability of FPT, mean FPT, and probability of successful escape through a particular exit for a general stochastic process obeying Equation 6.84. The drift coefficient $A(x)$ depends on the free energy landscape where the particle is undergoing drift–diffusion process (Equations 6.60 and 6.64). The diffusion coefficient $B(x)$ of the particle is taken to be uniform in most of the cases and it can in principle be position-dependent (Equation 6.61). By inputting the details of the free energy landscape for a specific system, the experimentally relevant quantities associated with the mobility of particles can be calculated from the above formulas.

6.7 PROPERTIES OF DRIFT–DIFFUSION PROCESS

We now illustrate the utility of the above equations by considering situations of stochastic processes where the coefficients $A(x)$ and $B(x)$ in Equations 6.56 and 6.57 are, respectively, constant velocity and constant diffusion coefficient,

$$A(x) = v, \quad B(x) = D. \tag{6.111}$$

The probability distribution function obeys the simple drift–diffusion equation (Equations 6.25 and 6.63),

$$\frac{\partial P(x,t)}{\partial t} = -v\frac{\partial P(x,t)}{\partial x} + D\frac{\partial^2 P(x,t)}{\partial x^2}. \tag{6.112}$$

Insight into the various physical aspects of the drift–diffusion process can be easily gained by considering the following situations. Let the initial position of the particle at $t = 0$ be at $x = x_0$. First, we treat the consequences of erecting one absorbing barrier at $x = b > x_0$, with the velocity of the particle being in the positive direction of x (Figure 6.7a). This situation is pertinent to the capture

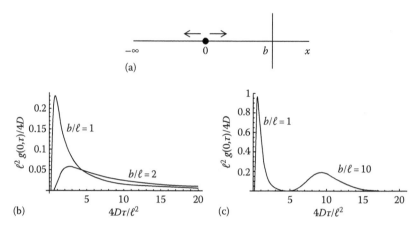

Figure 6.7 (a) A diffusion–drift process of a particle along the x-direction in the presence of an absorbing boundary at $x = b$. (b) Probability distribution function $g(0, \tau)\ell^2/4D$ of the FPT τ to the absorbing boundary at $b/\ell = 1$ and 2 with no drift. (c) Same as in (b) for $b/\ell = 1$ and 10 in the presence of drift, $v\ell/4D = 1$.

of a particle, undergoing a drift–diffusion process, by an absorbing trap. Next, we consider the effect of two barriers (Figure 6.6) at $x = 0$ and $x = L$, with the particle initially confined between these two boundaries ($0 < x < L$). Let the boundary at $x = L$ be absorbing. The other boundary at $x = 0$ satisfies one of the three boundary conditions ($BC1, BC2$, and $BC3$) given in Table 6.1. This situation is pertinent to the escape of a particle initially trapped inside a pore, out of the two pore exits.

6.7.1 One Absorbing Barrier

The solution of Equation 6.112, with the boundary condition that $P(x = b, t) = 0$, follows from standard techniques in solving partial differential equations (Sommerfeld 1949) as

$$P(x, t) = \frac{1}{\sqrt{4\pi Dt}} \left\{ \exp\left[-\frac{(x - vt)^2}{4Dt}\right] - \exp\left[\frac{vb}{D} - \frac{(x - 2b - vt)^2}{4Dt}\right] \right\}.$$
(6.113)

Substituting this result in Equation 6.86, the probability distribution function for the FPT is given by

$$g(0, \tau) = \frac{b}{\sqrt{4\pi D\tau^3}} \exp\left[-\frac{(b - v\tau)^2}{4D\tau}\right],$$
(6.114)

where the initial position of the particle is taken to be at the origin, $x = 0$. In the absence of drift, this equation reduces to

$$g(0, \tau) = \frac{b}{\sqrt{4\pi D\tau^3}} \exp\left[-\frac{b^2}{4D\tau}\right], \tag{6.115}$$

where b is the distance of the absorbing barrier from the starting point of the particle. Let ℓ be the length of each step of the particle. By defining the dimensionless time $\tilde{\tau} \equiv 4D\tau/\ell^2$, $\ell^2 g(0, \tau)/4D$ is plotted in Figure 6.7b for b/ℓ and 2, in the absence of drift. Thus, for a particle undergoing pure diffusion, the probability of the FPT is very small except when the absorbing barrier is very close to the starting point of the particle. The effect of drift is presented in Figure 6.7c for the dimensionless velocity $u \equiv v\ell/4D = 1$. In the presence of drift, the probability of the FPT along the direction of drift is nonnegligible even if the starting point of the particle is away from the boundary. Also, the probability to reach the boundary is higher in the presence of drift compared to the situation with only diffusion. Since $g(0, \tau)$ is proportional to $\tau^{-3/2}$ for $\tau \to \infty$, moments of the FPT cannot be defined for a biased random walk with one absorbing boundary.

6.7.2 Two Absorbing Barriers

We now consider a particle initially located at $x = x_0$ between the boundaries at $x = 0$ and $x = L$ (Figure 6.6a, with $a = 0$ and $b = L$). Let both boundaries be absorbing, that is, $P(0, t) = 0 = P(L, t)$. By solving Equations 6.86 and 6.87 with these boundary conditions, $g(x_0, \tau)$ can be derived as

$$g(x_0, \tau) = \frac{2}{L} \sum_{p=1}^{\infty} \beta_p \left[1 - \cos(\beta_p L) \exp\left(\frac{vL}{2D}\right)\right]$$

$$\times \sin(\beta_p x_0) \exp\left[-\frac{vx_0}{2D} - \left(\beta_p^2 + \frac{v^2}{4D^2}\right)D\tau\right], \tag{6.116}$$

where $\beta_p = \pi p/L$. The above infinite series can be approximated by

$$g(x_0, \tau) = \frac{1}{L^2\sqrt{4\pi\tilde{\tau}^3}} \exp\left(-\frac{\alpha\tilde{x}_0}{2} - \frac{\alpha^2\tilde{\tau}}{4}\right) \left\{\tilde{x}_0 \exp\left(-\frac{\tilde{x}_0^2}{4\tilde{\tau}}\right)\right.$$

$$-\frac{1}{2} \exp\left(\frac{\alpha}{2}\right) \left[(1 + \tilde{x}_0) \exp\left(-\frac{(1 + \tilde{x}_0)^2}{4\tilde{\tau}}\right)\right.$$

$$\left.\left. - (1 - \tilde{x}_0) \exp\left(-\frac{(1 - \tilde{x}_0)^2}{4\tilde{\tau}}\right)\right]\right\}, \tag{6.117}$$

with

$$\tilde{\tau} \equiv \frac{D\tau}{L^2}, \quad \tilde{x}_0 \equiv \frac{x_0}{L}, \quad \alpha \equiv \frac{vL}{D}. \tag{6.118}$$

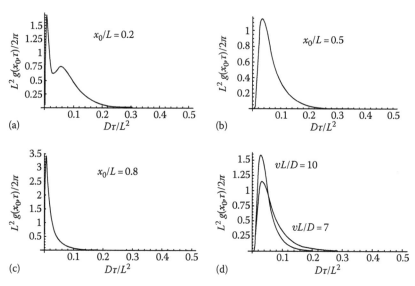

Figure 6.8 Probability distribution function $g(x_0, \tau)L^2/2\pi$ of the FPT versus $D\tau/L^2$. The initial position of the particle is at x_0 and the two absorbing boundaries are separated by L. (a), (b), and (c) correspond to $x_0/L = 0.2, 0.5$, and 0.8, respectively, for the drift strength $vL/D = 7$. (d) The role of $vL/D = 7$ and 10 for $x_0/L = 0.5$.

Some representative results from Equation 6.117 are given in Figure 6.8. The dependence of the probability distribution function of the FPT, $L^2 g(x_0, \tau)/2\pi$ on $D\tau/L^2$ is given in Figure 6.8a through c, for the initial positions at $x_0/L = 0.2, 0.5$, and 0.8 at $vL/D = 7$. It is seen from Figure 6.8a that there are two peaks in the distribution when the particle is close to the boundary upstream. The larger peak represents exit through $x = 0$ and the other peak represents the exit through $x = L$. For $x_0/L = 0.5$, there is only one peak (Figure 6.8b). The probability distribution function is asymmetric in shape; it rises sharply, reaches a maximum, and then decreases roughly algebraically. The same shape of the distribution function is seen for $x_0/L = 0.8$, that is, the initial position is close to the downstream boundary (Figure 6.8c). Now, the peak position occurs at a shorter time than in the case of Figure 6.8b. The effect of velocity is illustrated in Figure 6.8d, for $x_0/L = 0.5$. The distribution function for the FPT is compressed to shorter times at higher velocities.

The mean FPT is obtained from Equation 6.91 or Equations 6.99 through 6.101 as

$$\tau_1(x_0) = \frac{1}{v} \frac{L(1 - e^{-vx_0/D}) - x_0(1 - e^{-vL/D})}{(1 - e^{-vL/D})}. \tag{6.119}$$

The limits of this equation for $v \to 0$ and $v \to \infty$ are

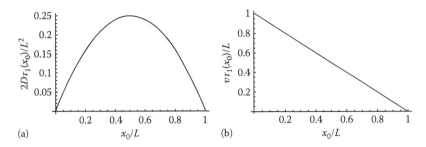

Figure 6.9 Dependence of the mean FPT $\tau_1(x_0)$ on the initial particle position for two absorbing boundaries separated by distance L: (a) diffusion-limited regime and (b) drift-limited regime.

$$\tau_1(x_0) = \begin{cases} \frac{x_0(L-x_0)}{2D}, & v/D \to 0 \\ \frac{L-x_0}{v}, & v/D \to \infty. \end{cases} \qquad (6.120)$$

The diffusion-limited result for the mean FPT $(2D\tau_1(x_0)/L^2)$ is plotted in Figure 6.9a as a function of x_0/L. The drift-limited result is given in Figure 6.9b as a plot of $v\tau_1(x_0)/L$ versus x_0/L. In this limit, the time to reach the boundary downstream is simply the distance to be traveled divided by the velocity. Equation 6.120 gives the crossover behavior between the two limits provided in Figure 6.9.

The conditional probability to exit the boundary at $x = L$, without ever reaching the other boundary, is

$$\pi_+(x_0) = \frac{(1 - e^{-vx_0/D})}{(1 - e^{-vL/D})}. \qquad (6.121)$$

The conditional probability to exit through the other boundary is $\pi_-(x_0) = 1 - \pi_+(x_0)$. The dependence of $\pi_+(x_0)$ on the strength of the drift (vL/D) is given in Figure 6.10.

The mean exit time $\tau_+(x_0)$ through the downstream boundary at $x = L$ follows from Equation 6.107 as

$$\tau_+(x_0) = \frac{L}{v}\frac{(1 + e^{-vL/D})}{(1 - e^{-vL/D})} - \frac{x_0}{v}\frac{(1 + e^{-vx_0/D})}{(1 - e^{-vx_0/D})}. \qquad (6.122)$$

Similarly, Equation 6.109 yields the mean exit time through the upstream boundary as

$$\tau_-(x_0) = \frac{1}{v}\left\{ x_0\frac{(1 + e^{-v(L-x_0)/D})}{(1 - e^{-v(L-x_0)/D})} - \frac{2L}{(1 - e^{-v(L-x_0)/D})}\frac{(e^{vx_0/D} - 1)}{(e^{vL/D} - 1)} \right\}. \qquad (6.123)$$

The role of drift on the x_0-dependence of τ_+ and τ_- is given in Figure 6.11.

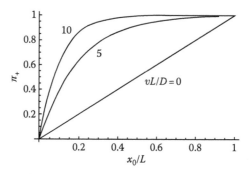

Figure 6.10   Dependence of the conditional probability $\pi_+(x_0)$ to reach the downstream absorbing boundary at $x = L$ on the initial particle position at x_0, for the drift strength values $vL/D = 0, 5$, and 10.

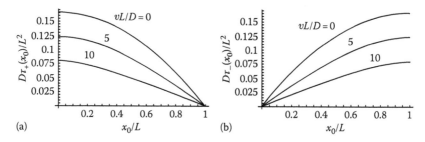

Figure 6.11   (a) Dependence of the mean exit time $\tau_+(x_0)$ to reach the downstream absorbing boundary at $x = L$ on the initial particle position at x_0, for the drift strength values $vL/D = 0, 5$, and 10. (b) Same as in (a), but for reaching the upstream boundary at $x = 0$.

6.7.3 Reflecting Boundary

For the combination $(BC2)$ of a reflecting boundary at $x = 0$ and an absorbing boundary at $x = L$, the same procedure as above can be carried out to find the following results.

The unconditional probability distribution function for the FPT is

$$g(x_0, \tau) = 2 \sum_{p=1}^{\infty} \frac{\beta_p(\beta_p^2 + v^2/4D^2) e^{v(L-x_0)/2D - (\beta_p^2 + v^2/4D^2)D\tau} \sin(\beta_p(L - x_0))}{[L(\beta_p^2 + v^2/4D^2) + v/2D]},$$

(6.124)

with

$$\beta_p \cot(\beta_p L) = -v/2D.$$

(6.125)

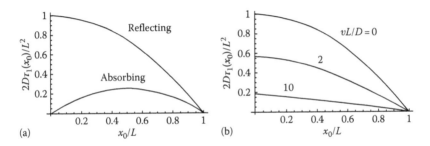

Figure 6.12 (a) Comparison of the dependence of the mean FPT $\tau_1(x_0)$ on the initial particle position, between a reflecting boundary and an absorbing boundary at $x = 0$, with an absorbing boundary at $x = L$, and with no drift. (b) Effect of drift on the mean FPT for a reflecting boundary condition at $x = 0$, with $vL/D = 0, 2$, and 10.

The mean FPT is obtained by substituting $A(x) = v$ and $B(x) = D$ in Equations 6.99, 6.100, and 6.102 to get

$$\tau_1(x_0) = \frac{1}{v}\left[L - x_0 + \frac{D}{v}(e^{-vL/D} - e^{-vx_0/D})\right]. \tag{6.126}$$

The asymptotic limits of this result for the weak and strong strengths of the drift with respect to diffusion follow as

$$\tau_1(x_0) = \begin{cases} \frac{L^2 - x_0^2}{2D}, & v/D \to 0 \\ \frac{L - x_0}{v}, & v/D \to \infty. \end{cases} \tag{6.127}$$

This limiting behavior in the diffusion-dominated regime is plotted in Figure 6.12a and compared with the result of Equation 6.120 for the boundary condition $BC1$. For the case of reflecting barrier, the mean FPT decreases monotonically with the decreasing distance of the absorbing barrier downstream from the initial position of the particle. In the drift-dominated regime, the time taken to reach the boundary downstream is simply the distance divided by the velocity of the particle, as in the case of two absorbing barriers (Figure 6.9b). The crossover behavior for intermediate values of the drift strength ($\alpha \equiv vL/D$) given by Equation 6.126 is illustrated in Figure 6.12b.

6.7.4 Radiation Boundary Condition

For the combination ($BC3$) of an absorbing boundary at $x = L$ and a mixed boundary condition at $x = 0$, the same procedure as above can be carried out to find the various quantities associated with the FPT. For the sake of completeness with respect to different boundary conditions, we provide the formulas for the probability distribution function and the unconditional mean FPT. Other quantities can also be calculated in an analogous manner.

With the mixed boundary condition for the probability at $x = 0$,

$$\left.\frac{\partial P(x, t)}{\partial x}\right|_{x=0} - \left(\frac{v}{D} + h\right) P(x = 0, t) = 0, \qquad (6.128)$$

where h is the mixing parameter, the probability distribution function for the FPT is given by

$$g(x_0, \tau) = 2 \sum_{p=1}^{\infty} \left\{ [\beta_p^2 + (v/2D + h)^2][\beta_p + h\sin(\beta_p L)e^{-vL/2D}] \right.$$

$$\times e^{v(L-x_0)/2D - (\beta_p^2 + v^2/4D^2)D\tau}$$

$$\times \left. \frac{\sin(\beta_p(L - x_0))}{[L(\beta_p^2 + (v/2D + h)^2) + v/2D + h]} \right\}, \qquad (6.129)$$

with

$$\beta_p \cot(\beta_p L) = -\left(h + \frac{v}{2D}\right). \qquad (6.130)$$

The mean FPT is obtained by substituting $A(x) = v$ and $B(x) = D$ in Equations 6.99, 6.100, and 6.103 and using the boundary condition Equation 6.98 at $a = 0$, to get

$$\tau_1(x_0) = \frac{1}{v}$$

$$\times \frac{\left\{ L - x_0 + \frac{D}{v}(e^{-vL/D} - e^{-vx_0/D}) + \frac{Dh}{v}[L(1 - e^{-vx_0/D}) - x_0(1 - e^{-vL/D})] \right\}}{\left[1 + \frac{Dh}{v}(1 - e^{-vL/D})\right]}.$$

$$(6.131)$$

The above results for mixed boundary condition reduce to those of the absorbing and reflecting boundary conditions discussed above, for $h \to \infty$ and $h = 0$, respectively. Other moments of the FPT can be similarly written down for the general boundary condition.

6.8 SUMMARY

The motion of polymer molecules in a solution under chemical potential gradients or externally imposed electric fields is an example of the "drift–diffusion" stochastic processes. We have introduced several equivalent formalisms for studying these processes: biased random walk, master equation, and Langevin equation of motion. In each of these lines of arguments, we have arrived at the Fokker–Planck–Smoluchowski equation.

By using the Fokker–Planck methodology for a general drift–diffusion process, we have provided key formulas for several quantities of experimental interest. These quantities include the average time for a particle to approach a boundary and the probability of success in reaching one boundary against another possible boundary. We have presented the dependencies of these quantities on the initial distance of the particle from the boundary, its velocity, and its diffusion coefficient.

The utility of the general formulas is illustrated by treating a simple drift–diffusion process with a constant diffusion coefficient and a constant velocity. The limits of drift–domination and diffusion–domination, and the crossover behavior between these limits are discussed for various quantities of interest in the context of translocation. We shall use the key formulas provided here in later chapters dealing with polymer capture by a pore and polymer translocation through a pore.

7

POLYELECTROLYTE DYNAMICS

Polymer translocation through a nanopore has several rich molecular attributes unlike that of electrolyte ions. Whereas the transport of an ion through a channel can be envisaged as shuttling of structureless marble-like particles, the capacity of the polymer molecule to adopt many conformations and to deform under geometrical restrictions and external forces necessitates an entirely different approach. Even away from the pore, the dynamics of polymer molecules are complex. Every segment of the polymer is subjected to the Brownian forces arising from incessant collisions by the solvent molecules. As a result, each segment would have the tendency to undergo diffusion on its own right, but now modulated by the fact that the segments are correlated by chain connectivity and not independent. We may therefore expect an anomalous diffusion for a labeled monomer of the polymer. Nevertheless, for very long times and large distances compared to the polymer size, we expect the center of mass of the polymer to obey the Einsteinian dynamics. There are several characteristic timescales associated with the dynamics of the individual segments and the collective behavior of the whole chain. The description of the crossover between the behavior of polymers at short times and at long times can be cumbersome, but tremendous progress has been made for uncharged polymers (Doi and Edwards 1986).

When the polymer chain bears charges, the dynamics exhibits many intriguing phenomena, unknown for uncharged polymers. For example, the collective diffusion coefficient of a polyelectrolyte molecule, as measured in dynamic light scattering, can be several orders of magnitude larger than expected for its uncharged counterpart, and it can even be as high as that of an electrolyte ion. When driven by a constant electric field, polyelectrolyte molecules move with the same average speed, independent of their molecular weight. Features of this nature arise from the coupled dynamics of counterions and the polymer. In order to maintain local electroneutrality, the dynamics of the polymer and the counterions are strongly coupled. The polymer molecule cannot be treated as a separate entity without a simultaneous consideration of the counterion cloud. Furthermore, if some flow fields are imposed, for example, by sucking the fluid at the pore mouth, the molecules can deform and their dynamics are modified rather dramatically.

In this chapter, we shall summarize only the salient concepts behind the dynamics of polyelectrolyte molecules by relegating the technical details to the original papers. After introducing the concept of hydrodynamic interaction, we shall present its consequences on the polymer dynamics. First, we shall consider uncharged polymers in dilute solutions, followed by a treatment of coupling between counterion dynamics and polyelectrolyte dynamics. The laws of polymer deformation and polyelectrolyte mobility under externally imposed fields will be considered next. The main results presented here will form the basis for understanding the process of delivering the polymer molecules to the pore, which is the first necessary step for translocation.

7.1 SOLVENT CONTINUUM AND HYDRODYNAMIC INTERACTION

When a polymer chain is dispersed in a solvent, the characteristic size and time for describing the collective motion of the polymer molecule are many orders of magnitude large in comparison with those for individual solvent molecules. As a result, the solvent background can be assumed to be a hydrodynamic continuum. Let us consider a small volume element around the spatial location $\mathbf{r}$, which is big enough to consider the solvent as a continuum, at time t. The time dependence of the velocity field $\mathbf{v}(\mathbf{r}, t)$ at $\mathbf{r}$ and t is obtained from the Newton's second law of motion, namely the rate of change of momentum of a fluid element is equal to the net force acting on it, as (Landau and Lifshitz 1959)

$$\rho_0 \frac{d\mathbf{v}(\mathbf{r}, t)}{dt} + \mathbf{v} \cdot \nabla \mathbf{v}(\mathbf{r}, t) - \eta_0 \nabla^2 \mathbf{v}(\mathbf{r}, t) + \nabla p(\mathbf{r}, t) = \mathbf{F}(\mathbf{r}, t), \qquad (7.1)$$

where $p(\mathbf{r}, t)$ is the local pressure field, ρ_0 is the mass density of the fluid, and η_0 is the shear viscosity of the fluid. $\mathbf{F}(\mathbf{r}, t)$ is an external force, such as gravity or a suction force by a capillary or a shear force, and it specifies the particular experimental setup.

As the experimental situations of our current interest generally lie in the weak velocity fields and the inertial forces become negligible for relatively long timescales relevant to polymer molecules, the first two terms on the left-hand side of the above equation may be ignored resulting in the "zero-frequency" linearized Navier–Stokes equation (Landau and Lifshitz 1959),

$$-\eta_0 \nabla^2 \mathbf{v}(\mathbf{r}, t) + \nabla p(\mathbf{r}, t) = \mathbf{F}(\mathbf{r}, t). \qquad (7.2)$$

In addition, we assume that the fluid is incompressible. This condition is given as

$$\nabla \cdot \mathbf{v}(\mathbf{r}, t) = 0. \qquad (7.3)$$

The pressure field in Equation 7.2 can be eliminated by using Equation 7.3, so that the velocity field can be expressed in terms of any externally imposed force field $\mathbf{F}$.

The fundamental property of the background fluid is how it transmits a force field $\mathbf{F}(\mathbf{r}', t)$ at the location $\mathbf{r}'$ to another location $\mathbf{r}$. Combining Equations 7.2 and 7.3, the velocity field at $\mathbf{r}$ can be written in terms of $\mathbf{F}$ as

$$\mathbf{v}(\mathbf{r}) = \int d\mathbf{r}' \mathbf{G}(\mathbf{r} - \mathbf{r}') \cdot \mathbf{F}(\mathbf{r}'), \tag{7.4}$$

where $\mathbf{G}$ is called the Oseen tensor,

$$\mathbf{G}(\mathbf{r} - \mathbf{r}') = \frac{1}{8\pi \eta_0 |\mathbf{r} - \mathbf{r}'|} \left[\mathbf{1} + \frac{(\mathbf{r} - \mathbf{r}')(\mathbf{r} - \mathbf{r}')}{|\mathbf{r} - \mathbf{r}'|^2} \right], \tag{7.5}$$

where $\mathbf{1}$ is the unit tensor. The operator $\mathbf{G}$ may be taken as a switch that converts a force at $\mathbf{r}'$ to a velocity at a different location $\mathbf{r}$ (Figure 7.1a). If we were to perform an angular average of the Oseen tensor, the tensor inside the square brackets reduces to the factor $4/3$ so that

$$G_{ang.av}(\mathbf{r} - \mathbf{r}') = \frac{1}{6\pi \eta_0 |\mathbf{r} - \mathbf{r}'|}. \tag{7.6}$$

It is important to recognize that $\mathbf{G}$ is inversely proportional to the separation distance between the point of source of a force and the point where velocity perturbation is monitored. This long-ranged correlation in the solvent continuum is called the *hydrodynamic interaction*.

The long-ranged nature of the hydrodynamic interaction in the solvent background leads to correlations among the various monomers of a dispersed polymer although these monomers could be at large separation distances (Figure 7.1b). The velocity of a specified monomer is affected by all sources of forces where some of which could even be spatially far away. These sources can be other monomers of the same chain or other chains, charge distributions from counterions, and externally generated flow fields. Before we consider the hydrodynamic coupling in polymer dynamics in the following sections, let us

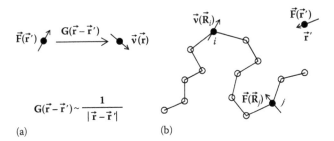

(a)   (b)

Figure 7.1   (a) The hydrodynamic interaction between two fluid elements at $\mathbf{r}$ and $\mathbf{r}'$ is long-ranged. The Oseen tensor $\mathbf{G}$ transmits a force $\mathbf{F}$ at $\mathbf{r}'$ to a velocity $\mathbf{v}$ at $\mathbf{r}$. (b) Each monomer is coupled hydrodynamically to all other monomers. The velocity of the ith monomer depends on the force generated by the jth monomer as well as the force field at $\mathbf{r}'$ in the solvent.

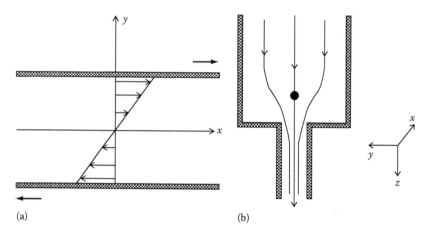

Figure 7.2 Geometry of flow for (a) simple shear and (b) uniaxial elongation.

consider a few examples of the external force **F**, in terms of the velocity fields due only to **F** in the absence of polymer molecules.

Let us first consider a simple shear flow. When a fluid is confined between two parallel plates, both perpendicular to the y-axis (Figure 7.2a), and one is moved with a constant velocity in the direction of the x-axis relative to the other, the velocity field of the fluid is given by

$$v_x = \dot{\gamma}y, \quad v_y = 0, \quad v_z = 0, \tag{7.7}$$

where $\dot{\gamma}$ is the shear rate or velocity gradient. Such flows have both stretching and rotational components. The x-, y-, and z-components of the stretching part of the velocity are $\frac{1}{2}\dot{\gamma}y$, $\frac{1}{2}\dot{\gamma}x$, and 0, respectively. The corresponding components for the rotational part are $\frac{1}{2}\dot{\gamma}y$, $-\frac{1}{2}\dot{\gamma}x$, and 0. The flow field corresponding to these parts given by Equation 7.7 is depicted in Figure 7.3a with the streamlines. The tangent at any point of a streamline is the direction of velocity at that point. If a flexible polymer molecule is placed in this flow field, it undergoes rotation with periodic expansion along the stretching direction and contraction along the direction orthogonal to the stretching direction.

Another example of a flow field of common interest, such as an effective suction of the fluid at the pore mouth (Figure 7.2b), is the uniaxial elongational flow along the central axis (Morrison 2001),

$$v_x = -\frac{1}{2}\dot{\epsilon}x, \quad v_y = -\frac{1}{2}\dot{\epsilon}y, \quad v_z = \dot{\epsilon}z, \tag{7.8}$$

where $\dot{\epsilon}$ is the elongational flow rate (strain rate). The streamlines for this flow along the central axis are sketched in Figure 7.3b. The maximum elongational rate occurs at a distance comparable to the pore diameter, in front of the pore.

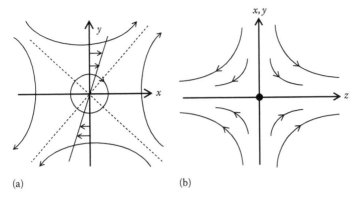

Figure 7.3 Streamlines for a (a) simple shear flow and (b) uniaxial elongational flow.

7.2 UNCHARGED POLYMER

The movement of a polymer coil in a solution is a net effect of motions of all monomers constituting the molecule. The dynamics of monomers are highly correlated due to chain connectivity and inter-monomer interactions. Nevertheless, for sufficiently long times, to be described below, the whole polymer chain is expected to move like a Brownian particle as discussed in Section 6.2. On the other hand, if we were to look at length and timescales that would probe the internal dynamics of the polymer, new laws are expected to emerge. Before introducing a few successful models of polymer dynamics to describe these features, we first outline the general expectations based on the universal Einsteinian dynamics.

7.2.1 General Expectations

In a quiescent solution, the center of mass of the polymer $\mathbf{R}_{cm}(t)$ is expected to undergo diffusion just as a rigid spherical Brownian particle (see Equations 6.42 through 6.45),

$$\langle [\mathbf{R}_{cm}(t) - \mathbf{R}_{cm}(0)]^2 \rangle = 6Dt, \quad \text{(Einsteinian)} \tag{7.9}$$

where the center-of-mass diffusion coefficient is

$$D = \frac{k_B T}{f_t}. \tag{7.10}$$

The friction coefficient f_t of the whole polymer molecule is expected to follow the Stokes law of proportionality to η_0 times the radius of the polymer coil. By choosing the same prefactor of 6π as in the relation for a rigid sphere (Equation 6.44), we define the "hydrodynamic radius" R_h as

$$f_t = 6\pi \eta_0 R_h, \tag{7.11}$$

where R_h is proportional to the radius of gyration R_g (Section 2.1). Since R_g is related to the chain length via the size exponent, $R_g \sim N^\nu$ (where N is the number of Kuhn segments per chain, which is proportional to the molecular weight), it follows from Equations 7.10 and 7.11 that

$$D \sim N^{-\nu}. \tag{7.12}$$

Therefore, we anticipate the center-of-mass diffusion coefficient to be inversely proportional to the square root of the molecular weight of a polymer obeying Gaussian chain statistics, etc.

While the suspended particles undergo diffusion in a solvent, the shear viscosity η_0 of the solvent is also modified by the particles to the shear viscosity η of the whole solution. As was originally addressed by Einstein, the increment in viscosity is directly proportional to the volume fraction of the suspended particles, if the interparticle interactions are ignored. For a solution containing n particles of radius R in volume V, the result is

$$\frac{\eta - \eta_0}{\eta_0} = \frac{5}{2} \frac{n}{V} \frac{4\pi R^3}{3}. \tag{7.13}$$

For a dilute solution containing n polymer chains of radius of gyration R_g, we anticipate from the above Einsteinian law for the change in viscosity that

$$\frac{\eta - \eta_0}{\eta_0} \sim \frac{nR_g^3}{V}. \tag{7.14}$$

Rewriting the above result in terms of the monomer concentration $c \equiv nN/V$, we get

$$\frac{\eta - \eta_0}{\eta_0} \sim c\frac{R_g^3}{N}. \tag{7.15}$$

Since $R_g \sim N^\nu$, the "reduced viscosity" of the solution is given by

$$\frac{\eta - \eta_0}{\eta_0 c} \sim N^{3\nu - 1}. \tag{7.16}$$

In the limit of $c \to 0$, the reduced viscosity is called the intrinsic viscosity. Therefore, we expect the intrinsic viscosity to scale as $N^{1/2}$ and $N^{4/5}$ in polymer solutions under ideal ($\nu = 1/2$) and good solvent ($\nu = 3/5$) conditions, respectively.

According to Equation 6.1 and Section 6.2, the Einsteinian dynamics can be generally paraphrased as

$$\langle (\text{distance})^2 \rangle \sim (\text{diffusion constant}) \times (\text{time}) \tag{7.17}$$

Let us consider a polymer chain of radius of gyration R_g. If the center-of-mass of the chain were to move a distance comparable to its radius of gyration, the

new polymer conformation is expected to be uncorrelated to its initial conformation. Without worrying about the numerical coefficients, we argue that the
correlation time τ over which the conformations of a single chain are correlated
is the time taken by the polymer to explore a distance comparable to R_g by the
diffusion process. Since the diffusion coefficient is inversely proportional to
R_g, we get from Equation 7.17,

$$R_g^2 \sim \left(\frac{1}{R_g}\right)\tau. \qquad (7.18)$$

Therefore, we expect the scaling law for the correlation time τ as being proportional to the molecular volume of the polymer coil, and in view of $R_g \sim N^\nu$,

$$\tau \sim N^{3\nu}. \qquad (7.19)$$

The general expectations embodied in Equations 7.12, 7.16, and 7.19 are
borne out to be valid as shown by experiments in dilute solutions of uncharged
polymers. Depending on the experimental conditions, the value of the size
exponent changes and this change is directly manifest in D, η, and τ in terms
of their dependencies on the molecular weight of the polymer and solvent conditions. In order to obtain the numerical prefactors for the above scaling laws
and to understand the internal dynamics of the polymer molecules, it is necessary to build polymer models that explicitly account for the chain connectivity.
The two basic models of polymer dynamics are the Rouse and Zimm models
(Rouse 1953, Kirkwood and Riseman 1948, Zimm 1956), which are discussed
next.

7.2.2 Rouse Model

Consider a chain of $N+1$ Kuhn segments, each of length ℓ and mass m, placed
inside a solvent continuum of shear viscosity η_0. The equation of motion for
the position vector $\mathbf{R}_i$ of the ith segment at time t is given by the Langevin
equation,

$$m\frac{d^2\mathbf{R}_i}{dt^2} = \mathbf{F}_{connectivity} + \mathbf{F}_{pot} + \mathbf{F}_{fric} + \mathbf{F}_{ext} + \mathbf{F}_{ran}, \qquad (7.20)$$

where the various forces on the right-hand side are given below based on intuitive arguments instead of formal derivations. The force arising from the chain
connectivity is related to the probability of realizing particular chain conformations. Since the ith segment is connected to only its neighbors ($i+1$ and $i-1$),
and ignoring the end effects, the connectivity force acting on the ith segment is
given by

$$\mathbf{F}_{connectivity}(\mathbf{R}_i) = -\frac{3k_BT}{\ell^2}\sum_{j=0}^{N}(2\delta_{i,j} - \delta_{i,j+1} - \delta_{i,j-1})\mathbf{R}_j. \qquad (7.21)$$

The connectivity matrix $(2\delta_{i,j} - \delta_{i,j+1} - \delta_{i,j-1})$ is called the Rouse matrix. If $V(\mathbf{R}_i - \mathbf{R}_j)$ is the intersegment potential energy due to excluded volume interactions (and electrostatic interactions if charges are present), the force on the ith segment is

$$\mathbf{F}_{pot}(\mathbf{R}_i) = -\nabla_{\mathbf{R}_i} \sum_{j=0}^{N} V(\mathbf{R}_i - \mathbf{R}_j). \qquad (7.22)$$

Since every segment is in principle bathed by the solvent molecules, the frictional force on each of the segments is written as if it itself is a Brownian particle (Section 6.2),

$$\mathbf{F}_{fric}(\mathbf{R}_i) = -\zeta_b \frac{d\mathbf{R}_i}{dt}, \qquad (7.23)$$

where ζ_b is a phenomenological parameter, called the bead friction coefficient. ζ_b represents the microscopic details of a segment against the background solvent and is related to the Kuhn length. $\mathbf{F}_{ext}(\mathbf{R}_i)$ is the external force such as the shear or electric force on the segment. $\mathbf{F}_{ran}(\mathbf{R}_i)$ is the fluctuating force arising from collisions by the solvent molecules. This random force is related to ζ_b by the fluctuation-dissipation theorem as in Section 6.2.

For reasonable long times that are typically explored experimentally in polymer solutions, the inertial term in Equation 7.20 can be ignored. Combining Equations 7.20 through 7.23, we get the equation of motion of the ith segment as

$$\zeta_b \left(\frac{d\mathbf{R}_i}{dt} - \mathbf{v}_{ext} \right) + \frac{3k_B T}{\ell^2} \sum_{j=0}^{N} (2\delta_{i,j} - \delta_{i,j+1} - \delta_{i,j-1})\mathbf{R}_j + \nabla_{\mathbf{R}_i} \sum_{j=0}^{N} V(\mathbf{R}_i - \mathbf{R}_j)$$

$$= \mathbf{F}_{ran}(\mathbf{R}_i). \qquad (7.24)$$

Here, $\mathbf{v}_{ext}$ is the velocity field at $\mathbf{R}_i$ in the absence of the polymer chain.

The above equation is known as the Rouse equation (Rouse 1953), when there are no intersegment potential interactions (Gaussian chain). Now the dynamics is equivalent to that of a vibrating string. By adopting the continuous chain representation, $i\ell \equiv s, 0 \leq s \leq L$, where $L = N\ell$ is the contour length of the chain, the Rouse equation for a Gaussian chain becomes

$$\zeta_b \frac{d\mathbf{R}(s)}{dt} - 3k_B T \frac{\partial^2 \mathbf{R}(s)}{ds^2} = \mathbf{F}_{ran}(s). \qquad (7.25)$$

This equation is solved by resolving $\mathbf{R}(s)$ in terms of its normal modes $2\pi p/N$, where the normal mode variable p is called the Rouse mode variable, and using the condition that the chain obeys the Gaussian statistics in equilibrium. Lower the value of p, longer is the length along the polymer that is being probed. The $p = 1, 2, \ldots$ Rouse modes represent the behavior of the whole chain, half of the chain along the chain contour, etc. The $p = 0$ mode corresponds to the center-of-mass of the chain. The major results of the Rouse chain dynamics are the following.

The mean square displacement of the center-of-mass of the chain obeys the Einsteinian dynamics at longer times compared to the Rouse time τ_{Rouse} (given below),

$$\langle [\mathbf{R}_{cm}(t) - \mathbf{R}_{cm}(0)]^2 \rangle = 6Dt, \tag{7.26}$$

with the diffusion coefficient D given by Equation 7.10. The key result of the Rouse dynamics is that the various segments contribute to the friction coefficient of the chain independently so that

$$f_t = \zeta_b N, \qquad D = \frac{k_B T}{\zeta_b N}. \tag{7.27}$$

Therefore, for a Rouse chain, the diffusion coefficient of the center-of-mass of the polymer is always inversely proportional to the chain length (independent of the solvent quality),

$$D \sim \frac{1}{N}. \tag{7.28}$$

In contrast to the diffusive behavior of the center-of-mass of the chain, the dynamics of a labeled monomer is even slower due to the fact that the tendency of the monomer to undergo diffusion is curtailed by the chain connectivity. It can be shown that the mean square displacement of the labeled ith segment of a Rouse chain is given by

$$\langle [\mathbf{R}_i(t) - \mathbf{R}_i(0)]^2 \rangle \sim \sqrt{t}. \tag{7.29}$$

Furthermore, the relaxation time τ_p of the Rouse mode p, namely the time taken by the block of segments representing the pth mode to relax if the chain were to be allowed to go back to equilibrium after deformation, depends inversely on p^2,

$$\tau_p = \frac{\zeta_b N^2}{12\pi^2 k_B T p^2} \sim \left(\frac{N}{p}\right)^2. \tag{7.30}$$

Therefore, the longest relaxation time τ_{Rouse} of the Rouse model, corresponding to $p = 1$, is proportional to the square of the chain length,

$$\tau_{Rouse} \sim N^2. \tag{7.31}$$

Also, it can be shown that the viscosity change of the solution due to n Rouse chains is proportional to chain length,

$$\frac{\eta - \eta_0}{\eta_0} \sim cN, \tag{7.32}$$

where c is the monomer concentration.

The exact results of Equations 7.28, 7.31, and 7.32 for the Rouse model are in disagreement with the general expectations of Section 7.2.1 and are never observed experimentally in dilute solutions of polymers, where the model was intended. The reason for this discrepancy is the absence of hydrodynamic coupling between the segments in the Rouse model. Nevertheless, due to the general complexity of the problem of polymer dynamics, there have been considerable activities in simulating the dynamics of polymer chains in silico, usually without hydrodynamic interactions. For this artificial situation of a Rouse chain (i.e., without hydrodynamic interaction) with the size exponent v, the various theoretical results can be summarized as

$$\tau_{Rouse} \sim N^{2v+1}, \tag{7.33}$$

$$D \sim \frac{1}{N}, \tag{7.34}$$

$$\langle [\mathbf{R}_i(t) - \mathbf{R}_i(0)]^2 \rangle \sim t^{2v/(1+2v)}, \tag{7.35}$$

and

$$\frac{\eta - \eta_0}{\eta_0} \sim cN^{2v}. \tag{7.36}$$

Although the above results of the Rouse dynamics are not pertinent to experimental situations of dilute solutions, they are relevant to computer simulations of translocation of single uncharged polymer molecules through nanopores without any consideration of hydrodynamic interaction.

7.2.3 Zimm Dynamics

The long-ranged hydrodynamic interaction between any two points in the fluid, as described by Equations 7.4 through 7.6, couples the dynamics of all monomers. For the ith segment, the hydrodynamic interaction between i and all other segments ($j \neq i$) must be taken into account (Figure 7.1b) in addition to the forces described in the Rouse model. Let σ_j be the force at every other segment $j(\neq i)$ arising from connectivity and potential interactions,

$$\sigma_j = -\frac{3k_BT}{\ell^2} \sum_{p=0}^{N} (2\delta_{j,p} - \delta_{j,p+1} - \delta_{j,p-1})\mathbf{R}_p - \nabla_{\mathbf{R}_j} \sum_{p=0}^{N} V(\mathbf{R}_j - \mathbf{R}_p). \tag{7.37}$$

σ_j is then transmitted to the ith segment by the solvent via the Oseen tensor of Equation 7.4,

$$\mathbf{v}_{hyd}(\mathbf{R}_i) = \sum_{j \neq i}^{N} \mathbf{G}(\mathbf{R}_i - \mathbf{R}_j) \cdot \sigma_j. \tag{7.38}$$

This velocity at the ith segment is converted to the local force by multiplying by ζ_b. Adding this additional hydrodynamic force to the Rouse equation (Equation 7.24), we get

$$\zeta_b \left(\frac{d\mathbf{R}_i}{dt} - \mathbf{v}_{ext} \right) - \sigma_i - \zeta_b \sum_{j \neq i}^{N} \mathbf{G}(\mathbf{R}_i - \mathbf{R}_j) \cdot \sigma_j = \mathbf{F}_{ran}(\mathbf{R}_i), \qquad (7.39)$$

where σ_i is defined through Equation 7.37 as the force exerted by the ith segment on the fluid due to chain connectivity and potential interactions. The dynamics of polymer segments with intersegment hydrodynamic correlations given by Equation 7.39 is called the Kirkwood–Riseman–Zimm dynamics or simply the Zimm dynamics (Kirkwood and Riseman 1948, Zimm 1956, Yamakawa 1971, Doi and Edwards 1986). Upon averaging over chain conformations and taking the average of the random force to be zero, the above equation becomes

$$\langle \sigma_i \rangle = \zeta_b \left(\left\langle \frac{d\mathbf{R}_i}{dt} - \mathbf{v}_{ext}(\mathbf{R}_i) \right\rangle \right) - \zeta_b \sum_{j \neq i} \langle \mathbf{G}(\mathbf{R}_i - \mathbf{R}_j) \cdot \sigma_j \rangle. \qquad (7.40)$$

The translational friction coefficient f_t of the polymer chain is calculated by defining the net force acting on the chain $- \sum_i \langle \sigma_i \rangle$ as the net frictional force on the polymer,

$$-f_t \left(\dot{\mathbf{R}}_{cm} - \mathbf{v}_{ext}^0 \right) = - \sum_i \langle \sigma_i \rangle, \qquad (7.41)$$

where $\dot{\mathbf{R}}_{cm}$ is the average velocity of the center-of-mass of the chain and $\mathbf{v}_{ext}^0$ is the velocity field due to external forces that would be present in the location of the center-of-mass of the polymer.

An analytical solution of Equation 7.39 for $\mathbf{R}_i$ is impossible due to the nonlinearity arising from the presence of $|\mathbf{R}_i - \mathbf{R}_j|$ in the denominator of the Oseen tensor (Equation 7.5). Therefore, a useful approximation, known as the *preaveraging* approximation, was invoked by Kirkwood and Riseman, where the prefactor term $\mathbf{G}(\mathbf{R}_i - \mathbf{R}_j)$ is averaged over chain conformations, namely $\langle \mathbf{G}(\mathbf{R}_i - \mathbf{R}_j) \cdot \sigma_j \rangle \simeq \langle \mathbf{G}(\mathbf{R}_i - \mathbf{R}_j) \rangle \cdot \langle \sigma_j \rangle$. The physical implication of this approximation is that the hydrodynamic coupling between segments is instantaneous in comparison with all other motions of the segments. The consequence of the preaveraging approximation is that the Oseen tensor now depends only on $|i - j|$ and not on $|\mathbf{R}_i - \mathbf{R}_j|$. For example, for chains with Gaussian statistics, the conformational average of $\mathbf{G}(\mathbf{R}_i - \mathbf{R}_j)$ is

$$\langle \mathbf{G}(\mathbf{R}_i - \mathbf{R}_j) \rangle = \frac{1}{6\pi \eta_0} \left\langle \frac{1}{|\mathbf{R}_i - \mathbf{R}_j|} \right\rangle = \frac{1}{\eta_0} \left(\frac{1}{6\pi^3 \ell^2 |i - j|} \right)^{1/2}. \qquad (7.42)$$

Substituting the preaveraged result of the Oseen tensor and performing the normal mode analysis of the Zimm equation for the various Rouse modes, we can calculate the mean square displacement of the center-of-mass of the chain, mean square displacement of a labeled monomer, translational friction coefficient of the chain, and the relaxation times of the various Rouse modes with the Zimm dynamics (Doi and Edwards 1986). The main results of these calculations are the following.

The mean square displacement of the center-of-mass of the chain obeying Zimm dynamics is the same as the Einsteinian dynamics for sufficiently long times as in Equations 7.9 and 7.10 with the translational friction coefficient given by

$$f_t = \frac{\zeta_b N}{\left[1 + \frac{4\zeta_b N}{9\pi^{3/2}\eta_0 R_g}\right]}. \tag{7.43}$$

This result follows from Equations 2.3, 7.11, and 7.42. When the hydrodynamic interaction among segments dominates over the frictional contribution from the segments, the second term in the denominator is much greater than unity so that f_t for the Zimm dynamics is

$$f_t = \frac{3\sqrt{\pi}}{8}(6\pi \eta_0 R_g). \tag{7.44}$$

Comparison of Equations 7.11, 7.42, and 7.44 provides the reason for defining the hydrodynamic radius according to Equation 2.3 and the proportionality between R_h and R_g.

Although Equation 7.44 is derived for a Gaussian chain with hydrodynamic interaction, the proportionality of f_t with R_g is valid for non-Gaussian chains as well. In fact, with the assumption of uniform chain expansion due to excluded volume interactions, the result of Equation 7.44 can be shown to be valid even in good solutions. Thus, generally,

$$f_t \sim \eta_0 R_g \sim \eta_0 N^\nu, \tag{7.45}$$

and from the Einsteinian relation $D = k_B T/f_t$, we obtain the generalized Stokes–Einstein law for polymer chains in dilute solutions as

$$D \sim \frac{T}{\eta_0 R_g} \sim \frac{T}{\eta_0 N^\nu}. \tag{7.46}$$

Furthermore, the same derivation given above can be repeated for rigid rod-like molecules, where conformational averages are unnecessary. Only orientational averages are required. The net result is that there is a logarithmic correction in N, with the diffusion coefficient proportional to $\ln(N)/N$ (Riseman and Kirkwood 1950, Muthukumar and Edwards 1983). This is consistent with the scaling law of Equation 7.46, with $\nu = 1$ for rod-like conformations, apart from the logarithmic correction.

The above results are valid only if the time of measurement is longer than the characteristic time for the relaxation of the various Rouse modes of the Zimm chain. The longest relaxation time for a chain with the Zimm dynamics (corresponding to the Rouse mode $p = 1$), where the hydrodynamic interaction dominates, is called the Zimm time given by

$$\tau_{Zimm} = \frac{\sqrt{3}\eta_0}{\pi^{3/2}k_BT}R_g^3. \qquad (7.47)$$

Therefore, the Zimm time is proportional to the molecular volume of the polymer and thus depends on the chain length according to

$$\tau_{Zimm} \sim R_g^3 \sim N^{3\nu}. \qquad (7.48)$$

It can also be shown from the preaveraged Kirkwood–Riseman–Zimm equation that the mean square displacement of a labeled monomer i is proportional to the $2/3$ power of time for times shorter than τ_{Zimm},

$$\langle[\mathbf{R}_i(t) - \mathbf{R}_i(0)]^2\rangle \sim t^{2/3}. \qquad (7.49)$$

The reduced viscosity of a dilute solution of chains with the Zimm dynamics is

$$\frac{\eta - \eta_0}{\eta_0 c} = 12\frac{R_g^3}{N} \sim N^{3\nu-1}, \qquad (7.50)$$

where c is the monomer concentration.

The results of Equations 7.46, 7.48, and 7.50 for the Zimm dynamics are entirely consistent with the universal laws expected in Section 7.2.1 and are fully supported by experimental data in dilute solutions. If the hydrodynamic interaction among segments is suppressed in the Kirkwood–Riseman–Zimm equation, then the problem reduces to the Rouse dynamics and all results of Section 7.2.2 are recovered.

7.2.4 Semidilute Solutions and Hydrodynamic Screening

As the polymer concentration is increased, the chains begin to overlap at about the overlap concentration $c^\star$, which depends (Section 2.7) on the chain length according to

$$c^\star \sim \frac{1}{N^{3\nu-1}}. \qquad (7.51)$$

For polymer concentrations higher than the overlap concentration, two regimes may be identified. For very high monomer concentrations, the interpenetrating chains entangle with each other and their dynamics is slowed down considerably. This concentration regime is the entangled regime. In the intermediate

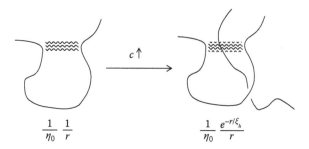

Figure 7.4 In dilute solutions, intrachain hydrodynamic interaction dominates (Zimm dynamics). At higher polymer concentrations, chain interpenetration
screens the hydrodynamic interaction, resulting in an apparent Rouse dynamics.

concentration regime, where c is larger than $c^\star$, but not large enough to have
significant entanglement effects, we have the unentangled concentration regime
called the semidilute solutions.

The main dynamical consequence of the interpenetration of the chains in
the semidilute regime is the screening of the hydrodynamic coupling among
the intrachain segments of a chain, as sketched in Figure 7.4. The long-ranged
hydrodynamic interaction given in Equations 7.5 and 7.6 is modified into the
"Debye–Hückel" form (Equation 3.18),

$$\frac{1}{\eta_0|\mathbf{r} - \mathbf{r}'|} \to \frac{e^{-|\mathbf{r}-\mathbf{r}'|/\xi_h}}{\eta_0|\mathbf{r} - \mathbf{r}'|}, \tag{7.52}$$

where ξ_h is the hydrodynamic screening length. The intrachain hydrodynamic
coupling is strong only if the separation distance is shorter than ξ_h. At larger distances, the hydrodynamic effect is essentially absent. The value of ξ_h depends
on the polymer concentration, and as expected it is a decreasing function
of the polymer concentration. It can be shown (Muthukumar and Edwards
1982b) that the hydrodynamic screening length is proportional to the correlation length ξ for the monomer density fluctuations in semidilute solutions
(Section 2.7),

$$\xi_h \sim \xi \sim c^{-\nu/(3\nu-1)}. \tag{7.53}$$

The main results of the theory of collective dynamics of chains in semidilute
solutions with hydrodynamic screening are the following. The mean square
displacement of a labeled chain follows the diffusion law:

$$\langle[\mathbf{R}_{cm}(t) - \mathbf{R}_{cm}(0)]^2\rangle = 6D_t t, \tag{7.54}$$

where the diffusion constant is

$$D_t \sim \frac{T}{c^{(1-\nu)/(3\nu-1)}N}. \tag{7.55}$$

This diffusion constant describing the diffusion of a labeled chain is called the tracer diffusion constant. It decreases with N and the monomer concentration c. The sharpness of the concentration dependence of D_t depends on the size exponent. For example, in good solutions ($v = 3/5$), $D_t \sim c^{-1/2}$, and in theta solutions ($v = 1/2$), $D_t \sim c^{-1}$. The longest relaxation time τ is

$$\tau \sim c^{(2-3v)/(3v-1)}N^2, \tag{7.56}$$

and the change in the shear viscosity of the solution is

$$\frac{\eta - \eta_0}{\eta_0} \sim c^{1/(3v-1)}N. \tag{7.57}$$

One of the convenient ways of measuring the diffusion coefficient is by monitoring the dynamical correlations of the local density of the scattering entities in experiments such as the dynamic light scattering. In dilute solutions, the number density of polymer chains can be monitored. In semidilute solutions or within polymer chains in dilute solutions, monomer density can be monitored. As an example, let us consider the time-dependent correlation of fluctuations in local concentration. Let the local monomer concentration, in number of monomers per unit volume, at the spatial location $\mathbf{r}$ and time t be

$$c_p(\mathbf{r}, t) = \sum_{\alpha}^{n} \sum_{i=0}^{N} \delta(\mathbf{r} - \mathbf{R}_{\alpha,i}(t)), \tag{7.58}$$

where there are n chains each of N Kuhn steps in the total volume, and $\mathbf{R}_{\alpha,i}(t)$ refers to the location of the ith segment of the αth chain at time t. Analogous to Equations 6.66 and 6.67, the monomer density obeys the continuity equation

$$\frac{\partial c_p(\mathbf{r}, t)}{\partial t} = -\nabla \cdot \mathbf{J}, \tag{7.59}$$

where the flux is given by

$$\mathbf{J} = -D_c \nabla c_p(\mathbf{r}, t). \tag{7.60}$$

D_c is called the cooperative or mutual diffusion coefficient, as thermodynamic correlations among all chains are used in writing the above equations. In contrast, the tracer diffusion coefficient D_t measures the diffusion of a tagged polymer. Defining the fluctuation in the local monomer density $\delta c_p(\mathbf{r}, t)$ as the deviation of $c_p(\mathbf{r}, t)$ from its average value $\bar{c}_p = c$, combining Equations 7.59 and 7.60, and taking the Fourier transforms yield

$$\frac{\partial \delta c_p(\mathbf{k}, t)}{\partial t} = -D_c k^2 \delta c_p(\mathbf{k}, t). \tag{7.61}$$

This equation gives the time evolution of density fluctuations resolved in their Fourier modes. The Fourier variable $\mathbf{k}$ is actually the scattering wave vector

introduced in Section 2.1. The cooperative diffusion coefficient can therefore be conveniently measured by constructing the dynamic structure factor
$\langle \delta c_p(\mathbf{k}, t) \delta c_p(\mathbf{k}, 0) \rangle$ (which is related to the scattered intensity correlations
at the scattering wave vector $\mathbf{k}$ and correlation time t), which by solving
Equation 7.61 is proportional to $\exp(-D_c k^2 t)$,

$$\langle \delta c_p(\mathbf{k}, t) \delta c_p(\mathbf{k}, 0) \rangle = \langle \delta c_p(\mathbf{k}, 0) \delta c_p(\mathbf{k}, 0) \rangle e^{-D_c k^2 t}. \qquad (7.62)$$

For infinitely dilute solutions, the same prescription as above is used in measuring the diffusion coefficient, except that correlations of the fluctuations in the
polymer number density are now monitored instead of monomer density fluctuations (in the $kR_g \ll 1$ limit). In dilute solutions, D_c is identical to D_t, and
so both diffusion coefficients are assigned the symbol D. However, at higher
concentrations, D_c is qualitatively different from D_t. In fact, the cooperative
diffusion coefficient in semidilute solutions, measured using light scattering,
increases with polymer concentration, whereas the tracer diffusion coefficient
decreases. The general result for the cooperative diffusion coefficient is

$$D_c = \frac{k_B T}{6\pi \eta_0 \xi}, \qquad (7.63)$$

where ξ is the correlation length for the monomer density described in
Section 2.7. The appearance of Equation 7.63 in the format of the Stokes–
Einstein law is due to the fact that the cooperative diffusion coefficient in
semidilute solutions corresponds to the collective diffusion of all monomers
within a correlation length ξ, independent of which chains the monomers
belong to. Therefore, the radius of an isolated particle in dilute solutions is
replaced by the linear size of the volume over which the segments are correlated both thermodynamically and dynamically. Since $\xi \sim c^{-\nu/(3\nu-1)}$, the
cooperative diffusion coefficient depends on the polymer concentration as

$$D_c \sim \frac{T}{\eta_0} c^{\nu/(3\nu-1)}. \qquad (7.64)$$

For good solutions ($\nu = 3/5$), the cooperative diffusion coefficient increases
with monomer concentration as $c^{3/4}$ and is independent of N. On the other hand,
the tracer diffusion coefficient decreases as $c^{-1/2} N^{-1}$. These results have been
thoroughly validated experimentally (Doi and Edwards 1986).

As the hydrodynamic interaction is screened in semidilute solutions, the
molecular weight dependencies of the diffusion coefficient, the longest relaxation time, and the viscosity change in the semidilute solutions are exactly the
same as in the Rouse model. However, since the Rouse model was originally
designed for an isolated chain, the concentration dependencies of these quantities are not captured by the Rouse model. Nevertheless, we shall refer to the
correct description of polymer dynamics in semidilute solutions as the Rouse
regime. A summary of the main results for the Zimm model in dilute solutions

TABLE 7.1

Polymer Dynamics in Dilute and Semidilute Solutions

Quantity	Dilute (Zimm)	Semidilute (Rouse)	Dilute (Rouse)
Diffusion coefficient of			
center-of-mass	$D \sim N^{-\nu}$	$1/\left(c^{(1-\nu)/(3\nu-1)}N\right)$	$1/N$
Cooperative diffusion coefficient	$D \sim N^{-\nu}$	$c^{\nu/(3\nu-1)}$	$1/N$
Longest relaxation time	$N^{3\nu}$	$c^{(2-3\nu)/(3\nu-1)}N^2$	$N^{2\nu+1}$
$(\eta - \eta_0)/\eta_0$	$cN^{3\nu-1}$	$c^{1/(3\nu-1)}N$	$cN^{2\nu}$

and semidilute solutions with hydrodynamic interaction screened (labeled as the Rouse regime) is provided in Table 7.1. There is no such a practical situation as the Rouse model in dilute solutions. However, this model is popular in computer simulations toward a fundamental understanding of simpler models of polymer dynamics. In view of this, the results for this model are also included in Table 7.1, as the last column.

7.2.5 Entanglement Regime

At very high polymer concentrations of long polymer chains, the entanglement effect arising from the impenetrability of chains across their contours leads to dramatic consequences on the viscoelastic properties of such systems. The simplest and perhaps the most elegant way of describing a collection of entangled chains is to imagine that any labeled chain is essentially localized inside a tube-like region around its chain contour, and the tube is a representation of the entanglement effects arising from all other chains. The time taken by all chains contributing to the tube to move and make a new tube-like region is expected to be much larger than the time required by the labeled chain to move out of its original tube. As a result, the dynamics of a labeled chain can be approximated as that of a chain trapped inside a tube, with the mobilities of the two ends dominating over those of the intermediate monomers. This model dynamics is called reptation (de Gennes 1971, 1979, Doi and Edwards 1986).

According to the reptation model, the tracer diffusion coefficient and the longest relaxation time are proportional to N^{-2} and N^3, respectively,

$$D_t \sim \frac{1}{N^2}, \quad \tau \sim N^3. \tag{7.65}$$

The dynamics of the polymer chain is naturally slowed down by entanglements, in comparison with the results for dilute and semidilute conditions (Table 7.1). There is a vast literature on the subject of viscoelasticity of polymer melts and how close the experimental results are with the predictions of the reptation model (Doi and Edwards 1986). Since conditions such as polymer melts are not of common place in the experiments exploring polymer translocation, we do not dwell on this subject here.

7.3 DIFFUSION OF POLYELECTROLYTE CHAINS

Although the direct consequences of the Einsteinian dynamics of the center-of-mass diffusion of a polyelectrolyte chain in dilute solutions,

$$\langle [\mathbf{R}_{cm}(t) - \mathbf{R}_{cm}(0)]^2 \rangle = 6Dt, \tag{7.66}$$

are observed experimentally, the simple identification of D as given by the Stokes–Einstein law

$$D = \frac{k_B T}{6\pi \eta_0 R_h}, \tag{7.67}$$

with R_h being proportional to the radius of gyration of the molecule is generally inapplicable for polyelectrolytes (Forster and Schmidt 1995). For example, in salt-free solutions containing the polyelectrolyte chains and their counterions, where R_h is proportional to N, we expect D to be proportional to $1/N$. However, the experimentally measured diffusion coefficient is independent of N and also of the polymer concentration. This is shown in Figure 7.5a where the measured diffusion coefficient in salt-free aqueous solutions of sodium poly(styrene sulfonate) is plotted against the polymer concentration for different molecular weights (Sedlak and Amis 1992a,b). Furthermore, we expect the diffusion coefficient given by the Stokes–Einstein law to increase with the concentration of added salt due to the shrinkage of R_g in the presence of salt. However, the experimental data show the opposite trend of a decrease in the diffusion coefficient upon addition of salt. This is shown in Figure 7.5b, where the measured diffusion coefficient is plotted against the ratio of monomer concentration to the salt concentration for aqueous solutions of quarternized poly(vinyl pyridinium) salt (Forster et al. 1990). The upper branch is called the fast mode and the lower branch is called the slow mode. The slow mode is not yet understood except for the recognition that it represents the diffusive behavior of large clusters much bigger than the R_g of single chains. We shall treat only the fast mode here. The resolution of the discrepancies between the experimentally observed polyelectrolyte behavior for the fast mode and the expectations from the Stokes–Einstein law lies in the role played by the counterions (Muthukumar 1997).

The behavior of polyelectrolyte molecules cannot be interpreted as arising solely from the polymer. The coupled behavior of the chain and its counterion cloud must be treated together. The counterions exert an electric field, even in the absence of any externally imposed electric field, in the immediate environment of a charged segment. This field in turn modifies the collective diffusion coefficient of the polyelectrolyte. The polyelectrolyte molecule must be treated not as an individual particle but as a quasiparticle along with its counterion cloud.

Let $\rho_p(\mathbf{r}, t)$ be the local number density of polymer chains in the volume element $\mathbf{r}$ at time t. By addressing the induced electric field $\mathbf{E}_{ind}(\mathbf{r})$ arising from the counterion cloud in the volume element $\mathbf{r}$ that acts on the polymer chains,

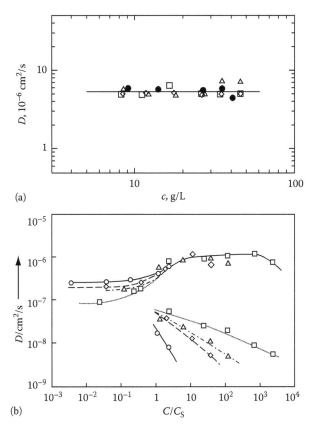

Figure 7.5 (a) Diffusion coefficient measured in salt-free dilute solutions of sodium poly(styrene sulfonate) is independent of the concentration and molecular weight (different symbols, in the range between 5000 and 200,000 g/mol) of the polymer. (From Sedlak, M. and Amis, E.J., *J. Chem. Phys.*, 96, 817, 1992b. With permission.) (b) Diffusion coefficient decreases with salt concentration for the upper branch (fast mode). The lower branch is the slow mode. The data are for quarternized poly(vinyl pyridinium) salt with the molecular weight of 200,000 g/mol at different salt concentrations (squares, $\approx 10^{-5}$ M; triangles, 10^{-3} M; diamonds, 10^{-2} M; and open circles, 0.1 M). (From Forster, S. et al., *Polymer*, 31, 781, 1990. With permission.)

the continuity equation and the net flux of polyelectrolyte chains follow from Equations 6.66 and 6.67 as

$$\frac{\partial \rho_p(\mathbf{r}, t)}{dt} = -\nabla \cdot \mathbf{J} \tag{7.68}$$

$$\mathbf{J} = -D_c \nabla \rho_p(\mathbf{r}, t) + \rho_p(\mathbf{r}, t) \mu \mathbf{E}_{ind}. \tag{7.69}$$

Here, D_c is the cooperative diffusion coefficient of the chain if the coupling to the counterion cloud were to be absent. In dilute solutions, D_c is inversely proportional to the radius of gyration. $\mu \mathbf{E}_{ind}$ is the velocity of the polyelectrolyte chains and μ is the electrophoretic mobility of the polyelectrolyte chain (discussed in Section 7.4). From the Poisson equation (Equation 3.9), we get

$$-\nabla^2 \psi(\mathbf{r}) = \nabla \cdot \mathbf{E}_{ind}(\mathbf{r}) = \frac{\rho(\mathbf{r})}{\epsilon_0 \epsilon}, \tag{7.70}$$

where $\rho(\mathbf{r})$ is the net charge density at $\mathbf{r}$ and $\epsilon_0 \epsilon$ is the permittivity of the solution. Focusing on the polyelectrolyte and its counterions and ignoring the presence of salt ions for now, we get

$$\nabla \cdot \mathbf{E}_{ind}(\mathbf{r}, t) = \frac{1}{\epsilon_0 \epsilon} [Q \rho_p(\mathbf{r}, t) + z_c e \rho_c(\mathbf{r}, t)], \tag{7.71}$$

where Q is the polymer charge, $z_c e$ is the charge of the counterion, and $\rho_c(\mathbf{r}, t)$ is the local concentration of counterions. It is thus evident from Equations 7.68 through 7.71 that the time evolution of local polyelectrolyte concentration is coupled to that of the counterion concentration. Analogous to the continuity equation for the polyelectrolyte concentration, the continuity equation for the counterion concentration is

$$\frac{\partial \rho_c(\mathbf{r}, t)}{dt} = D' \nabla^2 \rho_c(\mathbf{r}, t) - \nabla \cdot \rho_c(\mathbf{r}, t) \mu_c \mathbf{E}_{ind}, \tag{7.72}$$

where D' and μ_c are, respectively, the diffusion coefficient and electrophoretic mobility of the counterion related by (Equation 7.91)

$$\mu_c = \frac{z_c e D'}{k_B T}. \tag{7.73}$$

Defining the fluctuations in the concentrations of the polyelectrolyte chains and counterions,

$$\rho_p(\mathbf{r}, t) = \overline{\rho}_p + \delta \rho_p(\mathbf{r}, t) \tag{7.74}$$

$$\rho_c(\mathbf{r}, t) = \overline{\rho}_c + \delta \rho_c(\mathbf{r}, t), \tag{7.75}$$

where $\overline{\rho}_p$ and $\overline{\rho}_c$ are, respectively, the average concentrations of the polyelectrolyte and counterions, and performing Fourier transforms, we get the coupled equations

$$\frac{\partial \delta \rho_{p,\mathbf{k}}}{\partial t} = -D_c k^2 \delta \rho_{p,\mathbf{k}} - \frac{1}{\epsilon_0 \epsilon} \overline{\rho}_p \mu (Q \delta \rho_{p,\mathbf{k}} + z_c e \delta \rho_{c,\mathbf{k}}) \tag{7.76}$$

$$\frac{\partial \delta \rho_{c,\mathbf{k}}}{\partial t} = -D' k^2 \delta \rho_{c,\mathbf{k}} - \frac{1}{\epsilon_0 \epsilon} \overline{\rho}_c \mu_c (Q \delta \rho_{p,\mathbf{k}} + z_c e \delta \rho_{c,\mathbf{k}}) \tag{7.77}$$

to leading orders in $\delta\rho_{p,\mathbf{k}}$ and $\delta\rho_{c,\mathbf{k}}$. In general, a solution of these two coupled equations results in two collective modes for the relaxation of the local polymer concentration. We shall call these modes as the fast and superfast modes. We consider only the fast mode in the following discussion, as it is experimentally difficult to monitor the superfast mode. Since a fluctuation in polymer concentration relaxes in the omnipresent counterion cloud, we assume that $\partial\delta\rho_{c,\mathbf{k}}/\partial t = 0$. This allows an analytical expression for $\delta\rho_{c,\mathbf{k}}$ in terms of $\delta\rho_{p,\mathbf{k}}$, which in turn, upon substitution into Equation 7.76, yields

$$\frac{\partial\delta\rho_{p,\mathbf{k}}}{\partial t} = -D_f k^2 \delta\rho_{p,\mathbf{k}}, \tag{7.78}$$

where the *coupled* diffusion coefficient is (Muthukumar 1997)

$$D_f = D_c + \frac{1}{\epsilon_0\epsilon} \frac{\overline{\rho}_p \mu Q}{\left(k^2 + \kappa_c^2\right)}, \tag{7.79}$$

with

$$\kappa_c^2 = \frac{z_c^2 e^2 \overline{\rho}_c}{\epsilon_0\epsilon k_B T}. \tag{7.80}$$

The coupled diffusion coefficient D_f is the diffusion coefficient corresponding to the fast mode.

For a polyelectrolyte chain of degree of ionization α,

$$\overline{\rho}_c = \alpha N \overline{\rho}_p. \tag{7.81}$$

As will be shown in Section 7.4, for salt-free solutions,

$$\mu = \frac{QD_c}{k_B T}. \tag{7.82}$$

Combining Equations 7.79 through 7.82 in the limit of $k \to 0$ (corresponding to very large probe lengths), the coupled diffusion coefficient of the polyelectrolyte chain in dilute solutions becomes

$$D_f = D_c\left[1 + \frac{1}{\alpha N}\left(\frac{Q}{z_c e}\right)^2\right]. \tag{7.83}$$

Since the polymer charge $|Q|$ is αNe, and $|z_c| = 1$ for monovalent counterions, Equation 7.83 gives

$$D_f = D_c(1 + \alpha N). \tag{7.84}$$

Therefore, the coupling of polyelectrolyte chain to the counterion cloud, which is directly responsible for the αN term in the above equation, dominates the experimentally observed diffusion coefficient. Since $D_c \sim 1/R_g \sim 1/N$, for salt-free solutions, the coupled fast diffusion coefficient D_f turns out to be independent of N.

The above derivation can readily be extended to the presence of salt ions (Berne and Pecora 1976, Muthukumar 1997). The final formula for D_f is the same as Equation 7.79, except that κ_c^2 is replaced by $\kappa_c^2 + \kappa_s^2$ with κ_s^2 for $z{:}z$ electrolytes at number concentration ρ_s given by

$$\kappa_s^2 = \frac{2z^2e^2\rho_s}{\epsilon_0\epsilon k_B T}. \tag{7.85}$$

For high salt concentrations, this term is much larger than κ_c^2 and the second term of Equation 7.79 becomes negligible. Therefore, D_f is simply D_c at high salt concentrations as the electrostatic coupling between the polymer and the counterion cloud is broken by the Debye screening due to the salt. Thus, at high salt concentrations, the experimentally observed diffusion coefficient must approach the value appropriate for uncharged polymers in good solutions. Therefore, the diffusion coefficient is expected to decrease from a large N-independent value in salt-free solutions to a lower value, proportional to $N^{-3/5}$, as the salt concentration is increased. This behavior is indeed observed experimentally as shown in Figure 7.5b, where the diffusion coefficient measured in dynamic light scattering is plotted against c_p/c_s, with c_p and c_s being, respectively, the polymer concentration and salt concentration in molarity.

The coupling theory described above is pertinent to only the fast mode where D_f decreases with an increase in c_s. Furthermore, the counterion coupling theory (Muthukumar 1997) predicts that D_f is independent of c_p in salt-free dilute and semidilute polyelectrolyte solutions, consistent with Figure 7.5a. The superfast mode mentioned above is not yet measurable in experiments with polyelectrolytes. The slow mode appearing for $c_p/c_s \geq 1$ may be attributed to temporary clustering of polyelectrolyte chains and its full understanding is yet to be reached.

A brief summary of the above results for the diffusional behavior of polyelectrolytes in dilute solutions is presented in Table 7.2. Results for semidilute solutions are also included for comparison.

7.4 ELECTROPHORETIC MOBILITY

Electrophoresis is the movement of a charged molecule relative to a stationary liquid by an applied electric field. For weak enough electric fields, the average velocity $\mathbf{u}$ of the molecule is proportional to the electric field $\mathbf{E}$,

$$\mathbf{u} = \mu\mathbf{E}, \tag{7.86}$$

where μ is the electrophoretic mobility of the molecule.

TABLE 7.2
Diffusion Coefficients of Polyelectrolyte Chains in Dilute and Semidilute
Solutions

Quantity	Salt Level	Dilute	Semidilute
Tracer diffusion coefficient, D	High	$N^{-3/5}$	$c_p^{-1/2}N^{-1}$
Tracer diffusion coefficient, D	Low	N^{-1}	$c_p^0 N^{-1}$
Cooperative diffusion coefficient, D_c	High	$N^{-3/5}$	$c_p^{3/4} N^0$
Cooperative diffusion coefficient, D_c	Low	N^{-1}	$c_p^{1/2} N^0$
Coupled diffusion coefficient, D_f	High	$N^{-3/5}$	$\sqrt{c_p}/(c_p + 2c_s)$
Coupled diffusion coefficient, D_f	Low	N^0	$c_p^0 N^0$

Source: Muthukumar, M., Dynamics of polyelectrolyte solutions, J. Chem.
   Phys., 107, 2619, 1997. With permission.
Note: c_p and c_s are, respectively, the concentrations of the polymer and
   monovalent added salt.

Let us consider a rigid spherical particle of radius R, bearing a net charge Q. Let $\mathbf{E}$ be the externally applied uniform electric field and η_0 be the shear viscosity of the solution. The charged particle experiences an electric force $\mathbf{F}_{el}$ directed toward the oppositely charged electrode,

$$\mathbf{F}_{el} = Q\mathbf{E}. \tag{7.87}$$

This force is opposed by the viscous resistance of the liquid given by the frictional force

$$\mathbf{F}_{friction} = -f_t\mathbf{u}, \tag{7.88}$$

where $\mathbf{u}$ is the velocity of the particle, and f_t is the translational friction coefficient. The stationary state velocity is attained when the net force from these contributions is zero, so that

$$\mathbf{u} = \frac{Q}{f_t}\mathbf{E}. \tag{7.89}$$

The electrophoretic mobility defined in Equation 7.86 follows from Equation 7.89 as

$$\mu = \frac{Q}{f_t}. \tag{7.90}$$

Using the Einsteinian law of Equation 6.43, μ can be rewritten as

$$\mu = \frac{QD}{k_BT}, \tag{7.91}$$

where D is the diffusion coefficient of the particle. This equation is also referred to as the Einsteinian relation for electrophoretic mobility. Substituting the Stokes result (Equation 6.44) for the translational friction coefficient, we get

$$\mu = \frac{Q}{6\pi \eta_0 R}. \tag{7.92}$$

A straightforward extension of Equation 7.92, derived above for a small ion or a colloidal particle, to polyelectrolyte molecules fails badly. Since the total charge of a polyelectrolyte molecule is proportional to the number of Kuhn segments N in the chain (which is proportional to the degree of polymerization), and the chain radius is proportional to N^{ν}, we expect from Equation 7.92 that

$$\mu \sim N^{1-\nu}. \tag{7.93}$$

For salt-free conditions, the size exponent ν can be unity, corresponding to rod-like conformations. In this limit, μ is independent of N. However, as the concentration of added salt increases, the size exponent decreases toward 3/5. Therefore, we expect from Equation 7.92 that the electrophoretic mobility to actually increase with molecular weight of the polyelectrolyte at high salt concentrations, which is not observed in solutions.

These expectations based on the Einsteinian relation are erroneous. It is well known (Olivera et al. 1964, Hoagland et al. 1999, Nkodo et al. 2001, Stellwagen et al. 2003) experimentally that the electrophoretic mobility of long polyelectrolyte molecules in solution electrophoresis is independent of molecular weight at *all* salt concentrations,

$$\mu \sim N^0. \tag{7.94}$$

This result is illustrated in Figure 7.6 where the electrophoretic mobility of sodium poly(styrene sulfonate) in solutions containing sodium chloride in the range of 10^{-4}–3 M is plotted against the degree of polymerization (Hoagland et al. 1999). Furthermore, the diffusion coefficient and the electrophoretic mobility were measured directly for ssDNA and dsDNA at many values of the number of repeat units, in order to check the validity of the Einsteinian relation (Stellwagen et al. 2003). The diffusion coefficient was found to follow the Zimm dynamics (with hydrodynamic interaction dominating), $D \sim N^{-\nu}$, with $\nu \approx 0.6$, whereas μ was found to be independent of N. By writing the polymer charge Q as $Q \equiv ze$, where z is the number of charged residues in the polymer, the Einsteinian relation of Equation 7.91 gives

$$\frac{\mu}{zD} = \frac{e}{k_B T} = 39.6 \, \text{V}^{-1} \quad \text{at } 20°\text{C}. \tag{7.95}$$

Plots of the experimentally determined values of the left-hand side of the above equation against the number of repeat units are given in Figure 7.7 for ssDNA and dsDNA (Stellwagen et al. 2003). The constant value of $39.6 \, \text{V}^{-1}$

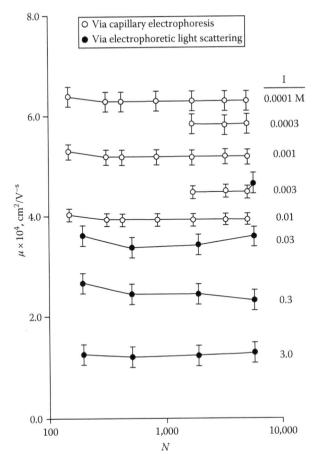

Figure 7.6 Electrophoretic mobility in dilute solutions is independent of molecular weight at all salt concentrations. The ionic strength I is indicated for each data set. Data are from capillary electrophoresis and electrophoretic light scattering. (From Hoagland, D. et al., *Macromolecules*, 32, 6180, 1999. With permission.)

is not observed, demonstrating the inapplicability of the Einsteinian relation of electrophoretic mobility for polyelectrolyte molecules.

The strong discrepancy between the Einsteinian relation for the electrophoretic mobility and the experimental results for polyelectrolytes suggests that the chain cannot be treated as an isolated entity. As in Section 7.3, the polyelectrolyte molecule and its counterion cloud must be treated together. The coupling between the charged monomers and their counterion clouds must be addressed, in addition to the forces arising from chain connectivity, potential interaction and hydrodynamic interaction among segments. In the presence of an electric field, the counterion cloud around each segment generates a force, and this gets transmitted to another location in the fluid by the hydrodynamic

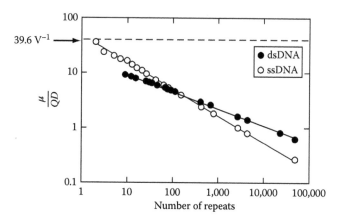

Figure 7.7 Inapplicability of the Einsteinian relation for electrophoretic mobility to polyelectrolyte chains. (From Stellwagen, E. et al., *Biochemistry*, 42, 11745, 2003. With permission.)

interaction, as sketched in Figure 7.8. Let $\rho_{jc}(\mathbf{r}')$ be the local charge density at $\mathbf{r}'$ due to the counterion surrounding the jth segment of the chain. The electric force arising from all counterions at $\mathbf{r}'$ in the presence of a uniform electric field is

$$\mathbf{F}_{counterion}(\mathbf{r}') = \sum_{j=0}^{N} \rho_{jc}(\mathbf{r}')\mathbf{E}. \qquad (7.96)$$

Therefore, a part of the external velocity $\mathbf{v}_{ext}$ in Equation 7.39 arises from the counterions as

$$\mathbf{v}_{ext,counterions}(\mathbf{R}_i) = \int d\mathbf{r}'\mathbf{G}(\mathbf{R}_i - \mathbf{r}') \cdot \sum_{j=0}^{N} \rho_{jc}(\mathbf{r}')\mathbf{E}, \qquad (7.97)$$

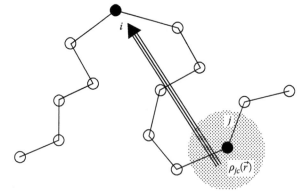

Figure 7.8 The electric force from the counterion cloud around the jth segment is transmitted to the position of the ith segment by the long-range hydrodynamic interaction.

where $\mathbf{G}$ is the Oseen tensor. By performing the preaveraging approximation as in the Kirkwood–Riseman–Zimm model, the net induced force from the chain $\sum_i \langle \sigma_i \rangle$, due to chain connectivity, intersegment potential interactions, and hydrodynamic correlations among all segments and counterions, is balanced with the net electric force $Q\mathbf{E}$ acting on the polymer,

$$\sum_{i=0}^{N} \langle \sigma_i \rangle = Q\mathbf{E}. \tag{7.98}$$

By repeating the same calculation as in the derivation of Zimm dynamics, it can be shown (Muthukumar 1996b) that the electrophoretic mobility follows as

$$\mu = Q \left(\frac{1}{f_t} + \tilde{A} \right). \tag{7.99}$$

The first term on the right-hand side is the hydrodynamic part as in the Einsteinian relation. The second term is due to the counterions and can be shown, within the Debye–Hückel approximation for the potential interaction among segments, as

$$\tilde{A} = \frac{1}{6\pi \eta_0 (N+1)^2} \sum_{i=0}^{N} \sum_{j \neq i} \left\langle \frac{1}{|\mathbf{R}_i - \mathbf{R}_j|} (e^{-\kappa |\mathbf{R}_i - \mathbf{R}_j|} - 1) \right\rangle. \tag{7.100}$$

It should be noted that the second term of Equation 7.100 is the negative of $1/f_t$ appearing in Equation 7.99.

Substitution of Equation 7.100 into Equation 7.99 yields (Muthukumar 1996b)

$$\mu = \frac{QD}{k_B T} \Theta \left(\frac{R_g}{\xi_D} \right), \tag{7.101}$$

where $\Theta(R_g/\xi_D)$ is the factor arising from the coupling of the dynamics of the polyelectrolyte and the counterions. This is a function of the ratio of the radius of gyration of the polyelectrolyte to the Debye length ξ_D, which is proportional to the inverse square root of the salt concentration $(\xi_D \sim 1/\sqrt{c_s})$. The term in front of Θ is exactly the same result as in the Einsteinian relation.

In the limit of $\kappa \to 0$ corresponding to the limit of salt-free conditions, $\tilde{A}$ is zero. As a result, the Einsteinian result

$$\mu = \frac{Q}{f_t} \tag{7.102}$$

is recovered in the $\kappa \to 0$ limit. Since $Q \sim N$ and $f_t \sim R_g \sim N$ in the salt-free limit, the electrophoretic mobility is independent of N,

$$\mu \sim N^0. \tag{7.103}$$

In the high salt limit, $\kappa R_g \gg 1$, it can be shown (Muthukumar 1996a) from Equations 7.99 and 7.100 that

$$\mu = \frac{Q}{N}\left[\frac{1}{\zeta_b} + \frac{\beta}{\eta_0 \ell (\kappa \ell)^{(1/\nu)-1}}\right], \qquad (7.104)$$

where β is a known numerical coefficient. Since $Q \sim N$, the electrophoretic mobility is independent of N in the high salt limit also. Furthermore, since $\nu = 3/5$ for flexible polyelectrolytes in solutions with high salt concentration, we expect

$$\mu \sim \frac{1}{\eta_0 \kappa^{2/3}} \sim \frac{1}{\eta_0 c_s^{1/3}}, \qquad (7.105)$$

where Equation 3.26 is used.

The crossover function $\Theta(R_g/\xi_D)$ in Equation 7.101 depends on N, at any salt concentration, according to

$$\Theta\left(\frac{R_g}{\xi_D}\right) \sim N^{\nu-1}, \qquad (7.106)$$

where ν is the effective size exponent and its value depends on the salt concentration. Since $QD \sim QN^{-\nu} \sim N^{1-\nu}$, Equations 7.101 and 7.106 lead to the general result,

$$\mu \sim N^0, \quad \text{at all salt concentrations.} \qquad (7.107)$$

It must be recognized that the independence of the electrophoretic mobility on N occurs at all salt concentrations within the Zimm model where the hydrodynamic interaction is unscreened. A popular conjecture in the literature (Long et al. 1996) is that the hydrodynamic interaction is screened and that the Rouse result of $f_t \sim N$ is suitable in the Einsteinian relation. The experimental results shown in Figure 7.7 clearly demonstrate that the hydrodynamic interaction among the segments is not screened. The above analysis of counterion coupling explains simultaneously the behaviors of the diffusion coefficient and electrophoretic mobility with hydrodynamics. However, only the qualitative aspects of the dependence of μ on N are addressed above. More accurate calculations accounting for the relaxation effect of deformed counterion clouds are still lacking to enable quantitative comparisons between theory and experiments.

7.5 COIL STRETCH UNDER FLOW

When a polymer chain is subjected to a shear or an extensional flow field, polymer conformations undergo deformation, resulting in many technologically relevant rheological properties such as shear thinning. In the context of polymer translocation through a pore with strong suction forces, substantial deformation of the polymer chain can occur in front of the pore, which can significantly influence the translocation kinetics. If the elongational component of the flow field is strong, the polymer can undergo a coil-stretch phase transition (de Gennes 1974). Experiments have been carried out in simple shear and elongational flow fields to follow the coil-stretch transition by monitoring scattered light intensity, birefringence, and video fluorescence microscopy (Odell and Keller 1986, Menasveta and Hoagland 1992, Perkins et al. 1997, Smith et al. 1999, Babcock et al. 2003, Atkins and Taylor 2004, Larson 2005).

The strength of the flow field (shear rate for shear flows and strain rate for elongational flows) required to significantly deform the chain is related to the characteristic relaxation time of the chain under quiescent conditions. For dilute solutions, the longest relaxation time of the chain is the Zimm time τ_{Zimm} given by Equation 7.48. By combining the terms arising from the externally imposed flow field and the net velocity field from the intersegment hydrodynamic interactions in the Kirkwood–Riseman–Zimm model, the dimensional analysis of timescales shows that the dimensionless flow field parameter is the Weissenberg number Wi,

$$Wi = \begin{cases} \dot{\gamma}\,\tau_{Zimm}, & \text{shear} \\ \dot{\epsilon}\,\tau_{Zimm}, & \text{elongational} \end{cases} \tag{7.108}$$

where $\dot{\gamma}$ and $\dot{\epsilon}$ are defined in Equations 7.7 and 7.8. The coil-stretch transition is predicted to occur at $Wi = 1$, that is, the critical shear rate $\dot{\gamma}_c$ and strain rate $\dot{\epsilon}_c$ are the reciprocal of the Zimm time,

$$\dot{\gamma}_c = 1/\tau_{zimm}, \tag{7.109}$$

$$\dot{\epsilon}_c = 1/\tau_{zimm}. \tag{7.110}$$

Since $\tau_{Zimm} \sim N^{3\nu}$, both $\dot{\gamma}_c$ and $\dot{\epsilon}_c$ are expected to decrease with N as $N^{-3\nu}$. The exponent is therefore expected to depend on the quality of the solvent ($\nu = 1/2$ for theta solutions and $\nu = 3/5$ for good solutions). However, most of the experiments point in the direction of an exponent of ≈ 1.5, independent of the solvent quality. This question continues to remain unresolved, and many issues such as the broadly distributed residence time of the chain near the stagnation point in elongational flows and nonuniform unraveling of the coil into extended conformations are under discussion. Nevertheless, the coil-stretch transition occurs when Wi is about unity. For $Wi \gg 1$, the chains are stretched. The mean fractional extension (the ratio of the mean span of the chain in the dominant stretching direction to the contour length of the chain) observed for λDNA with video fluorescence microscopy versus Wi is given in Figure 7.9 (Smith et al. 1999). As expected, the chain stretching is more prominent for elongational flows (which have no rotational component) than for the shear flows (which have a rotational component, Figure 7.3a).

In a shear flow field, the chain tends to stretch in the elongational direction as sketched in Figure 7.3a. However, due to the rotational component, the chain

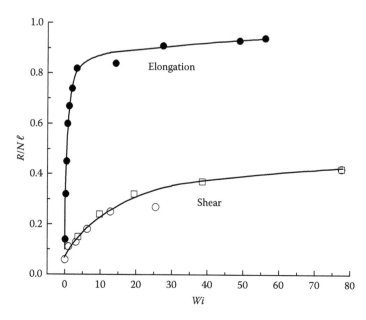

Figure 7.9 The dependence of the mean fractional extension of λDNA on the Weissenberg number *Wi*. The chain stretches more readily in an elongational flow in comparison with a shear flow. In both kinds of flows, the polymer undergoes coil-stretch transition at *Wi* ≈ 1. (Adapted from Smith, D. et al., *Science*, 283, 1724, 1999.)

rotates (clockwise) simultaneous to the stretch and is now subjected to chain contraction. Further along in time, the rotational component of the flow field puts the chain in a condition of stretch again, and so on. Consequently, the chain is subjected to a periodic stretching and contraction. As a result, complete extension of the chain cannot be reached in shear flows in contrast with elongational flows, as is evident from Figure 7.9. Computer simulations of an isolated bead-rod chain under flow fields have also revealed vivid details of how chain conformations are coupled to flow fields (Petera and Muthukumar 1999, Jendrejack et al. 2002, Liu et al. 2004, Larson 2005).

7.6 SUMMARY

In dilute solutions of uncharged macromolecules, experiments show that the diffusion coefficient of a chain obeys the Stokes–Einstein law,

$$D = \frac{k_B T}{6\pi \eta_0 R_h}, \tag{7.111}$$

where the hydrodynamic radius R_h depends on the molecular weight of the macromolecule as

$$R_h \sim N^\nu. \tag{7.112}$$

This law can only be understood by considering long-ranged hydrodynamic interactions among polymer segments.

The longest relaxation time of the chain with hydrodynamic interactions, called the Zimm time, is proportional to the volume of the macromolecule,

$$\tau_{Zimm} \sim N^{3\nu}. \tag{7.113}$$

In the presence of an externally imposed flow, a flexible polymer molecule can undergo the coil-stretch transition at a critical value of shear rate (for shear flows) or strain rate (for elongational flows), which is proportional to the reciprocal of the Zimm time.

Dynamic light scattering experiments show for charged polymers that the Stokes–Einstein law breaks down. This is due to the dynamical coupling between the polyelectrolyte and the counterions. Decreasing salt concentration increases polymer size due to a decreased screening of intrachain electrostatic interaction; but, the coupled diffusion constant increases instead of the expected decrease according to the Stokes–Einstein law. The observed diffusion constant in salt-free solutions can be almost as high as that of an electrolyte ion, independent of molar mass of the polyelectrolyte. At high salt concentrations, the dynamical coupling between the polyelectrolyte and counterions is broken due to electrostatic screening, and only under these conditions, the diffusion coefficient obeys the Stokes–Einstein law.

The Einsteinian law of electrophoretic mobility of ions and colloidal particles breaks down for polyelectrolyte molecules, once again due to the coupling between the polymer and the counterions. In general, the electrophoretic mobility in dilute solutions is given by the formula,

$$\mu \sim N^0, \tag{7.114}$$

where the prefactor is independent of N and depends only on the salt concentration and the value of the size exponent. μ is a decreasing function of salt concentration.

At higher polymer concentrations with chains interpenetrating, the intrachain hydrodynamic interaction is screened. Under these conditions, the model chain dynamics, called the Rouse model, may be used as far as the molecular weight dependence is concerned, and not for the dependence of the polymer concentration. The Rouse model is inapplicable for describing any dynamical properties of polyelectrolyte chains or uncharged macromolecules in dilute solutions.

8

ION FLOW IN SINGLE PORES

The single-molecule electrophysiology procedure of monitoring the passage of macromolecules through a single pore is to measure the ionic current through the pore due to the transport of small electrolyte ions in the system under a voltage gradient. In general, when the macromolecules are either in the vicinity of the pore or inside the pore, the flow of small electrolyte ions through the pore is significantly modified, resulting in changes in the measured ionic current. When some monomers of the polymer are inside the pore, they exclude a certain number of the small ions that would have otherwise been present there, thus resulting in a reduction in the number of charge carriers in the electric field, as sketched in Figure 8.1. In this chapter, we describe the basic theoretical framework, called the Poisson–Nernst–Planck (PNP) formalism, to understand the key features of the measured open pore current and its modifications arising from the presence of macromolecules. Two kinds of situations, namely the equilibrium and the steady state, will be considered. In equilibrium, there is no net current flow, and the pore embedded inside a membrane develops an equilibrium membrane potential, called the Nernst potential, for a prescribed gradient in the electrolyte concentration across the pore, and vice versa. In the steady state, the net ionic current is a constant and time-independent, and we shall discuss this situation using the PNP formalism. For uniform electric fields across the pore, the steady-state situation in the PNP formalism reduces to analytical results known as the Goldman–Hodgkin–Katz (GHK) equations. After equipping with these classical theoretical procedures, we shall consider the effect of a polymer molecule on the ionic current through a pore and the role of fluctuations in the ionic current. Finally, we shall describe the electroosmotic flow (EOF) arising from the hydrodynamic flow of accumulated counterions near charge-bearing pore walls, which can play a major role in generating strong velocity fields in nanopores. These various components (PNP, GHK, and EOF) are required for interpreting the measured ionic currents in terms of molecular mechanisms.

8.1 A GENERAL SCENARIO

Let us consider the flow of ions through a cylindrical pore of length L, which is embedded in a membrane of lower dielectric constant (Figure 8.2). The pore and the membrane separate the donor ($cis \equiv inside$) and the receiver ($trans \equiv outside$) chambers. These chambers contain controllable

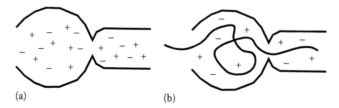

(a)   (b)

Figure 8.1 The presence of a polymer molecule inside a pore effectively reduces the number of electrolyte ions resulting in a reduced net ionic current.

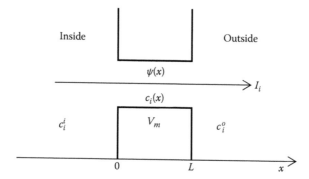

Figure 8.2 A cylindrical pore of length L with the membrane potential $V_m = \psi(0) - \psi(L)$. The concentration of the i-type ions is c_i^i and c_i^o in the inside and outside compartments. The spatial variations of the electric potential $\psi(x)$ and the concentration $c_i(x)$, and the ionic current I_i are calculated using the PNP formalism.

concentrations of electrolytes, and the polyelectrolyte molecules along with their counterions are initially in the inside chamber. Let the number concentration of the ith ion in the inside chamber be c_i^i and that in the outside chamber be c_i^o. A gradient in the electric potential ψ is created across the pore with a pair of electrodes, one in each chamber. We seek the spatial variations of the concentrations of the electrolyte ions and the electric potential, and the net ionic current through the pore.

As described in Section 6.4, the local concentration of the i-type ions in the spatial region $\mathbf{r}$ at time t obeys the law of conservation of the total number of i-type ions in the system according to

$$\frac{\partial c_i(\mathbf{r}, t)}{\partial t} = -\nabla \cdot \mathbf{J}_i(\mathbf{r}, t), \tag{8.1}$$

where the flux of the number of ions into a volume element around $\mathbf{r}$ is

$$\mathbf{J}_i(\mathbf{r}, t) = -D_i \nabla c_i(\mathbf{r}, t) + \mathbf{u}_i(\mathbf{r}, t) c_i(\mathbf{r}, t). \tag{8.2}$$

The first term on the right-hand side is the diffusive part in accordance with the Fick's law. The second term is the convective part with the velocity $\mathbf{u}_i$ of the

i-type ions arising from the gradient in the electric potential (Section 7.4),

$$\mathbf{v}_i(\mathbf{r}, t) = -\mu_i \nabla \psi(\mathbf{r}, t), \tag{8.3}$$

where μ_i is the electrophoretic mobility of the i-type ions (Section 7.4), and $-\nabla \psi(\mathbf{r})$ is the electric field $\mathbf{E}(\mathbf{r})$. Substituting Equation 8.3 into Equation 8.2, we get

$$\mathbf{J}_i(\mathbf{r}, t) = -[D_i \nabla c_i(\mathbf{r}, t) + \mu_i c_i(\mathbf{r}, t) \nabla \psi(\mathbf{r}, t)]. \tag{8.4}$$

The electric potential at every location must satisfy the Poisson equation (Section 3.1.2, Griffiths 1999),

$$\epsilon_0 \nabla \cdot [\epsilon(\mathbf{r}) \nabla \psi(\mathbf{r}, t)] = -\sum_i z_i e c_i(\mathbf{r}, t) - \rho_{pore}(\mathbf{r}, t) - \rho_{poly}(\mathbf{r}, t). \tag{8.5}$$

Here, $\epsilon(\mathbf{r})$ is the effective dielectric constant at the spatial location $\mathbf{r}$ accounting for the dielectric heterogeneity in the medium. For example, the dielectric constant in the region occupied by the membrane is in the range of 2–5, whereas it is ≈ 80 inside the chambers and the pore containing aqueous electrolyte solutions. ϵ_0 and e are the permittivity in the vacuum and the magnitude of the electronic charge, respectively. z_i is the valency of the i-type ions. $\rho_{pore}(\mathbf{r})$ is the charge density on the internal surfaces of the pore. The charge decoration on the pore, which can be tuned either with pH or by deliberate chemical modifications, contributes to the electric potential variations across the pore. Similarly, when polyelectrolyte monomers are present at the location $\mathbf{r}$, their charge density given by $\rho_{poly}(\mathbf{r}, t)$ needs to be included in the Poisson equation (Equation 8.5). The label i indicates any of the charged species in the system, namely the salt ions, polymer segments, the counterions of the pore and the polymer. Although the Einsteinian law for the electrophoretic mobility as given by Equation 7.91,

$$\mu_i = \frac{z_i e D_i}{k_B T}, \tag{8.6}$$

is valid for small electrolyte ions, care must be exercised in using the appropriate expression for the electrophoretic mobility of macromolecules (Section 7.4).

The contribution of the i-type ions to the ionic current is the charge of these ions times the flux in the number density given as

$$\mathbf{I}_i(\mathbf{r}, t) = z_i e \mathbf{J}_i(\mathbf{r}, t). \tag{8.7}$$

The net ionic current is the vectorial sum from all ions,

$$\mathbf{I}(\mathbf{r}, t) = \sum_i \mathbf{I}_i(\mathbf{r}, t). \tag{8.8}$$

The calculation of the net ionic current at any desired location and time requires a self-consistent computation of the local concentration of the ions, flux, and the electric potential given by the combination of Equations 8.1, 8.4, and 8.5, for a prescribed dielectric heterogeneity, charge decoration on the pore, and the presence of polymer molecules. The above coupled nonlinear equations constitute the PNP formalism. For general situations, and for monitoring the time evolution of transients, it is necessary to resort to numerical procedures in solving these nonlinear coupled equations.

The PNP equations become simpler under some special conditions allowing for physically transparent analytical results. For pores with diameters much larger than the diameter of the electrolyte ions, it is sufficient to consider the above equations in only one dimension along the direction of the central axis of the cylindrical pore, which is chosen as the x-axis (Figure 8.2). For this one-dimensional case, the closed form results from PNP for equilibrium and steady-state conditions are discussed in the following sections.

8.2 EQUILIBRIUM

Consider an electrolyte solution in the system without any polymer molecules. In equilibrium, there is no net flow of ions so that the current I_i is zero, and of course, the concentration c_i is independent of time in accordance with Equation 8.1. This requirement is valid individually for each type of electrolyte ions in the solution. In one-dimension along the x-axis, Equation 8.4 gives for $z_i e J_i = 0$,

$$D_i \frac{\partial}{\partial x} c_i(x) = -\mu_i c_i(x) \frac{\partial}{\partial x} \psi(x). \qquad (8.9)$$

Substituting the Einsteinian relation Equation 8.6 for the electrophoretic mobility of the electrolyte ion, we get

$$\frac{\partial}{\partial x} c_i(x) = -c_i(x) \frac{\partial}{\partial x} \left(\frac{z_i e \psi(x)}{k_B T} \right). \qquad (8.10)$$

The integration of this equation gives

$$c_i(x) = A e^{-(z_i e \psi(x)/k_B T)}, \qquad (8.11)$$

where A is the constant of integration. Thus, the local concentration of an ion in equilibrium is given by the Boltzmann distribution of the electrostatic energy of the ion in the potential acting on it in units of $k_B T$.

It is customary to investigate the interdependence of the electrolyte concentration and the electric potential by focusing on the entry and exit points of the pore. The concentration of the i-type ions at $x=0$ is usually different from that in the inside chamber due to confinement effects and specific interactions between the ion and the pore. Analogously, the concentration of the i-type ions at $x = L$ is usually different from that in the outside chamber. Due to the confinement inside the pore, there is an equilibrium of partitioning of the ions between a bulk solution and a pore. The equilibrium constants for the partitioning are defined as

$$\frac{c_i(x = 0)}{c_i^i} \equiv K_i; \qquad \frac{c_i(x = L)}{c_i^o} \equiv K_i', \tag{8.12}$$

where K_i and K_i' are called the partition coefficients. The integration of Equation 8.10 between $x = 0$ and $x = L$ yields

$$\ln\frac{c_i(x = L)}{c_i(x = 0)} = -\frac{z_i e}{k_B T}[\psi(x = L) - \psi(x = 0)]. \tag{8.13}$$

In view of Equation 8.12, the above equation rearranges to

$$\psi(x = 0) - \psi(x = L) = \frac{k_B T}{z_i e} \ln\frac{c_i^o}{c_i^i}, \tag{8.14}$$

where we have assumed that $K_i = K_i'$. Equation 8.14 is known as the Nernst equation. Thus, in equilibrium, a maintenance of a gradient in the concentration of the i-type ions across a pore generates a potential difference that is unique to that type of ions, provided that different types of ions behave independently. This potential is called the Nernst potential V_i for the particular i-type ions given by

$$V_i = \frac{k_B T}{z_i e} \ln\frac{c_i^o}{c_i^i}. \tag{8.15}$$

As noted in Equation 3.61, $k_B T / e$ is 25.7 mV at 25°C. The value of the Nernst potential depends on the valency of the ion and the ratio of the ion concentrations in the outside and inside chambers. The Nernst equation in the form of Equation 8.15 is one of the fundamental formulas used in understanding ionic transport in neurons (Kandel et al. 1991,Weiss 1996, Hille 2001).

8.3 STEADY STATE WITHOUT DIFFUSION

Consider an electrolyte solution with uniform ion concentrations in the system. Under this condition, there are no diffusive fluxes of ions. Therefore, only the convective term in Equation 8.2 contributes to the flux of the ions. We shall refer to this situation as the drift-dominated behavior. Furthermore, in the steady state, the ion concentration distribution is time-independent and the flux is a constant in accordance with Equation 8.1.

8.3.1 Ionic Current and Ohm's Law

Considering the one-dimensional situation, the constant flux of the i-type ions in the steady state (in the drift-dominated regime) is given by

$$J_i(x) = -\mu_i c_i \frac{\partial \psi}{\partial x}, \tag{8.16}$$

where $\psi(x)$ is due to the externally imposed field and other ions. In this section, we shall assume that the externally imposed electric field along the x-direction $E(x)$ is dominant and that the various electrolyte ions conduct the current independently. Therefore,

$$E(x) = -\frac{\partial \psi(x)}{\partial x}. \tag{8.17}$$

For uniform electric fields, $E(x) = E$, the steady-state flux of the number concentration of i-type ions in the drift-dominated regime is

$$J_i = \mu_i c_i E. \tag{8.18}$$

In general,

$$J_i(x) = -\frac{z_i e D_i}{k_B T} c_i \frac{\partial \psi}{\partial x}, \tag{8.19}$$

where Equation 8.6 is used. Substituting Equations 8.19 and 8.17 in Equations 8.7 and 8.8, we obtain the Ohm's law,

$$I(x) = \sum_i \sigma_i E(x), \tag{8.20}$$

where σ_i is the conductivity of the i-type ions

$$\sigma_i = \frac{z_i^2 e^2 D_i}{k_B T} c_i. \tag{8.21}$$

The net current through a pore of length L is obtained by integrating the current density of the i-type ions between $x=0$ and $x=L$ and summing over all kinds of electrolyte ions. It follows from Equations 8.7, 8.8, and 8.19 that

$$\int_0^L dx I(x) = \int_0^L dx \sum_i I_i(x) = -\sum_i \int_0^L dx \frac{z_i^2 e^2 D_i}{k_B T} c_i \frac{\partial \psi}{\partial x}. \tag{8.22}$$

For a uniform electric field across the pore, this equation gives the net uniform ionic current I as

$$I = G V_m, \tag{8.23}$$

where V_m is the membrane potential,

$$V_m \equiv \psi(x = 0) - \psi(x = L), \tag{8.24}$$

and G is the conductance,

$$G = \frac{1}{L} \sum_i \frac{z_i^2 e^2 c_i}{k_B T} D_i. \tag{8.25}$$

It is evident from Equation 8.21 that the conductance is the sum of conductivities of all ions per unit length of the pore. The inverse of the conductance is the resistance R of the pore,

$$R = \frac{1}{G}. \tag{8.26}$$

Basically, the PNP formalism in the steady state without any gradients in ion concentrations reduces to the Ohm's law.

8.3.2 Relaxation of Charge Density

Let us impose a temporary fluctuation on charge density, say $\rho(\mathbf{r}, t)$, at some location $\mathbf{r}$ in a uniform electrolyte solution. The time dependence of $\rho(\mathbf{r}, t)$ is given by the conservation equation (Equation 8.1) as

$$\frac{\partial \rho(\mathbf{r}, t)}{\partial t} = -\nabla \cdot \mathbf{I}(\mathbf{r}, t), \tag{8.27}$$

where $\mathbf{I}(\mathbf{r}, t)$ is the ionic current, namely the flux of net charge of all ions. Combining Equations 8.8, 8.7, 8.19, and 8.21, we get

$$\mathbf{I}(\mathbf{r}, t) = -\sigma \nabla \psi(\mathbf{r}, t), \tag{8.28}$$

where the conductivity σ of the electrolyte is

$$\sigma = \sum_i \frac{z_i^2 e^2 c_i}{k_B T} D_i. \tag{8.29}$$

Substituting Equation 8.28 into Equation 8.27, and with the use of the Poisson equation (Equation 3.9), we get

$$\frac{\partial \rho(\mathbf{r}, t)}{\partial t} = -\frac{\sigma}{\epsilon_0 \epsilon} \rho(\mathbf{r}, t). \tag{8.30}$$

The solution of this linear differential equation is that a fluctuation $\rho(\mathbf{r}, t)$ decays exponentially with time from its initial value $\rho(\mathbf{r}, o)$,

$$\rho(\mathbf{r}, t) = \rho(\mathbf{r}, o) e^{-t/\tau}, \tag{8.31}$$

with a relaxation time τ given by

$$\tau = \frac{\epsilon_0 \epsilon}{\sigma}. \tag{8.32}$$

Furthermore, assuming that the diffusion coefficients of the cation and the anion are the same, $D_i = D$, the relaxation rate can be rewritten for a $z{:}z$-type electrolyte of concentration c as

$$\frac{1}{\tau} = \frac{2z^2 e^2 c}{\epsilon_0 \epsilon k_B T} D. \tag{8.33}$$

Using the definition of the electrostatic screening length in Section 3.1.3.1, the relaxation time τ is simply the ratio of the square of the Debye length to the diffusion coefficient of the ions,

$$\tau = \frac{\xi_D^2}{D}. \tag{8.34}$$

For typical experimental conditions involving simple strong electrolytes in water, $\xi_D \approx 10^{-9}$ m (see Table 3.1) and $D \approx 10^{-5}$ cm^2/s, so that $\tau \approx 10^{-9}$ s. Therefore, any imposed charge density fluctuation in a uniform electrolyte solution dies out quickly in comparison with other characteristic times in the system such as the Zimm time or the translocation time for the polymer.

8.4 STEADY STATE WITH DRIFT AND DIFFUSION

In general, the flux of the ions is controlled by both the diffusion and convection, as given by Equation 8.4. If the diffusive part dominates over the convective (drift) part, the situation is referred to as the diffusion-dominated regime. In general, the behavior of ion fluxes is in the crossover region between the drift-dominated and diffusion-dominated regimes.

Considering the one-dimensional case of ion transport along the pore (x-axis), the charge flux due to the i-type ions follows from Equations 8.4, 8.6, and 8.7 as

$$I_i(x) = -z_i e D_i \left[\frac{\partial c_i(x)}{\partial x} + \frac{z_i e c_i(x)}{k_B T} \frac{\partial \psi(x)}{\partial x} \right]. \tag{8.35}$$

This equation can be rearranged into two equivalent forms. By multiplying and dividing the first term on the right-hand side by $c_i(x)$, we get

$$I_i(x) = -\frac{z_i^2 e^2 D_i}{k_B T} c_i(x) \frac{\partial}{\partial x} \left[\frac{k_B T}{z_i e} \ln c_i(x) + \psi(x) \right]. \tag{8.36}$$

The other convenient form is obtained by performing the same rearrangement as in Section 6.5.2 as

$$I_i(x) = -z_i e D_i e^{-\frac{z_i e \psi(x)}{k_B T}} \frac{\partial}{\partial x} \left[e^{\frac{z_i e \psi(x)}{k_B T}} c_i(x) \right]. \tag{8.37}$$

In the steady state, the charge flux $I_i(x)$ is a constant independent of the variable x. In this limit, the above two equations can be readily integrated between $x = 0$ and $x = L$ to obtain the following useful formulas.

8.4.1 Analogy to Electrical Circuits

The integration of Equation 8.36 between the limits of $x = 0$ and $x = L$ by keeping I_i constant yields

$$I_i \int_0^L dx \left(\frac{k_B T}{z_i^2 e^2 D_i} \right) \frac{1}{c_i(x)} = \psi(0) - \psi(L) - \frac{k_B T}{z_i e} \ln\left(\frac{c_i(L)}{c_i(0)} \right). \qquad (8.38)$$

This expression is in the form,

$$I_i = g_i(V_m - V_i), \qquad (8.39)$$

where V_m is the membrane potential defined in Equation 8.24, and V_i is the Nernst potential for the i-type ions defined in Equation 8.15. The conductance g_i of the i-type ions follows from Equation 8.38 as

$$g_i = \left[\int_0^L dx \left(\frac{k_B T}{z_i^2 e^2 D_i} \right) \frac{1}{c_i(x)} \right]^{-1}. \qquad (8.40)$$

When the membrane potential is the same as the Nernst potential for a particular type of ions, the ion current due to these ions is zero, if all kinds of ions were to behave independently of each other. This situation corresponds to the equilibrium condition discussed in Section 8.2. When V_m differs from the Nernst potential, there is a net current. The result of Equation 8.39 can be conveniently represented as an equivalent electrical circuit. Consider a resistor of resistance $R_i = 1/g_i$ and a battery of the Nernst potential V_i placed in series between the *in* and *out* terminals (Figure 8.3a). Since the resistor and the battery are in series, the current I_i and the voltage difference between the *in* terminal and the *out* terminal are related as

$$V_m = V_{in} - V_{out} = I_i R_i + V_i, \qquad (8.41)$$

which is equivalent to Equation 8.39.

As described in Section 3.2, ion distributions near a charged interface lead to a net accumulation of one kind of ions. As a result, the two sides of the membrane can acquire a charge buildup upon imposition of electric potentials on them. The net charge Q_m on the membrane built up this way is

$$Q_m = C_m V_m, \qquad (8.42)$$

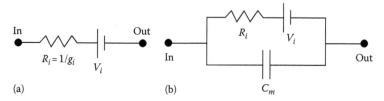

Figure 8.3 Electrical equivalent circuits for a pore. (a) The Nernst battery of potential V_i and a resistor of resistance R_i are in series. The voltage difference between the *in* and *out* terminals is the sum of the voltage drops from the resistor and the Nernst battery (Equation 8.41). (b) The charged membrane acts like a capacitor in parallel to the arrangement in (a).

where C_m is the capacitance of the membrane. In the electrical equivalent circuit, this is represented as a capacitor in parallel (Figure 8.3b). We shall return to this circuit in Section 8.7 when we discuss the fluctuations in ionic current through pores.

8.4.2 Solutions of PNP Equations

The integration of Equation 8.37 between $x=0$ and $x=L$ in the steady state, where I_i is a constant, gives

$$I_i \int_0^L dx e^{\tilde{z}_i \psi(x)} = -z_i e D_i \left[c_i(L) e^{\tilde{z}_i \psi(L)} - c_i(0) e^{\tilde{z}_i \psi(0)} \right], \qquad (8.43)$$

where

$$\tilde{z}_i \equiv \frac{z_i e}{k_B T}. \qquad (8.44)$$

Therefore, the steady-state charge flux due to the i-type ions is given as

$$I_i = -z_i e D_i \frac{[c_i(L) e^{\tilde{z}_i \psi(L)} - c_i(0) e^{\tilde{z}_i \psi(0)}]}{\int_0^L dx e^{\tilde{z}_i \psi(x)}}. \qquad (8.45)$$

The spatial dependence of the ion concentration $c_i(x)$ is obtained by integrating Equation 8.37 between $x=0$ and the desired location x as

$$-\int_0^x dx' \frac{I_i(x')}{z_i e D_i} e^{\tilde{z}_i \psi(x')} = c_i(x) e^{\tilde{z}_i \psi(x)} - c_i(0) e^{\tilde{z}_i \psi(0)}. \qquad (8.46)$$

Assuming that the diffusion coefficient D_i is independent of the spacial position along the pore and taking the charge flux in the steady state as given by Equation 8.45, Equation 8.46 rearranges to

$$c_i(x) = e^{-\tilde{z}_i\psi(x)}\left\{c_i(0)e^{\tilde{z}_i\psi(0)} + \frac{[c_i(L)e^{\tilde{z}_i\psi(L)} - c_i(0)e^{\tilde{z}_i\psi(0)}]\int_0^x dx' e^{\tilde{z}_i\psi(x')}}{\int_0^L dx e^{\tilde{z}_i\psi(x)}}\right\}.$$

$$(8.47)$$

The computation of the steady-state charge flux and the concentration profile requires the electric potential variation $\psi(x)$. This is given by the Poisson equation (Equation 8.5),

$$\epsilon_0\frac{d}{dx}\left[\epsilon(x)\frac{d}{dx}\psi(x)\right] = -\sum_i z_i e c_i(x), \qquad (8.48)$$

where we have considered only the electrolyte ions without any charges from the pore and in the absence of polymer chains. Equations 8.45, 8.47, and 8.48 constitute the solutions of the PNP equations for the transport of electrolyte ions under an applied voltage. Given the values of $\psi(0)$, $\psi(L)$, $c_i(0)$, and $c_i(L)$, and taking a suitable initial guess for the concentration profile, the electric potential $\psi(x)$ is computed from Equation 8.48, and then using this result, $c_i(x)$ is calculated from Equation 8.47, which is used as the next guess for the computation of $\psi(x)$. This procedure is repeated to obtain the final results for the current, concentration profile and the electric potential profile, within a desired numerical accuracy. There are several numerical procedures and theoretical discussions available in the literature (Cooper et al. 1985, Eisenberg 1996, Nonner et al. 1999, Kurnikova et al. 1999, Cardenas et al. 2000, Corry et al. 2000, Moy et al. 2000, Berezhkovskii et al. 2002, 2006, Kolomeisky 2007, Kolomeisky and Kotsev 2008, Ammenti et al. 2009) for accurately solving the PNP equations. There are also computer simulation results in these references in order to assess the validity of the mean field nature of the PNP formalism. In general, the PNP formalism is an excellent tool to capture the essentials of coupled transport of several kinds of ions through pores with prescribed charge decorations.

8.4.3 Constant Field Approximation

If the spatial dependence of the electric potential across the pore is linear with distance, the electric field $E(x) = -\partial\psi/\partial x$ is a constant. With a constant electric field, the PNP equations can be solved analytically (Weiss 1996, Jackson 2006). The closed-form solutions of the PNP equations with constant electric field are known as the GHK equations. For a constant electric field, given by V_m/L, across the pore, the electric potential $\psi(x)$ is

$$\psi(x) = \psi(0) - V_m\frac{x}{L}, \qquad (8.49)$$

where V_m is the membrane potential defined in Equation 8.24. The assumption of constant electric field inside the pore is equivalent to the assumption of electroneutrality at every location along the axis of the pore, as is evident from the Poisson equation.

8.4.3.1 GHK Current Equation

Substitution of Equation 8.49 in Equation 8.45 yields the steady-state charge flux of the i-type ions as

$$I_i = \frac{z_i^2 e^2 D_i V_m}{k_B T L} \frac{[c_i(L)e^{-\bar{z}_i V_m} - c_i(0)]}{(e^{-\bar{z}_i V_m} - 1)}. \tag{8.50}$$

Assuming the partition coefficients K_i and K_i' defined in Equation 8.12 to be the same, the above result is usually rewritten in terms of the ion concentrations in the inside and outside compartments as

$$I_i = \frac{z_i^2 e^2 P_i V_m}{k_B T} \frac{\left[c_i^o e^{-z_i e V_m / k_B T} - c_i^i\right]}{(e^{-z_i e V_m / k_B T} - 1)}, \tag{8.51}$$

where P_i is defined as the permeability of the i-type ions,

$$P_i \equiv \frac{K_i D_i}{L}. \tag{8.52}$$

Equation 8.51 is known as the GHK current equation.

The GHK current equation reduces to the Nernst equation and the Ohm's law in the appropriate limits. In equilibrium, I_i is zero so that the term inside the square brackets of Equation 8.51 results in the Nernst equation (Equation 8.15). If the ion concentrations are uniform in the system, $c_i(x=0) = c_i(x=L) = c_i$, Equation 8.50 gives the Ohm's law for the i-type ions

$$I_i = \frac{z_i^2 e^2 D_i c_i}{k_B T} \frac{V_m}{L} = \sigma_i E, \tag{8.53}$$

where σ_i is the conductivity of the i-type ions (Equation 8.21). When the ion concentrations are asymmetric in the inside and outside compartments, the Ohm's law emerges also in the asymptotic regimes of very large membrane potentials, where the drift dominates over diffusion. The two asymptotic limits of Equation 8.51 are

$$\frac{k_B T}{z_i^2 e^2 P_i} I_i = \begin{cases} V_m c_i^i, & z_i e V_m / k_B T \to \infty \\ V_m c_i^o, & z_i e V_m / k_B T \to -\infty. \end{cases} \tag{8.54}$$

The results of the GHK current equation (Equation 8.51) are given in Figure 8.4 as a graphical representation, where $I_i/(z_i e P_i)$ is plotted against $z_i e V_m / k_B T$ for different ratios of c_i^i / c_i^o. The asymptotic limits of Equation 8.54 are evident in Figure 8.4.

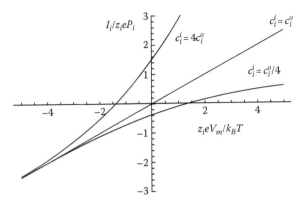

Figure 8.4 Plot of $I_i/z_i e P_i$ against $z_i e V_m/k_B T$ according to the GHK current equation. The asymptotic limits correspond to the Ohm's law. The curves are illustrated for $c_i^i/c_i^o = 4$, 1, and 1/4 (with the choice of $c_i^o = 1/2$).

8.4.3.2 Concentration Profile

Substitution of Equation 8.49 in Equation 8.47 gives the concentration profile as

$$c_i(x) = e^{\bar{z}_i V_m x/L} \left\{ c_i(0) + \frac{[c_i(L)e^{-\bar{z}_i V_m} - c_i(0)](e^{-\bar{z}_i V_m x/L} - 1)}{(e^{-\bar{z}_i V_m} - 1)} \right\}. \quad (8.55)$$

The spatial dependence is a function of only the ratio x/L. The concentration profiles given by Equation 8.55 are presented in Figure 8.5 as solid curves for monovalent cations ($z_i = +1$) and monovalent anions ($z_i = -1$). In Figure 8.5, $V_m = 100\,\text{mV}$, and the monovalent electrolyte concentrations at the pore ends are $c(x=0) = 0.5\,\text{M}$ and $c(x=L) = 0.1\,\text{M}$. The length L of the pore is taken as 2.5 nm. In fact, the length can be expressed in any unit of length as only the ratio x/L appears in Equation 8.55. The analytical results given by the constant field approximation are compared in Figure 8.5 with the full numerical results by solving the PNP equations in three dimensions (Corry et al. 2000). Although there are significant discrepancies in the quantitative details between the numerical results and the constant field approximation, the qualitative trends are adequately captured by the GHK equations (Chen et al. 1997, Syganow and von Kitzing 1999).

8.4.3.3 Resting Potential

When many electrolyte ions are present in the system, the net current due to all ions follows from Equations 8.8 and 8.51 as

$$I = \sum_i \frac{z_i^2 e^2 P_i V_m}{k_B T} \frac{[c_i^o e^{-z_i e V_m/k_B T} - c_i^i]}{(e^{-z_i e V_m/k_B T} - 1)}. \quad (8.56)$$

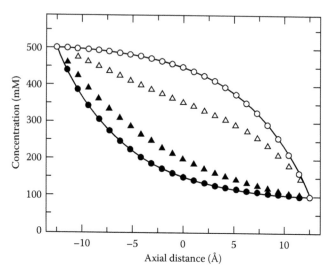

Figure 8.5 Concentration profile given by PNP for a concentration gradient of 0.4 M in monovalent salt concentration and a membrane potential of 100 mV. The constant field approximation gives the solid curves for anions (filled circles) and cations (open circles). The numerical PNP results are open triangles (cations) and filled triangles (anions). (From Corry, B. et al., *Biophys. J.*, 78, 2364, 2000. With permission.)

The value of the membrane potential at which the net current due to all ions is zero is called the resting membrane potential V_m^0. If the system contains only monovalent ions, the resting membrane potential can be calculated by taking the right-hand side of the above equation to be zero. The result is

$$\sum_{i\ cations} P_{i+} \frac{\left(c_{i+}^o e^{-eV_m^0/k_BT} - c_{i+}^i\right)}{\left(e^{-eV_m^0/k_BT} - 1\right)} + \sum_{i\ anions} P_{i-} \frac{\left(c_{i-}^o e^{eV_m^0/k_BT} - c_{i-}^i\right)}{\left(e^{eV_m^0/k_BT} - 1\right)} = 0.$$

(8.57)

A simple rearrangement of this equation gives an explicit expression for the resting membrane potential as

$$V_m^0 = \frac{k_BT}{e} \ln\left(\frac{\sum_{i\ cations} P_{i+}c_{i+}^o + \sum_{i\ anions} P_{i-}c_{i-}^i}{\sum_{i\ cations} P_{i+}c_{i+}^i + \sum_{i\ anions} P_{i-}c_{i-}^o}\right).$$

(8.58)

This equation is known as the GHK voltage equation. As an example, for a solution containing the electrolytes KCl and NaCl, the resting membrane potential is

$$V_m^0 = \frac{k_BT}{e} \ln\left(\frac{P_K c_K^o + P_{Na} c_{Na}^o + P_{Cl} c_{Cl}^i}{P_K c_K^i + P_{Na} c_{Na}^i + P_{Cl} c_{Cl}^o}\right).$$

(8.59)

When there is only one kind of ions controlling the membrane potential, Equations 8.58 and 8.59 reduce to the Nernst equation (Equation 8.15).

8.5 EFFECT OF BARRIERS

In the above treatment of charge fluxes through pores, we have been considering only the drift from the electrical forces and diffusion from local concentration gradients. However, since the pore walls are usually not inert for the ions, there can be multiple binding sites in the pore for the ion or the polymer segment that is undergoing translocation. A realistic pore can be quite rich in terms of the potential-energy landscape for the movement of an ion through the pore. Presence of such additional potential interaction between the pore and the ion can lead to significant modifications in the current–voltage curves given by the GHK current equation. We illustrate the nature of such modifications by considering the presence of a single potential barrier, as sketched in Figure 8.6a. Let the potential barrier $U(x)$ have the maximum value of U_m located at $x = L/2$,

$$U(x) = U_m - \frac{1}{2w^2}\left(x - \frac{L}{2}\right)^2, \tag{8.60}$$

where w is a measure of the width of the barrier.

As discussed in Section 6.4, there is an additional contribution from the potential barrier, $-(D_i/k_BT)\nabla U(\mathbf{r}, t)$, in the right-hand side of Equation 8.3. In view of Equation 8.6 for small ions, Equation 8.3 allows the simple modification of $z_i e\psi(\mathbf{r})$ into $z_i e\psi(\mathbf{r}) + U(\mathbf{r})$. For the one-dimensional situation described in Section 8.4, the charge flux due to the i-type ions given by Equation 8.35 is now modified to

$$I_i(x) = -z_i e D_i \left\{ \frac{\partial c_i(x)}{\partial x} + \frac{c_i(x)}{k_BT}\frac{\partial}{\partial x}[z_i e\psi(x) + U(x)] \right\}. \tag{8.61}$$

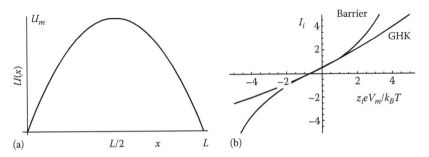

Figure 8.6 (a) Potential barrier with a maximum value U_m at $x = L/2$. (b) Plot of $I_i/z_i e P_{i,\mathrm{eff}}$ with the barrier against $z_i e V_m/k_BT$ is steeper than the corresponding plot of $I_i/z_i e P_i$ from GHK current equation ($c_i^i = 2c_i^o$; $c_i^o = 0.5$).

By repeating the same derivation as in Section 8.4.2, the analog of Equation 8.45 for the steady-state charge flux due to the i-type ions becomes

$$I_i = -z_i e D_i \frac{[c_i(L)e^{z_i e \psi(L)/k_B T} - c_i(0)e^{z_i e \psi(0)/k_B T}]}{\int_0^L dx e^{[z_i e \psi(x) + U(x)]/k_B T}}. \tag{8.62}$$

Taking the constant field approximation given by Equation 8.49, Equation 8.62 becomes

$$I_i \simeq -z_i e D_i \frac{e^{-U_m/k_B T}}{\sqrt{2\pi w^2 k_B T}} [c_i(L)e^{-z_i e V_m/2k_B T} - c_i(0)e^{z_i e V_m/2k_B T}]. \tag{8.63}$$

In obtaining the above result, we have assumed that the potential barrier is sufficiently large and narrow so that the integral in the denominator of Equation 8.62 can be evaluated accurately with the usual saddle point approximation (Arfken and Weber 2001). By defining an effective permeability,

$$P_{i,eff} = K_i D_i \frac{e^{-U_m/k_B T}}{\sqrt{2\pi w^2 k_B T}}, \tag{8.64}$$

Equation 8.63 becomes

$$I_i = -z_i e P_{i,eff} \left[c_i^o e^{-z_i e V_m/2k_B T} - c_i^i e^{z_i e V_m/2k_B T} \right]. \tag{8.65}$$

The Equation 8.64 containing the exponential term in $-U_m/k_B T$ is exactly the same as the general Kramers formula for barrier crossing. A comparison of Equations 8.52 and 8.64 shows that the pore length is essentially replaced by the width of the barrier accompanied by the exponential term in the barrier height. A plot of $I_i/(z_i e P_{i,eff})$ against $z_i e V_m/k_B T$ is given in Figure 8.6b. For comparison, the result from the GHK current equation (Equation 8.51) is included in the figure. It is evident that the presence of a potential barrier leads to steeper current–voltage curves. Occurrence of such steeper curves in experimental observations is an indicator of the possible presence of a potential barrier for the translocating ion through the pore. Although the GHK current equation is modified by the potential barrier, the GHK voltage equation and the Nernst equation are not affected by the barrier, as can be readily seen in Equation 8.65.

8.6 IONIC CURRENT THROUGH PROTEIN PORES

The experimentally observed current–voltage curves for protein channels can be quite complicated, instead of the simple Ohmic behavior (Aidley and Stanfield 1996, Hille 2001). These complications arise from the rich decoration of both positive and negative charges on the ionizable amino acid groups on the irregularly shaped internal interfaces of the pore. It is difficult to monitor the electric potential variations and the distributions of the mobile cations and anions along the pore, by using direct experimental probes. Only the net effect arising from all specific features of a particular pore and the mobile ions is measured as the pore current for an applied voltage. In view of this, computational modeling is desirable to develop an understanding of the molecular mechanisms behind the observed net current. However, the typical protein pores are very large in terms of their number of amino acids and water molecules. Even with modern computers, it is difficult to simulate the translocation of macromolecules for experimentally relevant timescales, by explicitly accounting for all amino acid residues, mobile ions, water molecules, and the polymer segments. There are also conceptual difficulties in performing reliable modeling due to insufficient understanding of the heterogeneous dielectric environment in the vicinity of pore and the membrane in which the pore is embedded. In contrast to such challenging detailed molecular dynamics simulations, the self-consistent PNP calculations are relatively easier.

Although the PNP formalism is only approximate, it has been successfully used to gain insight into the local variations of the electric potential and electrolyte concentrations along the pore. By comparing the calculated net pore current with experimental results, the correspondences between the pore structure and the pore current can be assessed. We illustrate the use of the PNP formalism by considering gramicidin A (GA) channel, which is permeable only to simple monovalent cations, and the protein pore made from α-hemolysin (αHL), which is permeable to single-stranded polynucleotides and synthetic flexible polyelectrolyte molecules.

8.6.1 Gramicidin A Channel

The GA channel is made of two identical small polypeptide molecules spanning a pore distance of about 2 nm across a membrane (Arseniev et al. 1986, Roux and Karplus 1994, Kurnikova et al. 1999) (Figure 8.7a). By prescribing the appropriate values for the atomic radii and partial charges of the various atoms on the inside of the pore, the ionic current has been calculated (Kurnikova et al. 1999, Cardenas et al. 2000) with the PNP equations. The calculated results are given in Figure 8.7b, by assuming that the dielectric constant in the membrane and protein region is 2 and that in the aqueous region inside the pore is 80. The monovalent electrolyte concentration is 0.5 M on the inside compartment and 0.04 M on the outside compartment. The steady-state current depends on the diffusion coefficients of the electrolyte ions. The value of the diffusion coefficient is adjusted to be 1.27×10^{-6} cm^2/s for both the cation and anion, although this value is smaller than the bulk value (Lynden-Bell and Rasaiah 1996). The computed current–voltage curve, with the appropriate charges on the pore, is nonlinear, in qualitative agreement with experimental data. If the charges were to be absent, the current–voltage relation is simply linear. The substantial difference in the current between the charged and uncharged GA channel demonstrates the significant role played by the charge decoration of the pore on the pore current.

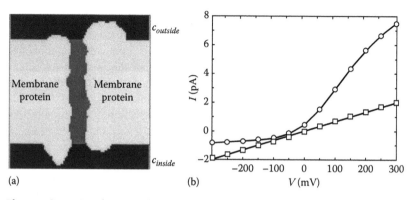

Figure 8.7 (a) The GA channel. (b) Current–voltage curves for charged GA (circles) and uncharged GA (squares). (From Kurnikova, M.G. et al., *Biophys. J.*, 76, 642, 1999. With permission.)

The variations of the electric potential and the ion concentrations along the pore axis are given in Figure 8.8 (Kurnikova et al. 1999). If the pore is uncharged, the electric potential drops essentially linearly across the pore. In Figure 8.8a, this is represented by the dot-dashed line for an external potential difference ($\psi_{inside} - \psi_{outside}$) of 300 mV. In contrast, when the pore is charged (circles in Figure 8.8a), the electric potential gradient is the steepest at the pore mouth, followed by a weaker electric field inside the pore. Finally, the potential reaches the value in the outside compartment with a negative electric field at the exit region of the pore. The ranges of distance from the pore, both in the inside and outside regions, where there is any significant spatial dependence of the electric potential, are merely in subnanometers for the electrolyte concentrations used in the calculation. If the polarity of the external potential difference is reversed to −300 mV, the curve with squares (Figure 8.8a) is obtained. The ion concentrations according to the PNP calculations are given in Figure 8.8b, for the electrolyte concentration being 0.5 and 0.04 M in the inside and outside regions, respectively, and for the external potential difference of 300 mV. The dot-dashed curve corresponds to the cations for an uncharged GA pore. Charging the GA pore results in substantial change in the ion distributions (circles). The triangles denote the distribution of anions along the pore axis for the charged GA pore. Thus, the PNP formalism offers a tool to explore the correlation between the measured pore current and the electrostatic environment inside the pore. Nevertheless, it must be recognized that the PNP methodology is only the zeroth-order description and the specificity of the ions cannot be addressed without parametrization.

8.6.2 α-Hemolysin Pore

The αHL protein pore is a heptameric assembly of αHL molecules with a mushroom shape (Figure 8.9a). In the single-molecule translocation experiments, a single αHL pore is incorporated into a membrane separating

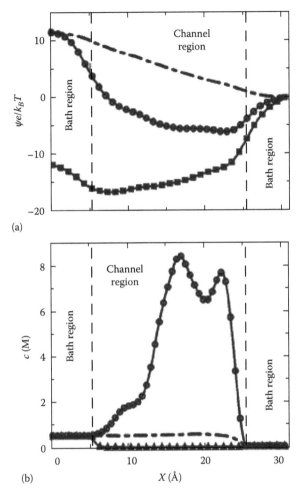

Figure 8.8 Variations of the electric potential and ion concentrations along the pore axis of the GA channel embedded inside a membrane. The inside and the outside regions are the bath regions on the left and right, respectively. The monovalent electrolyte concentration is 0.5 M (left) and 0.04 M (right). (a) Electrostatic potential for uncharged GA (dot-dashed) at $V_m = 300$ mV and charged GA at $V_m = 300$ mV (circles) and $V_m = -300$ mV (squares). (b) Concentrations at $V_m = 300$ mV: cations for uncharged GA (dot-dashed) and charged GA (circles), and anions for charged GA (triangles). (From Kurnikova, M.G. et al., *Biophys. J.*, 76, 642, 1999. With permission.)

the donor (*cis*) and receiver (*trans*) compartments, both of which contain a buffered electrolyte solution. The configuration of the αHL pore (Song et al. 1996) consists of a vestibule on the *cis* side and a transmembrane β-barrel on the *trans* side. The length of the pore is about 10 nm. The opening of the vestibule at the *cis* side is ≈2.9 nm and the diameter of the vestibule's cavity

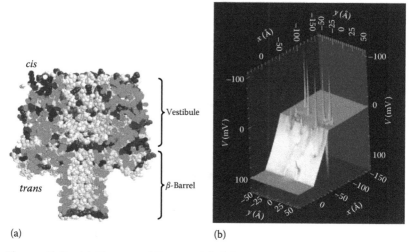

Figure 8.9 (a) The space-filling model of the αHL protein pore in pH = 7.5. Different shades denote the heterogeneities in charge and hydrophobicity. (From Wong, C.T.A. and Muthukumar, M., *J. Chem. Phys.*, 133, 045101, 2010. With permission.) (b) The variation of the electric potential along the axis of the αHL protein pore. (From Kong, C.Y. and Muthukumar, M., *J. Am. Chem. Soc.*, 127, 18252, 2005. With permission.)

is $\approx$4.1 nm. The average internal diameter of the β-barrel is $\approx$2 nm. The two domains of the pore's lumen are separated by a constriction of $\approx$1.4 nm. The inner surface of the pore is richly decorated with charged and hydrophobic amino acid residues. For the concentration of 1 M monovalent electrolytes, the PNP results (Kong and Muthukumar 2005) for the variation of the electric potential from the *cis* side ($\psi = 0$ mV) to the *trans* side ($\psi = 120$ mV) are presented in Figure 8.9b. The pore axis is along the x-axis and the y-axis is normal to the pore axis. The steep gradient in the electric potential is only across the pore, and the electric fields in the *cis* and *trans* chambers are very weak for the electrolyte concentrations used in these calculations. The sharp peaks in Figure 8.9b correspond to the locations of fixed charges of the amino acid residues.

The calculated value of the open pore current for the potential difference between the *cis* and *trans* of -120 mV for 1 M KCl-type electrolyte concentration is about 135 pA (Muthukumar and Kong 2006). The calculated open pore current is very close to the experimentally observed average value of about 120 pA. It must be stressed that the PNP value for the open pore current is a single number, because it is based on a mean field theory, whereas the experimental value fluctuates with time about an average value. Within the PNP formalism, the dynamics of the polyelectrolyte chain is modeled by the Langevin dynamics method (Section 4.3) by incorporating the forces from the pore, externally imposed potential gradient, and the dielectric heterogeneity in the system (Muthukumar and Kong 2006). As the polymer chain

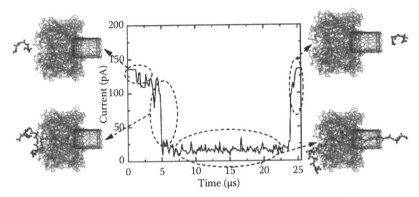

Figure 8.10 A typical ionic current trace, from computer simulations, as a polymer molecule undergoes translocation through the αHL protein pore. (From Muthukumar, M. and Kong, C.Y., *Proc. Natl. Acad. Sci. USA*, 103, 5273, 2006. With permission.)

translocates through the pore, the instantaneous steady-state ionic current is computed. A typical ionic current trace is presented in Figure 8.10 for the applied voltage of $-120\,\mathrm{mV}$ and 1 M KCl-type solutions.

Several important features of polymer translocation are evident from Figure 8.10. The presence of polymer monomers at the mouth of the vestibule is sufficient to elicit a response in the ionic current by providing a spatial blockade to the flow of electrolyte ions. As time progresses, the chain gets into the vestibule not necessarily as a single-file transport. Because of its relatively large vestibule volume in comparison with the volume of the polymer segments inside the vestibule, the segments reside in the cavity for some definite duration of time exploring their conformational entropy. As a result, the vestibule acts like an entropic trap. At this stage, the ionic current is $\approx 100\,\mathrm{pA}$. The fluctuations in the ionic current around its average value reflect the dynamics of the polymer rattling inside the vestibule. As time passes by, one end of the chain eventually enters the β-barrel. At this time, the ionic current drops sharply to about 17 pA. This low value of the ionic current is essentially due to the reduction in the cross-sectional area of β-barrel for flow of electrolyte ions in the presence of polymer segments. Furthermore, the fluctuations about 17 pA are solely due to the dynamics of the confined polymer inside the pore. The chain eventually gets out of the β-barrel and the ionic current returns to the open pore current. The translocation time is directly calculated as the time of ionic current blockade for every successful translocation event. By repeating the above calculation, which is a combination of the Langevin dynamics for the polymer and the PNP formalism for ion flow, the various measures of the translocation phenomenon such as the distribution functions of translocation times and blocked ionic currents can be obtained and interpreted in terms of molecular mechanisms. The advantage of the PNP methodology and the Langevin method for polymers is that the explored timescales are pertinent to the experimental scales in the range of microseconds to milliseconds. However, details of rearrangement of water molecules and orientational correlations of the bond vectors of the various bonds constituting the polynucleotides are not captured adequately. In the other extreme of molecular dynamics simulations (Mathe et al. 2005), where there is a potential to address these atomistic details, the timescales that can be explored currently are several orders of magnitude shorter than the typical values of the experimentally observed translocation times.

8.7 FLUCTUATIONS IN IONIC CURRENT

In general, the ionic current through an open pore is noisy, due to the thermal motion of the ions even with a drift along the electric field (Aidley and Stanfield 1996, Bezrukov 2000, Bezrukov et al. 2000, Hille 2001). As an example, a typical current trace for the open pore current in a single αHL pore is given in Figure 8.11a for an applied voltage of $-170\,$mV and 1 M KCl concentration, with a cutoff frequency of 10 kHz. The analogous traces when there are blockades of pore current are given in Figure 8.11b and c. These traces correspond to a single molecule of sodium (polystyrene sulfonate) undergoing translocation through the αHL pore. The trace in Figure 8.11b corresponds to the situation where the polymer resides inside the vestibule, and the trace in Figure 8.11c corresponds to the situation when the polymer is being threaded through the β-barrel. It is clear that the dynamics of the polymer influences the fluctuations of the ionic current through the pore. The interference between the polymer dynamics and the fluctuations in the ionic current through a pore is yet to be understood, and the analysis of the ionic current fluctuations in the presence of the polymer might be a quantitative way to monitor this coupling.

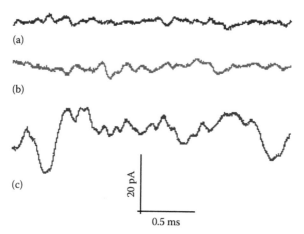

Figure 8.11 Typical ionic current traces for αHL protein pore. (a) Open pore current. In the presence of sodium poly(styrene sulfonate) in the vestibule (b) and the β-barrel (c), ionic current fluctuations are modified. The scales for the current and time are the same for all traces.

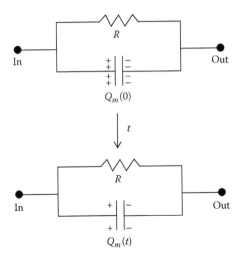

Figure 8.12 Decay of fluctuations in voltage with time.

The ionic current fluctuations in a pore are conveniently analyzed by constructing their power spectrum. We illustrate this procedure by recalling the electrical equivalent circuit of Figure 8.3b. Consider a simple equivalent electrical circuit consisting of a resistor (pore) and a capacitor (membrane) in parallel (Figure 8.12). Let there be a fluctuation δV in voltage across the leads representing the "in" and "out" terminals. This fluctuation arises from that in the ionic current through the pore. The charge buildup on the capacitor of capacitance C_m, due to δV, is (Young and Freedman 2000)

$$Q_m = C_m \delta V. \tag{8.66}$$

The energy associated with creating a change of δV in the capacitor is (Young and Freedman 2000)

$$E = \frac{1}{2} C_m (\delta V)^2. \tag{8.67}$$

Because the probability $W(\delta V)$ of realizing a fluctuation δV is given by the Boltzmann distribution function in $E/k_B T$,

$$W(\delta V) \sim e^{-C_m (\delta V)^2 / 2k_B T}, \tag{8.68}$$

the mean square fluctuation in voltage is

$$\langle (\delta V)^2 \rangle = \frac{k_B T}{C_m}. \tag{8.69}$$

The occurrence of δV is equivalent to a flow of current I. Since the rate of change of the charge is the current,

$$I = -\frac{dQ_m}{dt}, \tag{8.70}$$

where the negative sign is due to the charge on the capacitor being reduced when a positive current flows through a resistor. Combining Equations 8.66 and 8.70 and using the relation between the current and the resistance R, $I = \delta V / R$, we get

$$\frac{\partial(\delta V)}{\partial t} = -\frac{\delta V}{RC_m}, \tag{8.71}$$

with the solution,

$$\delta V(t) = \delta V(0)e^{-t/(RC_m)}. \tag{8.72}$$

This is a standard result for "RC-circuits" (Young and Freedman 2000).

In the presence of noise in the circuit, Equation 8.71 is modified to

$$\frac{\partial(\delta V)}{\partial t} = -\frac{1}{RC_m}\delta V(t) + \Gamma(t), \tag{8.73}$$

where $\Gamma(t)$ is the noise term with its average value assumed to be zero, $\langle \Gamma(t) \rangle = 0$. The above stochastic equation is solved by using the techniques described in Chapter 6. For now, the following simple argument would suffice. Rearranging Equation 8.73,

$$e^{-t/RC_m} \frac{d[(\delta V)e^{t/RC_m}]}{dt} = \Gamma(t). \tag{8.74}$$

Integration of this equation yields

$$\delta V(t) = \delta V(0)e^{-t/RC_m} + e^{-t/RC_m} \int_0^t dt' \Gamma(t')e^{t'/RC_m}. \tag{8.75}$$

By multiplying this equation by $\delta V(0)$ and performing the average over the fluctuating noise, we get

$$\langle \delta V(0)\delta V(t) \rangle = \langle (\delta V)^2 \rangle e^{-t/RC_m} = \frac{k_B T}{C_m}e^{-t/RC_m}, \tag{8.76}$$

where the property of the noise $\langle \Gamma(t) \rangle = 0$ and Equation 8.69 are used. Integrating the above equation over time, we obtain an expression for the resistance R as an integral over the correlation function in the fluctuations in voltage as

$$R = \frac{1}{k_B T} \int_0^\infty dt \langle \delta V(0)\delta V(t) \rangle. \tag{8.77}$$

In fact, this is another form of the fluctuation–dissipation theorem, as discussed generally in Section 6.2, now relating the resistance of the pore to the fluctuations in the pore current.

Going back to the traces of Figure 8.11, resolving their signals in terms of the Fourier transforms allows the determination of contributions from each frequency ω. The Fourier transform of the correlation function $\langle \delta V(0)\delta V(t)\rangle$, as defined through

$$P(\omega) = \frac{1}{2\pi} \int_{-\infty}^{\infty} dt\, e^{-i\omega t} \langle \delta V(0)\delta V(t)\rangle, \qquad (8.78)$$

is called the power spectrum or the spectral density. Substitution of Equation 8.76 into Equation 8.78 gives the spectral density as

$$P(\omega) = \frac{k_B TR}{\pi}\, \frac{1}{1 + (\omega/\omega_c)^2}, \qquad \omega_c \equiv 1/RC_m. \qquad (8.79)$$

The two limits of $\omega \ll \omega_c$ and $\omega \gg \omega_c$ are apparent from this equation,

$$P(\omega) = \begin{cases} \dfrac{k_B TR}{\pi} & \omega \ll \omega_c \\[2mm] \dfrac{k_B T}{\pi RC_m^2}\dfrac{1}{\omega^2} & \omega \gg \omega_c. \end{cases} \qquad (8.80)$$

The result in the limit of $\omega \to 0$ is known as the Nyquist theorem. The normalized spectral density $\pi P(\omega)/k_B TR$ is plotted against the frequency ω in Figure 8.13 as a double logarithmic plot for two different values of ω_c. Since ω_c gives the crossover between the plateau limit and the power-law regime, it is called the corner frequency. In the present case, ω_c is $1/RC_m$ and the slope in the large frequency regime is -2. The values of ω_c and the asymptotic slope are signatures of the particular pore and the particular molecule undergoing

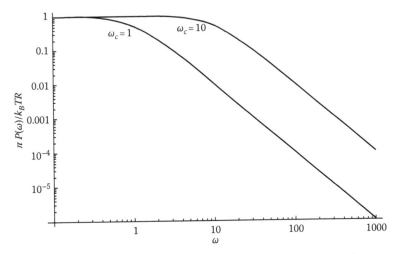

Figure 8.13 Plot of the normalized spectral density against the frequency. Two values of the corner frequency ω_c are illustrated.

translocation. More experimental work is needed to capitalize on the spectral analysis of the pore current in order to explore the dynamics of translocating polymer chains at intermediate and short timescales.

8.8 ELECTROOSMOTIC FLOW

The phenomenon of electroosmosis is the movement of a liquid relative to an immobile charged surface by an applied electric field. Due to the charge density on the surface, more number of counterions are present near the surface than in the bulk away from the surface (Section 3.2.2). These counterions being mobile will move collectively toward the oppositely charged electrode under an externally applied electric field, and at the same time carrying the solvent molecules with them. As a result, there is a net flow of the liquid, with its direction being dictated by the direction of the movement of counterions toward their oppositely charged electrode. For example, positively charged surface makes the negatively charged counterions move toward the (positively charged) anode, so that there is a net electroosmotic fluid flow toward the anode.

Useful expressions for the EOF velocity can be obtained as follows. By balancing the viscous forces of the fluid and the electrical forces due to the various ions in the solution, we get from the linearized Navier–Stokes equation (Equation 7.2),

$$-\eta_0 \nabla^2 \mathbf{v}(\mathbf{r}) = \rho(\mathbf{r})\mathbf{E}, \tag{8.81}$$

where η_0 is the zero shear rate solvent viscosity and $\mathbf{v}(\mathbf{r})$ is the local velocity. The right-hand side of the equation is the electrical force due to the local charge density $\rho(\mathbf{r})$ in the presence of the externally applied uniform electric field $\mathbf{E}$. The effects from the inertial and gravitational forces and any pressure gradients are ignored for now, which can be added on to the final formulas for EOF. We now consider the EOF velocity for two common situations, namely, near a planar surface and inside a cylindrical pore.

8.8.1 EOF near Planar Interfaces

Let the interface be an infinite plane along x-, z-directions (Figure 8.14) at $y = 0$ with a surface potential ψ_s, and the electric field be parallel to the surface in the positive x-direction, denoted as E_x. For infinitely long distances along x, the fluid velocity and the charge density are uniform so that their derivatives with respect to x are zero. The variations of the velocity and the charge density are only along the y-direction given by

$$-\eta_0 \frac{\partial^2 v_x}{\partial y^2} = \rho(y)E_x = -\epsilon_0\epsilon \frac{\partial^2 \psi}{\partial y^2} E_x, \tag{8.82}$$

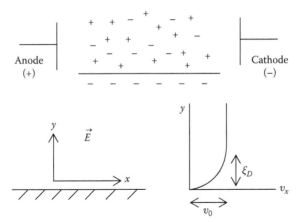

Figure 8.14 EOF near a charged planar interface.

where the Poisson equation for the charge density is used, with ψ being the local electric potential. Integration of the above equation gives

$$\eta_0 \frac{\partial v_x}{\partial y} = \epsilon_0 \epsilon \frac{\partial \psi}{\partial y} E_x + C_1, \tag{8.83}$$

where C_1 is the integration constant. At $y \to \infty$, far away from the charged surface, $\partial v_x / \partial y = 0 = \partial \psi / \partial y$ so that $C_1 = 0$. Integration of Equation 8.83 gives

$$\eta_0 v_x = \epsilon_0 \epsilon \psi E_x + C_2. \tag{8.84}$$

At the surface, the potential is ψ_s and the velocity of the fluid is taken to be zero with the assumed no-slip boundary condition. Using these boundary conditions, the integration constant C_2 becomes

$$C_2 = -\epsilon_0 \epsilon \psi_s E_x. \tag{8.85}$$

Substituting this result in Equation 8.84, we get

$$v_x(y) = \frac{\epsilon_0 \epsilon}{\eta_0}[\psi(y) - \psi_s]E_x. \tag{8.86}$$

The dependence of the electric potential on the distance from the planar interface is derived in Section 3.2.2. For example, with the Debye–Hückel approximation, $\psi(y)$ falls off exponentially with y and the characteristic decay length is the Debye length (Equation 3.62 and Figure 3.10). Therefore, at distances larger than the Debye length away from the surface, the potential vanishes and the velocity of the fluid acquires the terminal velocity v_0 given by

$$v_0 = -\frac{\epsilon_0 \epsilon \psi_s}{\eta_0}E_x. \tag{8.87}$$

This relation is known as the Helmholtz–Smoluchowski equation. A sketch of the result from Equation 8.86 is included in Figure 8.14. The thickness of the "skin" is essentially the Debye length, which is dictated by the salt concentration. For aqueous solutions at room temperatures with the surface potential of 0.1 V and an electric field of 10^5 V/m, the terminal EOF velocity is about 10^7 nm/s, which is substantial in nanofluidics.

8.8.2 EOF inside Cylindrical Pores

The generalization of the Helmholtz–Smoluchowski description of the EOF for cylindrical pores (Figure 8.15a) is carried out as follows. Let a and L be the radius and length of a uniform cylindrical pore with surface potential ψ_s. Let the electric field be along the x-direction and r be the radial distance from the axis of the cylinder in the yz-plane orthogonal to the x-axis. As in the case for the planar surfaces, the EOF velocity is calculated from Equation 8.81 by implementing the appropriate boundary conditions.

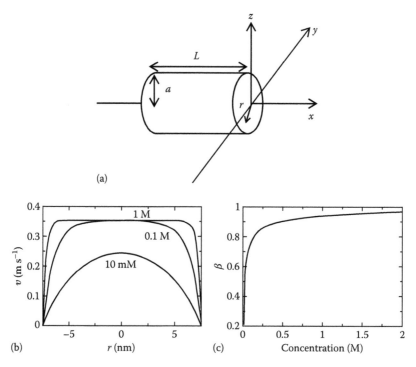

Figure 8.15 EOF inside a cylindrical pore. (a) Sketch of the pore. (b) Velocity profile at different monovalent salt concentrations. (c) Strength coefficient of the EOF flux as a function of salt concentration. ($\psi_s = 0.01$ V, $a = 7.5$ nm, $E_x = 5 \times 10^7$ V/m, $T = 300$ K, $\eta_0 = 1$ cP, and $\epsilon = 80$.). (From Wong, C.T.A. and Muthukumar, M., *J. Chem. Phys.*, 126, 164903, 2007. With permission.)

The charge density $\rho(r)$ at a radial distance r from the pore axis was derived for the geometry of Figure 8.15a in Section 3.2.3 as given by Equation 3.74, within the Debye–Hückel approximation. Substituting Equation 3.74 in Equation 8.81 and recognizing the cylindrical symmetry, we get

$$\eta_0 \frac{1}{r} \frac{\partial}{\partial r} \left[r \frac{\partial}{\partial r} v_x \right] = \epsilon_0 \epsilon \kappa^2 \psi_s E_x \frac{I_0(\kappa r)}{I_0(\kappa a)}, \tag{8.88}$$

where κ is the inverse Debye length, and $I_0(x)$ is the zeroth-order modified Bessel function of the first kind (Abramowitz and Stegun 1965). This equation can be solved for the velocity by using the no-slip boundary condition at the pore wall to yield

$$v_x(r) = -\frac{\epsilon_0 \epsilon \psi_s E_x}{\eta_0} \left[1 - \frac{I_0(\kappa r)}{I_0(\kappa a)} \right]. \tag{8.89}$$

The velocity profile is illustrated in Figure 8.15b for some typical values of the pore radius and salt concentration. At high salt concentrations, the velocity profile is flat except near the wall. The thickness of the narrow region with a gradient in the velocity is dictated by the Debye length. At low salt concentrations, the Debye layers overlap and the velocity profile is parabolic. In most of the experiments on translocation, the salt concentration is high and the velocity profile is essentially planar. For typical values of the parameters in aqueous solutions at room temperatures with $\psi_s = 0.01$ V, $a = 7.5$ nm, and $E_x = 5 \times 10^7$ V/m, the EOF velocity can be in the range of 0.3 m/s, which is a significantly large velocity.

The net flux J_0 of the fluid through the pore is given by the integral of $v_x(r)$ over the area of cross-section as

$$J_0 = \int_0^a dr \, (2\pi r) \, v_x(r) = -\frac{\epsilon_0 \epsilon \psi_s E_x}{\eta_0} (\pi a^2)\beta, \tag{8.90}$$

where β is a geometric numerical coefficient,

$$\beta \equiv 1 - \frac{2}{\kappa a} \frac{I_1(\kappa a)}{I_0(\kappa a)}. \tag{8.91}$$

Here $I_1(x)$ is the first-order modified Bessel function of the first kind (Abramowitz and Stegun 1965). The factor β is plotted in Figure 8.15b as a function of monovalent salt concentration. It approaches unity for large values of κa, corresponding to high salt concentrations, according to

$$\beta \simeq 1 - \frac{2}{\kappa a} + \cdots, \quad \kappa a \gg 1. \tag{8.92}$$

The negative sign in Equation 8.90 gives the correct direction of the flow. For positively charged cylindrical pore ($\psi_s > 0$), the flow is in the opposite direction to the electric field and toward the positive electrode (anode).

It must be noted that for high values of κa, that is, for the pore radius being larger than the Debye length, which can be realized at higher salt concentrations, the pore radius influences the EOF velocity only as the cross-sectional area of the pore. On the other hand, if there were a pressure drop $\Delta p/L$ across the pore, then the total flux would be

$$J_0 = J_{0,EOF} + \frac{\pi a^4}{8\eta_0} \frac{\Delta p}{L}, \tag{8.93}$$

where $J_{0,EOF}$ is the electroosmotic flux given by Equation 8.90. The second term on the right-hand side of the above equation is the Poiseuille's result for the flow of a viscous liquid through a cylindrical pipe under a pressure drop (Young and Freedman 2000). While the EOF contribution is proportional to a^2, the Poiseulle contribution is proportional to a^4. Therefore, the ratio of the EOF to the hydraulic flow, proportional to $1/a^2$, becomes higher as the pore radius decreases. In fact, controlling the extents of EOF and Poiseulle contributions with the surface charge and pressure drop offers opportunities to deliberately control the fluid flow and consequently the speed of the translocating macromolecules.

8.9 SUMMARY

In general, the flux of ions and charged macromolecules through a pore in the presence of an electric field is determined by their (a) diffusion due to concentration gradients, (b) drift due to the externally imposed force, and (c) penetration through potential barriers from the pore. Depending on the experimental conditions, three limiting regimes, namely, diffusion-dominated, drift-dominated, and barrier-dominated regimes, can emerge. A useful method to address these features is the PNP formalism. In equilibrium, the PNP equations reduce to the Nernst equation relating the potential difference and the electrolyte concentration gradient across the pore. In the steady state, explicit formulas for the variations of the electric potential and the concentrations of the electrolyte ions are derived. For situations where the electric field is uniform across the pore, the PNP equations yield the analytical formulas of GHK.

In the steady state, when the drift contribution dominates, the ionic current obeys the Ohm's law. In the absence of either drift or barriers, the behavior of ions is according to the Fick's law. The GHK equations offer a convenient way to describe the crossover behavior between the diffusion-, drift-, and barrier-dominated regimes. We have also shown the utility of the numerically solved results from the PNP equations for the ionic currents through the GA channel and the αHL protein pore. The PNP calculations show that the steepest gradient in the electrical potential is only very near and across the pore. We have also considered fluctuations in the local charge density and the ionic current through the pores. The fluctuations in charge density die out quickly in comparison with any relevant timescales for macromolecules. The coupling among the dynamics of the small ions and the translocating polymer inside a pore affects significantly the spectral density of the ionic current through the pore.

Finally, we have described the EOF, arising from the flow of the fluid near charged surfaces in the presence of an electric field. Explicit formulas are derived for the EOF in terms of the various experimental parameters of the pore. It turns out that the EOF velocity can be very high in charge-bearing nanopores and can affect the conformations and speed of a translocating polymer. The EOF velocity can be further tuned by imposing a hydrostatic pressure gradient across the pore. The derived formulas are useful for judiciously tuning the surface charge density and the diameter of the pore, electrolyte concentration, the electric field, and the pressure gradient in order to achieve the desired speed of the translocating polymer. This in turn will facilitate reliable interrogation of the polymer monomers as they undergo translocation through a nanopore.

9

POLYMER CAPTURE

Imagine a polyelectrolyte molecule of contour length $\sim 6\,\mu m$ and radius of gyration $\sim 500\,nm$ having to find a tiny hole of radius $\sim 2\,nm$ embedded in a wall that is essentially of infinite dimensions. This apparently daunting task of placing one end of the polymer at the pore entrance is the necessary first step for polymer translocation across a nanopore to occur. The same mechanisms discussed in the last chapter for the transport of small ions through pores, namely the diffusion, drift, and barriers, are operative for the transport of polyelectrolytes as well. The significant difference now is that the chain connectivity and the counterions of the polymer substantially modify these contributing factors. In addition, the electroosmotic flow (EOF) generated by charge-bearing pores in the presence of an electric field can lead to dramatic changes in the way the polymer is captured by the pore. There have been many experiments on the capture rate of the polymer by a nanopore, in terms of its dependence on the chain length, polymer concentration, salt concentration, and the strength of the applied voltage difference. We shall discuss the experimental findings in light of the various contributions arising from the diffusion, drift, barriers, and EOF, by assembling the various essential conceptual elements already described in the previous chapters. In the present chapter, we focus on the steady-state capture rate of the center-of-mass of the polymer and relegate the details of threading of the polymer through the pore to the next chapter.

9.1 REPRESENTATIVE EXPERIMENTAL RESULTS

As introduced in Section 1.2, there is a large body of phenomenology based on single-molecule nanopore-based translocation experiments. In most of these experiments, the events of polymer translocation are monitored by measuring the ionic current through a single nanopore. A typical trace of the measured ionic current as a function of time is given in Figure 1.6. An encounter between a polymer molecule arriving from the donor compartment and the pore is signaled by a temporary reduction in the ionic current through the pore. Let t_0 be the time duration of the interval between any two successive encounters between polymer molecules and the pore. During an encounter between the polymer and the pore, the polymer can either successfully translocate into the receiver compartment or unsuccessfully return to the donor compartment. Regardless of whether the translocation has actually occurred in an encounter or not, the frequency of the encounters between the polymer molecules and the pore can be recorded. This frequency of polymer–pore encounters, called the capture rate, is found to depend on many experimental variables such as the polymer concentration, polymer length, pore geometry, chemical decoration of the pore, ionic strength, applied voltage difference, and gradients in hydrostatic pressure and salt concentrations across the pore. As only a certain fraction of these encounters results in the eventual completion of the translocation process, the probability of successful translocation can also be monitored as a function of the various experimental variables. We briefly summarize below some of the salient features of the experimental findings on the capture rate.

Let J_c denote the rate of capture of the polymer of chain length N by the pore. J_c is the inverse of the average arrival time $\langle t_0 \rangle$ between any two successive events of ionic current blockades. Alternatively, a histogram of t_0 may be constructed, such as $\exp(-J_c t_0)$, and the rate J_c can then be extracted. The capture rate J_c was shown (Henrickson et al. 2000) to depend exponentially on the voltage difference (V_m) (Figure 9.1a), above a threshold voltage for biotinylated poly$(dC)_{30}$ through αHL protein pore in 1 M KCl. Furthermore, the dependence of J_c on the voltage bias depended on whether the polymer entered from the *cis* compartment or from the *trans* compartment, as seen in Figure 9.1a. The inset shows that J_c is directly proportional to the polymer concentration c_0 in the donor compartment. In contrast, two exponentials (Figure 9.1b) were reported (Meller and Branton 2002) for the capture of poly$(dC)_{40}$ through αHL in 1 M KCl, indicating two activated processes at low- and high-voltage difference regimes. Again, J_c is proportional to c_0. For the capture of dsDNA by a solid-state nanopore (3–4 nm in diameter), J_c was found (Wanunu et al. 2010) to increase with V_m first exponentially, and then linearly for higher values of V_m (Figure 9.1c). In this case, the chain length dependence of J_c showed an N-independent regime for large N and sharper increasing dependence on N for lower N values (Figure 9.1d). In all these experiments, J_c was found to be linearly proportional to the polymer concentration c_0.

Thus, the dependencies of the capture rate J_c on the applied voltage difference and the chain length are quite rich, although there is a uniform behavior in terms of the linear dependence on the polymer concentration. It is remarkable that the capture rate actually increases with chain length, which is counterintuitive because larger macromolecules are expected to undergo slower diffusion before getting captured. It seems from the experimental data that there are two threshold voltages for polymer capture, one for the capture to occur at all, and the other for the occurrence of a linear dependence on V_m. The above experimental results have attracted several theoretical efforts (Ambjornsson et al. 2002, Chen et al. 2004a, Muthukumar 2010, Wanunu et al. 2010).

In addition to the above features of polymer capture, several novel behaviors of polymer molecules have been observed in translocation experiments. As an example, time-resolved fluorescence monitoring (Chen et al. 2004a) of labeled DNA molecules as they translocated through a solid-state nanopore of 15 nm in diameter showed that there is a very large capture region of radius $\sim$3 μm in front of the nanopore (Figure 9.2). Far from the pore, the molecules

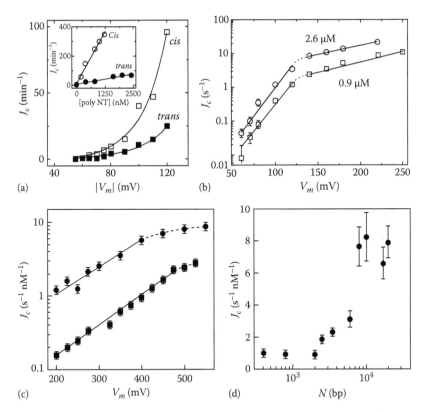

Figure 9.1 (a) Capture rate J_c depends exponentially on the voltage difference V_m for ssDNA through αHL. The rate of increase of J_c with V_m depends on the direction of translocation. Inset: J_c is proportional to the polymer concentration. (From Henrickson, S.E. et al., *Phys. Rev. Lett.*, 85, 3057, 2000. With permission.) (b) Double exponential dependence of J_c on V_m for ssDNA through αHL. (Adapted from Meller, A. and Branton, D., *Electrophoresis*, 23, 2583, 2002.) (c–d) Capture of dsDNA by 3–4 nm pore. (Adapted from Wanunu, M. et al., *Nat. Nanotechnol.*, 5, 160, 2010.) (c) Exponential and linear dependence of J_c on V_m (top, 3500 bp; bottom, 400 bp). (d) N-dependence for 300 mV and 1 M KCl.

appeared to undergo diffusion. But once the molecules reached a distance of about 3 μm from the pore, they were quickly pulled toward the nanopore for subsequent rapid disappearance into the pore. Inside the capture region, the molecule appeared to be elongated and the average polymer concentration is essentially zero. The magnitude of the capture region is unexpectedly several orders of magnitude larger than the pore radius. The literature on the phenomenology of polymer translocation continues to grow and many observed features are yet to be fully understood.

We describe below some of the basic principles behind the physics of polymer capture. Our primary goal is to delineate different regimes for the

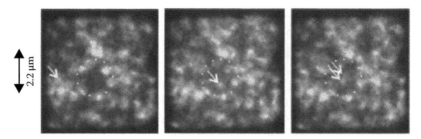

Figure 9.2 Occurrence of a large absorbing region in front of a solid-state nanopore, as seen in time-resolved fluorescence measurements. The arrow indicates the trajectory of a molecule. As soon as the molecule slowly enters the capture region, it gets rapidly captured, elongated, and translocated through the pore. (From Chen, P. et al., *Nano Lett.*, 4, 2293, 2004a. With permission.)

dependence of the rate of polymer capture on the various experimental variables and to understand the flow behavior in the vicinity of the nanopore.

9.2 GENERAL CONSIDERATIONS

As described in Chapter 1, there are three essential stages in the process of single-file polymer translocation through a nanopore. These are the (1) drift–diffusion of the polymer far away from the pore, (2) capture of the polymer at the pore entrance, and (3) crossing of the free energy barrier by the polymer at the pore entrance for the eventual successful translocation. These three stages are marked in Figure 9.3. In the first stage, away from the pore, the polymer undergoes a combination of electrophoretic drift due to the external electric field and diffusion due to thermal motion. Near the pore, strong flow fields can develop within a radial distance of r_c from the pore, primarily due to the EOF and pressure gradients across the pore. The range r_c and the magnitude of the velocity gradients inside this region in front of the pore depend on the experimental conditions. The behavior of the polymer inside and outside the region of r_c can be qualitatively different, depending on the nature of the velocity gradients. Within the range r_c, the polymer can undergo conformational changes such as the coil-stretch transition (Section 7.5). By experiencing such forces within r_c, the polymer approaches the pore. This is designated as stage 2 in Figure 9.3. When the polymer is driven to the pore mouth, it is generally in a jammed state, without any regard to the necessary condition for a single-file translocation where one of the chain ends must be at the pore entrance. The polymer must negotiate the free energy barrier, primarily arising from the entropy loss associated with the placement of one chain end at the pore entrance, for a successful translocation event to occur. This is designated as stage 3 in Figure 9.3. The barrier crossing in stage 3 is quite rich and will be discussed in the following chapter. Here we shall take the stage 3 only as a single-step passage of the chain through the free energy barrier.

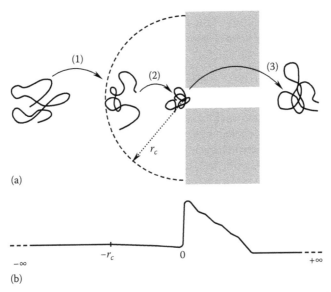

Figure 9.3 (a) Three main stages of polymer translocation process: (1) drift–diffusion, (2) capture, and (3) translocation through barrier crossing. (b) The free energy curve associated with these stages.

The time evolution of the number concentration of the polyelectrolyte chains is given by the same equations (Equations 8.1 and 8.2) as for small electrolyte ions, except that the values of the quantities such as the diffusion coefficient and the electrophoretic mobility must be taken to be the appropriate values for polyelectrolytes (Section 7.4). In addition, the role of convective flows and EOFs must be explicitly treated. Since we consider in this chapter the flux of only the number of molecules through the pore in describing the capture rate, it is sufficient to focus on the center-of-mass of a polymer chain as it gets caught by the pore and released into the receiver compartment. Furthermore, we first consider the flux along the x-direction, which is chosen along the pore axis from the donor compartment to the receiver compartment. The simpler mathematical details in the one-dimensional description enable to reach general physical conclusions as in the case of small electrolyte ions. However, we shall exercise care in the treatment of space-dimension–dependent, long-ranged correlations in the hydrodynamic flows and diffusion fields, which is necessary to properly consider the three dimensionality of the experimental situation.

Consider the time evolution of the number concentration of the polyelectrolyte chains $c(x, t)$ at the location x along the x-direction and at time t. Here we have taken the x-value for the polymer to be its center-of-mass value. According to the continuity equation (Section 6.4, Equations 8.1 and 8.2),

$$\frac{\partial c(x, t)}{\partial t} = -\frac{\partial}{\partial x} J(x, t), \tag{9.1}$$

where $J(x,t)$ is the net flux of the molecules at location x and time t. The flux has the diffusive and convective parts,

$$J(x,t) = -D\frac{\partial c(x,t)}{\partial x} + c(x,t)u(x,t), \qquad (9.2)$$

where D is the diffusion coefficient of the polymer molecule and $u(x,t)$ is the x-component of the net velocity of molecules at position x and time t. In writing the above equation, intermolecular interactions are ignored corresponding to an infinitely dilute polyelectrolyte solution in the donor compartment.

The velocity of the molecules arises from both the externally imposed force fields and the local gradients in the free energy landscape $F(x)$, as sketched in Figure 9.3b. Examples of the externally imposed force fields are an electric field E and a pressure gradient $\partial p/\partial x$. As described in Section 8.8, if the pore bears some residual charges, the externally imposed electric field creates an EOF as well. In general, the net velocity is

$$u(x) = u_{ext} + \mu E - \frac{1}{\zeta}\frac{\partial F(x)}{\partial x}, \qquad (9.3)$$

where u_{ext} is the velocity field generated by pressure gradients and EOF, μ is the electrophoretic mobility (Section 7.4), and ζ is the friction coefficient obeying the Einsteinian relation (Equation 6.43),

$$\zeta = \frac{k_B T}{D}. \qquad (9.4)$$

In writing Equation 9.3, we have assumed that the free energy landscape is time-independent.

Substitution of Equation 9.3 into Equation 9.2 gives

$$J(x,t) = -D\frac{\partial c}{\partial x} + c\mu E + cu_{ext} - cD\frac{\partial \tilde{F}(x)}{\partial x}, \qquad (9.5)$$

with

$$\tilde{F}(x) \equiv \frac{F(x)}{k_B T}. \qquad (9.6)$$

By writing E as the negative gradient of the electric potential $\psi(x)$ at x, the flux is given by

$$J(x,t) = -D\frac{\partial c}{\partial x} - c\mu\frac{\partial \psi(x)}{\partial x} + cu_{ext} - cD\frac{\partial \tilde{F}(x)}{\partial x}. \qquad (9.7)$$

The four terms on the right-hand side of Equation 9.7 represent the contributions from the diffusion, electrophoretic drift, convective drift from fluid velocity fields, and free energy barrier, respectively.

We shall treat each of these four contributions separately in the following sections, along with discussions of their effects in the context of experimental results. The same mathematical procedures described in Chapters 6 and 8, namely, the Fokker–Planck formalism, first passage times, Poisson–Nernst–Planck formalism, and the Goldman–Hodgkin–Katz equations, are implemented (Muthukumar 2010) to obtain the steady-state flux of the polymer chains and the probability of successful barrier crossing.

9.3 DIFFUSION-LIMITED CAPTURE

Consider the simplest situation of Figure 9.4a, where an absorbing spherical sink of radius R is at the center of the coordinate system and polymer chains are present in the solution around the sink. Following the classical theory of Smoluchowski (Chandrasekhar 1943), we assume that the sink absorbs the polymer chains as soon as the center-of-mass of a polymer chain approaches the surface of the sink. We identify the capture rate of polymer chains as the steady-state net flux of polymer chains into the absorbing sink. Let the initial number concentration of the polymer chains be c_0. The polymer concentration is continuously maintained as c_0 at distances far from the sink. The polymer chains undergo only diffusion and there are no other convective contributions. The continuity equation for the number concentration of polymer chains in three

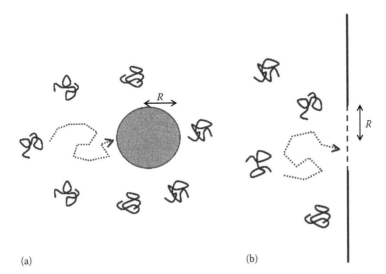

(a)   (b)

Figure 9.4 Diffusion-limited capture of polymer molecules by (a) an absorbing spherical sink of radius R and (b) an absorbing circular disk of radius R embedded on a thin infinite membrane.

dimensions follows from Equations 8.1, 8.2, 9.1, and 9.2 as the Fick's law,

$$\frac{\partial c(\mathbf{r}, t)}{\partial t} = D\nabla^2 c(\mathbf{r}, t). \tag{9.8}$$

In the spherical polar coordinate system, this equation is written as

$$\frac{\partial c(r, t)}{\partial t} = D\frac{1}{r^2}\frac{\partial}{\partial r}\left[r^2\frac{\partial c(r, t)}{\partial r}\right], \tag{9.9}$$

where r is the radial distance from the center of the sink. In the steady state, each term in the above equation is zero. Using the boundary conditions that $c(r = R, t) = 0$ and $c(r \to \infty) = c_0$ for all values of the time variable, the steady-state solution of Equation 9.9 is

$$c(r) = c_0\left(1 - \frac{R}{r}\right). \tag{9.10}$$

The net flux of the chains on the surface of the spherical sink is the surface area $(4\pi R^2)$ times the steady-state flux into the sink at the distance R, $(D\partial c(r)/\partial r)\,|_R$,

$$\bar{J}_{diffusion} = \left[(4\pi R^2)D\frac{\partial c}{\partial r}\right]_{r=R} = 4\pi DRc_0. \tag{9.11}$$

The time evolution of the polymer concentration near the absorbing sink is obtained by solving the time-dependent Equation 9.9 as

$$c(r, t) = c_0\left[1 - \frac{R}{r} + \frac{2R}{\sqrt{\pi}r}\int\limits_0^{(r-R)/2\sqrt{Dt}} dr'\, e^{-r'^2}\right]. \tag{9.12}$$

The integral in the above equation is the error function and is tabulated in mathematical handbooks (Abramowitz and Stegun 1965). The net flux of polymer chains arriving on the surface of the sink at time t follows as

$$\bar{J}_{diffusion} = \left[(4\pi R^2)D\frac{\partial c(r, t)}{\partial r}\right]_{r=R} = 4\pi DRc_0\left[1 + \frac{R}{\sqrt{\pi Dt}}\right]. \tag{9.13}$$

In the steady state, Equation 9.13 reduces to Equation 9.11.

The same procedure can be carried out if the sink is an absorbing circular disk of radius R embedded in a planar wall of infinite dimensions (Figure 9.2b), although the mathematical details are lengthier (Carslaw and Jaeger 1986). For this situation, which is directly pertinent to the capture of polymer chains from the donor compartment by the pore, the steady-state flux in the number concentration of polymer molecules is

$$\bar{J}_{diffusion} = 4DRc_0. \tag{9.14}$$

The key message of the above derivations is that the steady-state net flux of the polymer molecules into an absorbing region, namely the diffusion-limited capture rate, is simply proportional to the product of the linear size of the absorbing region, the diffusion coefficient of the polymer, and the average polymer concentration c_0 in the donor compartment,

$$\bar{J}_{diffusion} \sim DRc_0. \tag{9.15}$$

The various aspects of the polymer appear only through its diffusion coefficient. In dilute solutions, the diffusion coefficient of the polyelectrolyte in sufficiently salty solutions obeys the Stokes–Einstein law (Section 7.3),

$$D \sim \frac{T}{\eta_0 R_g}, \tag{9.16}$$

where R_g is the radius of gyration of the polymer, and η_0 is the viscosity of the solution. For a chain of contour length proportional to its number of segments N, the radius of gyration R_g is proportional to N^ν, where ν is the size exponent for the polymer. Therefore, the N-dependence of D is given as

$$D \sim \frac{1}{N^\nu}. \tag{9.17}$$

In solutions with high salt concentrations, the size exponent of a polyelectrolyte molecule is approximately the same as that of uncharged polymer molecules in good solvents, with $\nu \simeq 0.6$ (Section 4.4). Therefore,

$$D \sim \frac{1}{N^{0.6}}. \tag{9.18}$$

As discussed in Section 7.3, the above equations for D are valid only if the electrostatic interactions among the polymer segments are fully screened. If the concentration of the salt is not high enough, the dynamics of the polyelectrolyte molecule gets strongly coupled to that of its counterions, and care must be exercised in using a formula for D. Since almost all experiments dealing with single-molecule translocation processes have large amount of electrolytes, Equation 9.18 is sufficient for giving the molecular-weight dependence of the diffusion coefficient.

Therefore, when the first term in Equation 9.7 dominates over the other three terms, the steady-state capture rate of the molecules by the absorbing domain is given by the law

$$\bar{J}_{diffusion} \sim \frac{c_0}{N^{0.6}} \quad \text{(diffusion-limited)}, \tag{9.19}$$

where the proportionality factor depends on the geometry of the absorbing region. We call the regime of Equation 9.19, where the polymer diffusion dominates, as the diffusion-limited capture regime.

In the diffusion-limited regime, the capture rate decreases with chain length and is linear in the polymer concentration. It is, of course, independent of the electric field or convective velocity fields in this regime.

9.4 DRIFT-LIMITED REGIME

In this section, we consider the rate of capture of polymer chains arising solely from the second term on the right-hand side of Equation 9.7, due to the electrophoretic movement of the polyelectrolyte molecules. We call this limiting situation as the drift-dominated regime. In this limit, the steady-state flux in the number concentration is exactly the same as Equations 8.16 and 8.18, derived for small electrolyte ions,

$$\bar{J}_{drift} = -\mu c_0 \frac{\partial \psi}{\partial x} = \mu c_0 E. \tag{9.20}$$

If the moving species is an ion from the electrolyte or an isolated counterion, its electrophoretic mobility μ is given by the Einsteinian law,

$$\mu = \frac{zD}{k_B T}, \tag{9.21}$$

where z is the charge of the moving species. On the other hand, if the moving species is a polyelectrolyte molecule, the electrophoretic mobility is given by Equation 7.101 as

$$\mu = \frac{QD}{k_B T} \Theta\left(\frac{R_g}{\xi_D}\right), \tag{9.22}$$

where Q is the net polymer charge and $\Theta(R_g/\xi_D)$ is the factor arising from the coupling of the dynamics of the polyelectrolyte and the counterions (Section 7.4). This crossover factor is a function of the ratio of the radius of gyration of the polyelectrolyte to the Debye length ξ_D, which is proportional to the inverse square root of the salt concentration c_s ($\xi_D \sim 1/\sqrt{c_s}$). As described in Section 7.4, the crossover function $\Theta(R_g/\xi_D)$ depends on N according to

$$\Theta\left(\frac{R_g}{\xi_D}\right) \sim N^{\nu-1}, \tag{9.23}$$

where ν is the effective size exponent and its value depends on the salt concentration. Defining the function

$$\tilde{z} \equiv \frac{Q}{k_B T} \Theta\left(\frac{R_g}{\xi_D}\right), \tag{9.24}$$

the electrophoretic mobility of a polyelectrolyte chain is

$$\mu = \tilde{z}D = N^0 f(c_s), \tag{9.25}$$

where $f(c_s)$ is a decreasing function of the salt concentration, and the electrophoretic mobility is independent of the chain length at all salt concentrations.

In the drift-dominated regime, where the first, third, and fourth terms on the right-hand side of Equation 9.7 are negligible in comparison with the electrophoretic term, the constant flux in the steady state follows from Equations 9.1, 9.2, 9.20, and 9.25 as

$$\bar{J} \sim -c_0 N^0 \frac{\partial \psi}{\partial x}. \tag{9.26}$$

If V_m is the voltage difference that drives the negatively charged polymer from the donor compartment toward the positive electrode placed in the receiver compartment, it follows from Equation 9.26 that

$$\bar{J}_{drift} \sim c_0 N^0 V_m \quad \text{(drift-limited)}. \tag{9.27}$$

Therefore, in the drift-limited regime, where the effects of the diffusion, barrier, and convective flow may be ignored, the polymer flux is linearly proportional to the polymer concentration and the applied-voltage difference and is independent of the chain length.

9.5 EFFECT OF CONVECTIVE FLOW

In this section, we discuss the effect of flow by considering the first and third terms of Equation 9.7. The situation is sketched in Figure 9.5. As in Section 9.2, the polymer chains diffuse and get absorbed by a spherical sink of radius R, except that now the diffusion takes place in the presence of a fluid flow with the velocity $\mathbf{v}(\mathbf{r})$. The fluid flow is assumed to be directed along the x-axis as

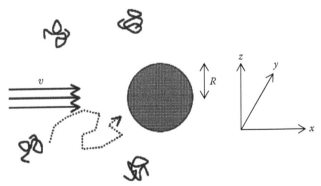

Figure 9.5 Polymer capture with convective diffusion.

shown in Figure 9.5. The fluid velocity at distances far away from the sink is taken to be uniform with a value v along the x-axis, and it is perturbed by the presence of the sink in its vicinity. As the polymer chains are carried by the fluid velocity, the average velocity of the polymer, $\mathbf{u}_{ext}$, is the same as the fluid velocity $\mathbf{v}$, as discussed in Section 7.2. Substituting the expression for the flux (Equation 6.67), into the equation for the conservation of number concentration (Equation 6.66), we get the following convective-diffusion equation,

$$\frac{\partial c(\mathbf{r},t)}{\partial t} + (\mathbf{v} \cdot \nabla)c(\mathbf{r},t) = D\nabla^2 c(\mathbf{r},t), \tag{9.28}$$

where the condition of the incompressibility of the fluid (Equation 7.3) is used. In the steady-state limit, the total flux of polymer chains into the absorbing spherical sink has been derived (Levich 1962, Probstein 1989) in the strong flow limit as

$$\bar{J}_{convective\ diffusion} = 7.84c_0 DR \left(\frac{vR}{D}\right)^{1/3}. \tag{9.29}$$

Thus, in addition to the diffusion-limited result, there emerges a multiplicative factor due to flow, which is proportional to the one-third power of the Peclet number $Pe \equiv vR/D$. In the flow-dominated regime, the capture rate is proportional to $D^{2/3}R^{4/3}v^{1/3}$. This is to be contrasted with the Smoluchowski result where the capture rate is proportional to DR. These two limits are sometimes combined as an approximate interpolation formula (Levich 1962) given by

$$\bar{J}_{convective\ diffusion} = 4\pi c_0 DR \left[1 + 0.624\left(\frac{vR}{D}\right)^{1/3}\right]. \tag{9.30}$$

Since $D \sim 1/R_g$, the two limits of the above formula are

$$\bar{J}_{convective\ diffusion} = \begin{cases} \frac{c_0 R}{R_g}, & weak\ flow \\ \frac{c_0 v^{1/3}R^{4/3}}{R_g^{2/3}}, & strong\ flow. \end{cases} \tag{9.31}$$

Therefore, the presence of strong flows can significantly influence the capture rate of the polymer molecules through a pore.

9.6 POLYMER CAPTURE WITH ELECTROOSMOTIC FLOW

We have seen in Section 8.8 that the EOF velocity generated by electric fields on the counterions adsorbed on immobile charged interfaces can be very significant. In this section, we address the effect of the EOF on the capture rate of the polymer chains in the steady state. As a specific example, consider a positively charged cylindrical pore (Figure 9.6a) with a surface electric potential

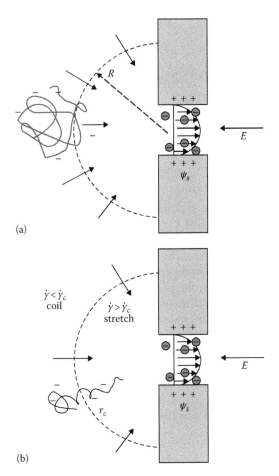

(a)

(b)

Figure 9.6 EOF assists polymer translocation if the surface charge of the pore is opposite to the polymer charge. (a) Sketch of the EOF effect. (b) When the polymer crosses a critical distance r_c from the pore, it is subjected to the coil-stretch transition and it passes rapidly through the pore.

ψ_s, which is generated by any one of the mechanisms described in Section 3.2. An external electric potential difference is applied across the pore so that the negatively charged polymer molecules are driven from the donor (*cis*) compartment to the receiver compartment (*trans*). With this arrangement, the direction of the EOF is the same as the direction of successful translocation events.

The steady-state net flux of the fluid, due to EOF, is given by Equation 8.90. Since the fluid flow must be continuous in the system, all the fluid elements in the *cis* compartment must flow into the pore, assuming that all nonlinear hydrodynamic effects such as the intermittency and turbulence are absent. This means that at a distance R away from the pore entrance, the surface area of the hemisphere $(2\pi R^2)$ times the velocity at that distance $(u(R))$ must match the flux J_0,

$$u(R) = -\frac{|J_0|}{2\pi R^2}, \tag{9.32}$$

where the negative sign is due to the choice of the origin being the entrance of the pore. Therefore, the local velocity gradient at R is $\dot{\gamma} = du(R)/dR$,

$$\dot{\gamma} = \frac{|J_0|}{\pi R^3}. \tag{9.33}$$

Thus, at distances near the pore, the velocity gradient is the strongest, and at large distances, it falls off like R^{-3}.

As described in Section 7.5, if the velocity gradient $\dot{\gamma}$ is greater than the critical value $\dot{\gamma}_c$ given as the reciprocal of the Zimm time (Equation 7.47), the polymer molecule undergoes the coil-stretch transition. This implies that there is a critical radius r_c from the pore, outside which $\dot{\gamma} < \dot{\gamma}_c$, and as a result, the polymer molecule is essentially in the coil state. For distances shorter than r_c, $\dot{\gamma} > \dot{\gamma}_c$ and the polymer chain is stretched. The threshold distance r_c for delineating the strong and weak velocity gradient regions is obtained from the criterion of the coil-stretch transition given by Equation 7.109,

$$\dot{\gamma}_c \tau_{Zimm} = 1. \tag{9.34}$$

Substituting Equation 9.33 into Equation 9.34 gives

$$r_c = \left(\frac{|J_0|\tau_{Zimm}}{\pi}\right)^{1/3}. \tag{9.35}$$

Combining Equations 8.90, 7.47, and 9.35, the critical capture radius follows as

$$r_c = \left[0.3\frac{\epsilon_0\epsilon\psi_s E\beta a^2}{k_B T}\right]^{1/3} R_g \equiv (0.3\alpha)^{1/3} R_g, \tag{9.36}$$

where a is the pore radius and β is the salt-concentration-dependent geometric factor given by Equations 8.91 and 8.92. All experimental parameters combine together as the single parameter α,

$$\alpha \equiv \frac{\epsilon_0\epsilon\psi_s E\beta a^2}{k_B T} = \left[1 - \frac{2}{\kappa a}\frac{I_1(\kappa a)}{I_0(\kappa a)}\right]\frac{\epsilon_0\epsilon\psi_s Ea^2}{k_B T}, \tag{9.37}$$

where Equation 8.91 is used. Therefore, the critical capture radius is proportional to the radius of gyration of the polymer times one-third power of the parameter α.

The number concentration of the polymer chains $c(R)$ at a distance R from the pore is obtained as follows. In the steady state, the divergence of the flux is zero. In the presence of diffusion and EOF, the divergence of the flux in the steady state is given in the spherical polar coordinate system as

$$D\frac{1}{R^2}\frac{\partial}{\partial R}\left[R^2\frac{\partial c(R)}{\partial R}\right] - \frac{1}{R^2}\frac{\partial}{\partial R}\left[R^2 c(R)u(R)\right] = 0, \quad (9.38)$$

where $u(R)$ is given by Equation 9.32. The solution of this equation with the boundary conditions $c(r_c) = 0$ and $c(R \to \infty) = c_0$ is

$$c(R) = c_0 \frac{\exp\left[-(|J_0|/2\pi D)(1/r_c - 1/R)\right] - 1}{\exp(-|J_0|/2\pi Dr_c) - 1}. \quad (9.39)$$

Near r_c, the polymer concentration decreases exponentially from the bulk value to zero with a decay length of $\sim 2\pi Dr_c^2/|J_0|$. Using Equations 7.10, 7.44, 8.90, and 9.36, we get this decay length as $0.07R_g/\alpha^{1/3}$, with α defined in Equation 9.37.

The steady-state capture rate of the polymer chains into the region at r_c is the surface area of the hemisphere at r_c times the flux given by

$$\bar{J}_{EOF} = -2\pi r_c^2 \left[-D\frac{\partial c(R)}{\partial R} + u(R)c(R)\right]_{R=r_c}. \quad (9.40)$$

Substituting Equation 9.39 into the above equation, we get

$$\bar{J}_{EOF} = \frac{c_0|J_0|}{1 - \exp(-|J_0|/2\pi Dr_c)} = \frac{\pi c_0 k_B T}{\eta_0}\frac{\alpha}{1 - \exp(-9.3\alpha^{2/3})}. \quad (9.41)$$

In writing the second equality, we have used Equations 7.10, 7.44, 8.90, and 9.36. For values of α larger than about 0.1, the exponential term in the denominator of the above equation is negligible. Under this condition, the capture rate becomes

$$\bar{J}_{EOF} \approx c_0|J_0| \sim c_0 E, \quad (9.42)$$

in view of Equation 9.37. Therefore, when the effects from diffusion and EOF are combined, the drift-dominated result of Equation 9.27 can emerge for large values of the parameter $\alpha (\geq 0.1)$, which is a combination of the surface potential, pore radius, salt concentration, and the electric field under certain experimental conditions.

We shall now return to the experimentally observed large capture region in the translocation experiments with dsDNA and solid-state nanopore (Figure 9.2). If the surface charge of the nanopore is opposite to the charge of the polymer, the EOF is in the same direction as the polymer translocation. The capture radius r_c is $(0.3\alpha)^{1/3}R_g$ (Equation 9.36). In typical nanopore experimental conditions, $(0.3\alpha)^{1/3}$ is of order unity. As a result, the capture radius of the absorbing region is comparable to the radius of gyration of the polymer. In the experiments of Figure 9.2, the radius of gyration of the DNA molecule is about 0.8 μm. As a result, the capture radius is in micrometers, consistent with the experimental observations. In view of this, the EOF has been attributed to the emergence of a large absorbing region in front of the pore (Wong and Muthukumar 2007).

When a pressure difference is present across the pore, Equation 8.93 must be used for J_0 in the above equations, instead of Equation 8.90. The control of capture rate of the polymer by the pore can be achieved by a combination of the EOF and the Poiseuille flow (Stein et al. 2006).

9.7 EFFECT OF BARRIERS ON CAPTURE RATE

When a polymer molecule approaches the pore entrance, it is usually in a jammed state with either of its ends not necessarily at the pore for further threading as a single file. As illustrated in Figure 1.9c, the chain end must unravel from a high-monomer-density jammed state in order to place itself at the pore entrance. This requires an entropic barrier. Let this entropic barrier for the localization of one chain end be $F_\ell^\dagger$. In addition to this part, there is a free energy barrier for squeezing the chain inside the pore. As discussed in Section 5.3, the latter arises from conformational changes of the polymer accompanying the confinement inside the pore, and let $F_c^\dagger$ denote this contribution. Adding these two parts, the total free energy barrier $F^\dagger$ is

$$F^\dagger = F_\ell^\dagger + F_c^\dagger. \tag{9.43}$$

For sufficiently large values of N and the monomer volume fraction in the jammed state higher than about 0.1, $F_\ell^\dagger$ decreases with N, according to Equation 5.31

$$\frac{F_\ell^\dagger}{k_B T} \sim \frac{1}{N^\alpha}, \quad \alpha \simeq 0.2 \pm 0.1. \tag{9.44}$$

The reduction in the free energy barrier with the chain length originates from the entropic pressure created by the confining region to push the chain end in the direction of the pore entrance (Kumar and Muthukumar 2009). This trend is opposite to that of $F_c^\dagger$, associated with chain confinement inside the pore. For example, for full confinement within a cylindrical pore of diameter d, $F_c^\dagger$ is extensive in N as given by Equation 5.44,

$$\frac{F_c^\dagger}{k_B T} \sim \frac{N}{d^{1/\nu}}. \tag{9.45}$$

In general, the free energy barrier sketched in Figure 9.1b is a complicated function of the characteristics of the pore and polymer. Nevertheless, in the simplest situation of a narrow hole embedded in a thin membrane, the free energy barrier is dominated by $F_\ell^\dagger$. If the chain is caught inside a small region in front of the pore mouth, where the electric potential gradient is the strongest, $F_\ell^\dagger$ decreases with N for sufficiently long chains (Muthukumar 2010).

The effect of entropic barrier on the rate of capture of the polymer by the pore is given by Equation 9.7. In the absence of external flow fields, the flux in the number concentration of the polymer chains is given by the contributions from the diffusion, electrophoretic drift, and the free energy barrier as

$$J(x) = -D\left[\frac{\partial c}{\partial x} + c\tilde{z}\frac{\partial \psi(x)}{\partial x} + c\frac{\partial \tilde{F}(x)}{\partial x}\right]. \tag{9.46}$$

In writing the electrophoretic drift contribution, we have used $\mu = \tilde{z}D$, where $\tilde{z}$ is defined in Equation 9.24, and $\tilde{F}(x)$ is the free energy in units of k_BT (Equation 9.6).

The generic aspects of the barrier contribution to the capture rate and the crossover behaviors among the diffusion-, drift-, and barrier-dominated regimes can be derived by the following analytically tractable model (Muthukumar 2010). Consider a cylindrical nanopore of length L along the x-direction (Figure 9.7a). Let the free energy barrier be ramp-like, as shown in Figure 9.7b, with a maximum value of $\tilde{F}_m$ located at a distance ηL from the pore boundary at the donor side. The length of pore is chosen to include a small distance in the donor compartment where the electric field becomes strong to trap the polymer. Alternatively, a distance a_0 in front of the pore may be identified where the polymer is initially trapped and the pore can be taken as a composite pore of effective length $L + a_0$. As the physical conclusions are not going to

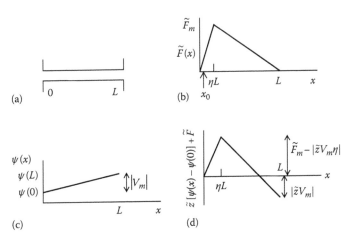

Figure 9.7 (a) A cylindrical pore of Length L. (b) Ramp-like profile for the free energy. (c) Uniform electric field along the pore. (d) Ramp-like profile by combining the free energy and electric potential gradient.

be changed by this microscopic parameter a_0, we assume that L is the effective length of the pore. Furthermore, we assume that $\tilde{F}(x)$ is ramp-like, given by

$$
\tilde{F}(x) = \begin{cases} \dfrac{\tilde{F}_m}{\eta} \dfrac{x}{L}, & \dfrac{x}{L} \leq \eta \\ \dfrac{\tilde{F}_m}{(1-\eta)} \left(1 - \dfrac{x}{L}\right), & \eta \leq \dfrac{x}{L} \leq 1. \end{cases} \tag{9.47}
$$

Here, $\tilde{F}_m$ and η are taken as parameters. Also, let x_0 be the initial position of the polymer in considering the probability of success or failure of capture events. The above choice of ramp-like profile is only for illustrative purposes and analytical tractability. In fact, as will be discussed in the following chapter, the gain in electrostatic energy by the translocating polymer inside a pore under an electric field is quadratic in x. This necessitates numerical calculations, except for steep barriers. If the barrier is steep, Kramers-like formula can be obtained as a saddle-point approximation (Section 8.5). For weak barriers, saddle-point approximation is insufficient. The present choice of ramp-like profile allows analytical formulas interpolating between the Kramers-like limit for steep barriers and the correct limit for weak barriers. A more complicated form that might be appropriate for a particular pore–polymer combination can of course be used instead, without losing the generality of the results given below.

For the variation of the electric potential across the pore, we assume that the electric field is uniform (Figure 9.7c), as in Section 8.4.3,

$$
\psi(x) = \psi(0) - V_m \frac{x}{L}, \quad 0 \leq \frac{x}{L} \leq 1, \tag{9.48}
$$

with

$$
V_m \equiv \psi(0) - \psi(L). \tag{9.49}
$$

Assuming that the polymer charge is negative, and combining the free energy associated with the encounter of polymer with the pore and the electric potential gradient, the net free energy curve is sketched in Figure 9.7d as given by

$$
\tilde{F}(x) + \tilde{z}[\psi(x) - \psi(0)] = \begin{cases} \left(\dfrac{\tilde{F}_m}{\eta} - |\tilde{z}V_m|\right) \dfrac{x}{L}, & 0 \leq \dfrac{x}{L} \leq \eta \\ \dfrac{\tilde{F}_m}{(1-\eta)} - \left(\dfrac{\tilde{F}_m}{(1-\eta)} + |\tilde{z}V_m|\right) \dfrac{x}{L}, & \eta \leq \dfrac{x}{L} \leq 1. \end{cases} \tag{9.50}
$$

The total free energy barrier is

$$
\tilde{F}^\dagger = \tilde{F}_m - |\tilde{z}V_m|\eta, \tag{9.51}
$$

and the free energy gain in the translocation process is

$$
\Delta\tilde{F} = |\tilde{z}V_m|. \tag{9.52}
$$

We now consider the consequences of the above barrier on the steady-state rate of capture of the polymer and probability of its successful transition across the barrier.

9.7.1 Capture Rate

The steady-state flux of the number concentration of polymer molecules is obtained from Equation 9.46, by rearranging and integrating between the limits of $x = 0$ and $x = L$ as in Section 8.4, as

$$\bar{J} = -D \frac{\left[c(L)e^{\tilde{F}(L)+\tilde{z}\psi(L)} - c(0)e^{\tilde{F}(0)+\tilde{z}\psi(0)} \right]}{\int_0^L dx e^{\tilde{F}(x)+\tilde{z}\psi(x)}}, \qquad (9.53)$$

where $c(L)$ and $c(0)$ are the number concentrations of the polymer at $x = L$ and $x = 0$, respectively. Substitution of Equation 9.50 into Equation 9.53 gives the steady-state capture rate in terms of the two parameters $\tilde{F}_m$ and η, and the net free energy gain $\Delta\tilde{F}$. We assume further that the polymer concentration in the receiver compartment is negligible, $c(L) \simeq 0$, as the polymer chains continuously get removed from the pore after successful translocation events. The other boundary condition is that $c(0)$ is the same as the average polymer concentration c_0 in the donor compartment. As a result, we get

$$\bar{J} = \frac{Dc_0}{L} \left[\frac{\eta(e^{\tilde{F}^\dagger} - 1)}{\tilde{F}^\dagger} + \frac{(1 - \eta)}{(\tilde{F}^\dagger + |\tilde{z}V_m|)}(e^{\tilde{F}^\dagger} - e^{-|\tilde{z}V_m|}) \right]^{-1}, \qquad (9.54)$$

where the free energy barrier is given in Equation 9.51.

The dependence of $\bar{J}$ on chain length, as given by Equation 9.54, can be addressed by incorporating the N-dependence of the free energy barrier. We take Equation 9.44 to describe the entropic barrier,

$$\tilde{F}_m \equiv \frac{\tilde{F}_0}{N^\alpha}. \qquad (9.55)$$

We use this choice to emphasize the importance of the entropic barrier and for the illustrative purpose of an explicit calculation. Naturally, any other form suitable for a particular pore–polymer arrangement can be readily used instead. Furthermore, $|\tilde{z}V_m|$ and D are N-dependent. The N-dependence of $|\tilde{z}V_m|$ is written as

$$|\tilde{z}V_m| \equiv z_0 N^\nu V_m \equiv v_0 N^\nu, \qquad (9.56)$$

where the coefficient z_0 is related to the function $\Theta(R_g/\xi_D)$ given in Equation 9.24. Also, we rewrite Equation 9.17 for D as

$$D \equiv \frac{D_0}{N^\nu}. \qquad (9.57)$$

Substituting Equations 9.55 through 9.57 into Equation 9.54, the N-dependence of the steady-state flux can be calculated.

A typical result given by Equation 9.54 is presented in Figure 9.8a and b, where $\bar{J}L/Dc$ is plotted against $|\tilde{z}V_m|$ for $\tilde{F}_m = 10$ and $\eta = 0.1$. The values

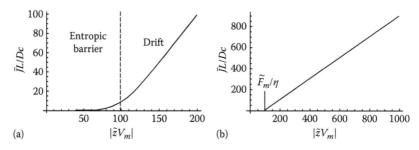

Figure 9.8 Plot of reduced polymer flux against the reduced voltage differ-ence. (From Muthukumar, M., *J. Chem. Phys.*, 132, 195101, 2010. With permission.)

of the parameters are typical of experimental conditions, essentially based on the fact that the applied-voltage difference of 25 mV corresponds to room tem-peratures and the entropic barriers are about $10k_BT$–$20k_BT$. It is evident from Figure 9.8b that the steady-state flux is linear in the electric potential differ-ence V_m for very large values of V_m, above an apparent threshold value of $|\tilde{z}V_m| \sim \tilde{F}_m/\eta$. This threshold value does not indicate an on/off process. There are always a finite number of capture events for voltage differences even below the threshold value, as is evident from Figure 9.8a. In the limit of $\eta|\tilde{z}V_m| \gg \tilde{F}_m$, Equation 9.54 reduces to

$$\bar{J} = \frac{Dc_0|\tilde{z}V_m|}{L}, \quad \eta|\tilde{z}V_m| \gg \tilde{F}_m. \tag{9.58}$$

In view of Equation 9.25, $\mu = \tilde{z}D$, the steady-state flux for high voltage differences becomes

$$\bar{J} = \frac{c_0\mu V_m}{L}, \tag{9.59}$$

which is exactly the same as the drift-limited result given by Equation 9.27. Therefore, the high voltage difference regime can be identified as the drift-dominated regime, whereas the lower voltage difference regime is identified as the entropic barrier regime. The crossover occurs at $|\tilde{z}V_m| \simeq \tilde{F}_m/\eta$.

In the absence of any voltage difference, where there is only diffusion and no drift, Equation 9.54 reduces to

$$\bar{J} = \frac{Dc_0\tilde{F}_m}{L}[\eta(e^{\tilde{F}_m} - 1) + (1 - \eta)(e^{-\tilde{F}_m} - 1)]^{-1}, \quad V_m = 0, \tag{9.60}$$

which for large barriers approaches the limit

$$\bar{J} = \frac{Dc_0\tilde{F}_m}{L\eta}e^{-\tilde{F}_m}, \quad \tilde{F}_m \gg 1. \tag{9.61}$$

In this limit, the flux is given by the Kramers-like formula (Equation 8.64). As in Equation 9.19, $\bar{J}$ is a decreasing function of chain length in the diffusion-limited regime, even when the barrier term brings an additional N-dependent factor.

The effects of barrier height and the chain length on the relation between $\bar{J}$ and V_m are illustrated in Figure 9.9a and b for $z_0 = 0.04$ and $\eta = 0.002$. V_m is given in mV. In Figure 9.9a, $N = 3000$ and $\tilde{F}_m = 8, 9$, and 10. As the barrier height increases, the capture rate decreases roughly exponentially for a given voltage difference. The result for $N = 3000$ is compared with that for $N = 2000$ in Figure 9.9b, for $\alpha = 0.2$, $z_0 = 0.04$, $\eta = 0.002$, and $\tilde{F}_0 = 23$. As the chain length increases, the flux increases for a fixed voltage difference.

The typical result on the dependence of $\bar{J}$ on N is shown in Figure 9.10, where $\bar{J}L/D_0c$ is plotted against $\ln N$ for $\tilde{F}_0 = 23$, $v_0 = 12$, $\alpha = 0.2$, and $v = 0.6$. The steady-state flux increases with chain length until it reaches the asymptotic drift-dominated regime of N-independence for very large values of N, as given by Equation 9.59. There is a crossover between the entropic barrier-dominated

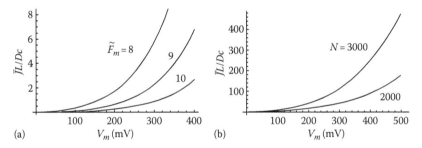

(a) (b)

Figure 9.9 Effects of (a) barrier height and (b) chain length on the relation between the flux and voltage difference. (From Muthukumar, M., *J. Chem. Phys.*, 132, 195101, 2010. With permission.)

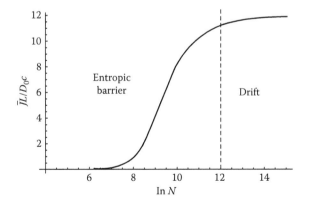

Figure 9.10 Chain length dependence of the reduced polymer flux. (From Muthukumar, M., *J. Chem. Phys.*, 132, 195101, 2010. With permission.)

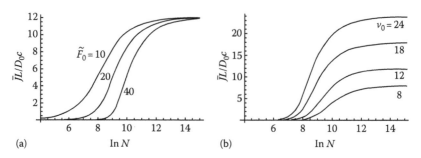

Figure 9.11 Effects of (a) barrier height and (b) voltage difference on the relation between the flux and chain length. (From Muthukumar, M., *J. Chem. Phys.*, 132, 195101, 2010. With permission.)

regime at lower chain lengths and the drift-dominated regime at higher chain lengths, as marked in Figure 9.10. The details of the crossover depend on the barrier height and the strength of the voltage difference. These effects are given in Figure 9.11a and b. In Figure 9.11a, $\bar{J}L/D_0c$ is plotted against $\ln N$ for $\tilde{F}_0 = 10, 20$, and 40, $v_0 = 12$, $\alpha = 0.2$, and $v = 0.6$. If the barrier due to the intrinsic nature of the pore is higher (i.e., $\tilde{F}_0$ is higher), the flux is lower. Also, it takes longer chains to have a reduced effective barrier so that the molecules can undergo translocation. This is seen in the upward shift in N for the apparent threshold for the polymer flux.

The analogous result for the dependence of the relation between the flux and chain length on the electric potential difference ($V_m = v_0/z_0$, Equation 9.56) is given in Figure 9.11b for $v_0 = 8, 12, 18$, and 24, $\tilde{F}_0 = 23$, $\alpha = 0.2$, and $v = 0.6$. As expected, for each value of N, higher the electric potential difference, higher is the flux.

The choice of the parameters used in constructing the Figures 9.8 through 9.11 is only for illustrative purposes in order to gain insight into the general trends on the dependencies of the steady-state capture rate on the features of the barrier, chain length, and applied voltage difference (Muthukumar 2010). The above predictions are in qualitative agreement with the experimental findings described in Section 9.1 and Figures 9.1 and 9.2. Naturally, different values of the parameters specific to a particular experimental system can be used in the above equations, and theoretical predictions can be made.

9.7.2 Probability of Successful Translocation

When the polymer gets caught at the pore entrance, it may subsequently penetrate through the barrier into the receiver compartment or it may return to the donor compartment. Let $\Pi_+(x_0)$ be the probability of successful transit through the barrier, if the polymer is initially at the location $x = x_0$. This problem is discussed in Section 6.6 as a general problem of a stochastic process with finite boundaries. With the polyelectrolyte concentration being very low in the translocation experiments, the polymer concentration $c(x,t)$ in Equation 9.1 can be equivalently taken as the probability $P(x,t)$ of finding a chain at x and time t. With this interpretation, the continuity equation is the Fokker–Planck equation (Equation 6.84),

$$\frac{\partial P(x,t)}{\partial t} = -\frac{\partial}{\partial x}J(x,t), \tag{9.62}$$

where

$$J(x,t) = \left[A(x) - \frac{\partial}{\partial x}B(x) \right] P(x,t), \tag{9.63}$$

with

$$A(x) = -D\frac{\partial}{\partial x}[\tilde{z}\,\psi(x) + \tilde{F}(x)], \tag{9.64}$$

and

$$B(x) = D. \tag{9.65}$$

The probability distribution of the first passage time and its moments can be obtained from the Fokker–Planck equation for general situations of $\tilde{F}(x)$ and $\psi(x)$, by following the standard procedures given in Section 6.6. In the present context, the average translocation time is the mean first passage time, which can be calculated by choosing the appropriate boundary conditions for $P(x,t)$.

Once the chain is caught at the pore mouth, the probability of successful translocation is obtained as follows. We imagine that the polymer is at some location x_0 very close to the entrance within the pore of length L (Figure 9.7b). Alternatively, we can imagine that the actual pore is extended a bit inside the donor region to include the jammed state of the polymer. Given the initial location of the captured polymer at x_0 ($0 \le x_0 \le L$), the probability of successful translocation is the probability that the stochastic process given by Equation 9.62 attains the value $x = L$ for the first time without ever reaching the other boundary at $x = 0$. This probability Π_+ follows from Equations 9.62 through 9.65 and 6.93 as

$$\frac{\partial^2 \Pi_+(x_0)}{\partial x_0^2} = \left(\frac{\partial}{\partial x_0}[\tilde{z}\,\psi(x_0) + \tilde{F}(x_0)] \right) \frac{\partial \Pi_+(x_0)}{\partial x_0}. \tag{9.66}$$

Using the absorbing boundary conditions, $\Pi_+(x_0 = 0) = 0$ and $\Pi_+(x_0 = L) = 1$, the solution of Equation 9.66 is (see Equation 6.104)

$$\Pi_+(x_0) = \frac{\Psi(x_0)}{\Psi(L)}, \tag{9.67}$$

where

$$\Psi(x) \equiv \int_0^x dy e^{\tilde{z}\psi(y)+\tilde{F}(y)}. \tag{9.68}$$

The above equations for the probability of successful translocation are general for the polymer transit, addressing the conformational attributes and the charged nature of the polyelectrolyte chains. The consequences of these equations for the specific model of Figure 9.7 are the following.

Substitution of Equations 9.50 and 9.51 into Equations 9.67 and 6.68 gives the expression for the probability of successful translocation in terms of the barrier height,

$$\Pi_+(x_0) = \frac{\left(e^{\frac{\tilde{F}^\dagger x_0}{\eta L}} - 1\right)}{\left[(e^{\tilde{F}^\dagger} - 1) + \frac{(1-\eta)}{\eta} \frac{\tilde{F}^\dagger}{(\tilde{F}^\dagger + |\tilde{z}V_m|)}(e^{\tilde{F}^\dagger} - e^{-|\tilde{z}V_m|})\right]}. \tag{9.69}$$

This result is derived for the initial position of the polymer on the left-hand side of the barrier peak. Analogous result can be derived if polymer is on the right-hand side of the barrier peak.

A typical result from the above equation is given in Figure 9.12, where $\Pi_+(x_0)$ is plotted against V_m (in mV) for $x_0 = 0.01L$, $\eta = 0.02$, and $\tilde{F}_m = 10$ and 20. In writing V_m in mV, we have taken z_0 in Equation 9.56 as $0.04(\text{mV})^{-1}$ and $N = 3000$ with $v = 0.6$. As is evident from this figure, higher the barrier, lower is the success rate for a fixed voltage difference. This expected result is given a quantitative form in Equation 9.69. Analogous to the results for the capture rate (Section 9.6.1), the dependence of the success rate on the chain length can be calculated from Equation 9.69. The dependence of Π_+ on $\ln N$ is given in Figure 9.13a for different values of the barrier height $\tilde{F}_0 = 5, 10, 20$, and 40. The other values of the parameters are $v_0 = 12$, $\alpha = 0.2$, $v = 0.6$, $\eta = 0.05$, and

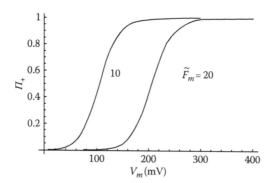

Figure 9.12 Dependence of the probability of successful translocation on voltage difference. (From Muthukumar, M., *J. Chem. Phys.*, 132, 195101, 2010. With permission.)

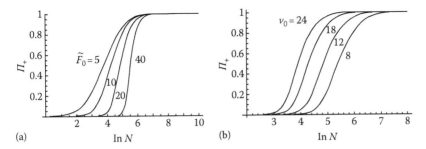

Figure 9.13 Effects of (a) barrier height and (b) voltage difference on the relation between the probability of successful translocation and chain length. (From Muthukumar, M., *J. Chem. Phys.*, 132, 195101, 2010. With permission.)

$x_0 = 0.01L$. As seen from this figure, the probability of success is lower for a given chain length if the barrier is higher. Also, the dependence of the success rate on chain length becomes more sharper as the barrier increases. The dependence of Π_+ on chain length at different values of voltage difference ($v_0 = 8, 12, 18$, and 24) is illustrated in Figure 9.13b. Here, $v_0 = z_0 V_m$ and we have chosen z_0 as $0.04(\text{mV})^{-1}$, and $\tilde{F}_0 = 20, \eta = 0.05, x_0 = 0.01L, \alpha = 0.2$, and $v = 0.6$. Again, Equation 9.69 provides physically expected result of the success rate of translocation being reduced at lower voltage bias for a fixed chain length.

The above predictions on the successful capture rate remain untested, except for some preliminary experimental results (Wong and Muthukumar 2010).

9.7.3 Phenomenological Chemical Kinetics Model

As an alternative phenomenological approach to the above Fokker–Planck formalism, which is common in the biochemistry literature, we now consider a model by drawing an analogy to chemical reactions (Lauger 1973, Hammes 1978). In this model, the molecular origins behind the capture and translocation of the polymer molecule through a pore are combined into a few phenomenological parameters. The whole translocation process is mapped into the format of a chemical reaction as sketched in Figure 9.14. Let the symbols P_c, C, and P_t

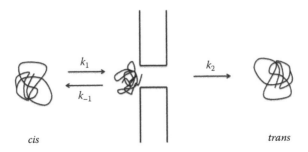

Figure 9.14 Chemical kinetics model of translocation.

denote the concentrations of the polymer in the *cis* chamber, the pore–polymer complex, and the polymer in the *trans* chamber, respectively. The rate constants for the capture of the polymer by the pore and release back into the *cis* chamber are taken as the phenomenological parameters k_1 and k_{-1}, respectively. Similarly, the rate constant for the release of the polymer from the pore–polymer complex into the *trans* chamber is k_2. We assume that the polymer in the *trans* chamber does not enter back into the pore, as the voltage difference pushes the polymer only in the direction of *cis* to *trans*. The rate of change of the concentration of the pore–polymer complex is given by

$$\frac{dC}{dt} = k_1 P_c - k_{-1}C - k_2 C. \tag{9.70}$$

In the steady state, the rate of formation of the complex is zero, so that C follows from Equation 9.70 as

$$C = \frac{k_1}{(k_{-1} + k_2)} P_c. \tag{9.71}$$

The flux into the C state from the *cis*, namely the number concentration passing across the boundary between the *cis* and the complex, is $k_1 P_c$,

$$J_{cis \rightarrow complex} = k_1 P_c, \tag{9.72}$$

and the flux into the *trans* from the complex is

$$J_{complex \rightarrow trans} = k_2 C = \frac{k_1 k_2}{(k_{-1} + k_2)} P_c, \tag{9.73}$$

where Equation 9.71 is used. The success ratio of the translocation events follows from the above two equations as

$$\Pi_+ = \frac{k_2}{(k_{-1} + k_2)}. \tag{9.74}$$

The above model can be readily generalized to the presence of two barriers. Consider an intermediate metastable state, labeled as m, in between the initial state (i) and the final state (f), as sketched in Figure 9.15a. Let the free energy barriers be $F_1 - F_m$ and $F_2 - F_m$ for the escape of the intermediate state into the initial state and the final state, respectively. Let the free energy of the initial state be the reference value $F_i = 0$ and the free energy of the final state be F_f. The chemical reaction scheme for this situation is given in Figure 9.15b, where $\alpha_1, \alpha_2, \beta_1,$ and β_2 are the rate constants for the corresponding reactions. These phenomenological constants can be taken as the Kramers-like rate constants given by the corresponding free energy differences (in units of $k_B T$),

$$\alpha_1 = k_0 e^{-\tilde{F}_1}, \quad \alpha_2 = k_0 e^{-(\tilde{F}_2 - \tilde{F}_m)}, \quad \beta_1 = k_0 e^{-(\tilde{F}_1 - \tilde{F}_m)}, \quad \beta_2 = k_0 e^{-(\tilde{F}_2 - \tilde{F}_f)}, \tag{9.75}$$

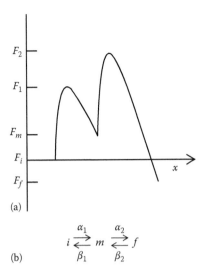

Figure 9.15 Chemical kinetics model of translocation with two barriers. (a) Free energy barriers. (b) Chemical reaction scheme.

where k_0 is a parameter. For crossing the first barrier, the steady-state flux is

$$\bar{J} = \alpha_1 p_i - \beta_1 p_m, \tag{9.76}$$

where p_i and p_m are the probability of finding the initial and intermediate states, respectively. It follows from the above equation,

$$p_m = \frac{\alpha_1}{\beta_1} p_i - \frac{\bar{J}}{\beta_1}. \tag{9.77}$$

Similarly, for the second barrier, the steady-state flux is

$$\bar{J} = \alpha_2 p_m - \beta_2 p_f, \tag{9.78}$$

so that

$$p_f = \frac{\alpha_2}{\beta_2} p_m - \frac{\bar{J}}{\beta_2}, \tag{9.79}$$

where p_f is the probability of finding the final state. Substituting Equation 9.77 into Equation 9.79 and using Equation 9.75, we obtain

$$p_f = p_i e^{-\tilde{F}_f} - \frac{\bar{J}}{k_0} \left(e^{\tilde{F}_2 - \tilde{F}_f} + e^{\tilde{F}_1 - \tilde{F}_f} \right), \tag{9.80}$$

so that the steady-state flux is given by

$$\bar{J} = k_0 \frac{\left(p_i - p_f e^{\tilde{F}_f} \right)}{\left(e^{\tilde{F}_1} + e^{\tilde{F}_2} \right)}. \tag{9.81}$$

If $p_f \simeq 0$ or if the free energy gain in reaching the final state is high $(-F_f \gg 1)$, we get

$$\bar{J} = k_0 \frac{p_i}{\left(e^{\tilde{F}_1} + e^{\tilde{F}_2}\right)}. \tag{9.82}$$

Since p_i is the polymer concentration in the donor compartment, the experimentally measured steady-state flux (after normalization with the factor of k_0) can be fitted in terms of two barrier heights as fitting parameters.

The rate constants appearing in the above equations are only phenomenological parameters. For treating the underlying mechanisms of the diffusion, electrophoretic drift, EOF effects, and the free energy barrier, we need to resort to the Fokker–Planck formalism.

9.8 SUMMARY

The rate of capture of polyelectrolyte chains by a nanopore is controlled by a combination of polymer diffusion, electrophoretic drift, convective flow, EOF, and free energy barriers associated with polymer–pore interactions. We have addressed the consequences of these contributing factors and presented explicit equations for use in interpreting experimental observations. Basically, there are the following four limiting regimes.

1. *Diffusion-limited regime*: When the diffusion of the macromolecule due to thermal motion dominates over the drift arising from any externally imposed flow fields and the barrier contributions, the steady-state flux is given by the law,

$$\bar{J}_{diffusion} \sim DRc_0 \sim \frac{Rc_0}{R_g} \sim \frac{Rc_0}{N^\nu}, \tag{9.83}$$

where D is the diffusion constant of the polymer, c_0 is the number concentration of the polymer, R is the linear dimension of the capturing object, R_g is the radius of gyration of the polymer, N is the number of Kuhn segments per chain, and ν is the size exponent of the polymer. In this regime, the capture rate is a decreasing function of the chain length and is independent of applied voltage.

2. *Drift-limited regime*: In this regime, the electrophoretic mobility of the polymer dominates over diffusion and barrier contributions. For large values of the applied voltage difference V_m, the steady-state flux is linear with the polymer concentration and V_m and is independent of the chain length

$$\bar{J}_{drift} \sim c_0 N^0 V_m. \tag{9.84}$$

3. *Flow-dominated regime*: In the presence of strong convective flows with velocity v, the above diffusion-limited law (Equation 9.83) is modified to

$$\bar{J}_{diffusion} \sim c_0 D^{2/3} R^{4/3} v^{1/3}. \tag{9.85}$$

When the pore wall bears charges, EOF is generated by the externally imposed electric field. The requirement of the continuity of fluid flow results in a large region of radius r_c in front of the pore, for capturing the polymer. Inside this region, the polymer can undergo the coil-stretch transition. The capture radius r_c is proportional to the radius of gyration of the polymer,

$$r_c \sim \alpha^{1/3} R_g, \tag{9.86}$$

where α is a parameter given by a unique combination of the surface charge density on the pore wall, pore radius, electric field, temperature, and salt concentration in a particular experimental system. The drift-dominated regime emerges readily in the presence of strong EOFs.

4. *Barrier-dominated regime*: A general formula (Equation 9.53) is derived for the capture rate of polyelectrolytes for any free energy profile and any electric potential variation during the polymer transport. For values of V_m not too high, the barrier-dominated regime emerges where the steady-state capture rate is an increasing function of both V_m and chain length. For larger values of V_m, the barrier effect is weak, and the capture rate follows the drift-limited behavior.

Overall, the capture rate falls either within the barrier-dominated regime or drift-dominated regime, and the diffusion-limited behavior is not dominant for polyelectrolyte transport under an applied voltage.

10

TRANSLOCATION KINETICS:
NUCLEATION AND THREADING

When a polymer molecule is initially directed to the nanopore, its chain ends are not necessarily at the entrance of the pore. As described in the preceding chapters and introduced in Figure 1.9c (Section 1.5.3), the entropic costs associated with the search by chain ends to be localized at the pore mouth and the reduction in chain conformations to be squeezed further into the pore are manifest as a free energy barrier. Once the free energy barrier is surmounted, subsequent transport of the chain along the nanopore is a favorable process, when the translocation is facilitated with a driving force. As a result, the combination of barrier crossing and the subsequent downhill process associated with the translocation event by a single polymer molecule can be treated as the classical nucleation and growth, which is of common occurrence in many phenomena such as the kinetics of phase transitions. We therefore import the technology of nucleation and growth mechanism to the translocation phenomenon. After introducing a few representative experimental results on the threading kinetics of polymers through nanopores and channels, and a few key ideas from computer simulations, we present the model of nucleation and growth for translocation. Based on the background materials developed in Chapters 5 and 6, in terms of entropic barriers and the Fokker–Planck formalism, we shall derive explicit formulas for the translocation time and its distribution function. We shall then return to the experimental contexts and discuss the progress that has been made with such theoretical concepts and the challenges that still remain for a better understanding.

10.1 REPRESENTATIVE EXPERIMENTAL RESULTS

Among the vast amount of experimental data on polymer translocation reported in the literature, we have selected only a few to emphasize the richness of the process and to recognize the most basic aspects that are common to all examples of this phenomenon. As described below, the experimental results are organized in terms of translocation processes through (a) α-hemolysin (αHL) protein pores, (b) solid-state nanopores, and (c) shallow channels.

10.1.1 Translocation through α-Hemolysin Pore

In the presence of an external voltage difference across a membrane containing the αHL pore, a temporary reduction in the pore ionic current occurs whenever the polymer creates a blockade in the pore. The duration of this blockade is called the dwell time τ_d or sometimes misleadingly the translocation time τ. Let I_b be the pore current during a blockade, which is lower than the open pore current I_0. One of the striking features of the experimental results is that there is a very broad distribution in the values of τ_d and I_b, although chemically identical molecules are undergoing translocation events. One of the early examples of such broad distributions is given in Figure 10.1a, where the distribution function $P(\tau_d)$ for the occurrence of a particular value of τ_d is plotted for the translocation of 210-nucleotide-long polyuridylic acid through αHL pore (Kasianowicz et al. 1996). This distribution exhibits three peaks. The first peak with the smallest dwell time can be attributed confidently to events in which the polymer only partially enters or collides with the pore. The average time associated with the first peak is independent of the chain length of the translocating polymer (Figure 10.1b). The second and third peaks represent full translocation events, and the dwell time is the translocation time. When the two peaks were deconvoluted, the average translocation time for each of these two peaks was found to be proportional to N and inversely proportional to V_m, where N is the number of bases in the polymer and V_m is the applied voltage difference (Figure 10.1c). The average obeys the expected law for the simplest physical situation of dragging a thread of length N, through the eye of a fixed needle, with uniform speed, under a constant force proportional to V_m (velocity = length/time; time = length/velocity $\sim$ length/force $\sim N/V_m$),

$$\langle \tau \rangle \sim \frac{N}{V_m}. \tag{10.1}$$

However, the breadths of the distributions are surprisingly large given the uniformity of the polymer molecules.

A more careful analysis of the ionic current traces revealed that a significant fraction of the blockade events is related to the trapping of the polymer in the vestibule of the αHL pore (Butler et al. 2006, 2007, 2008, Henrickson et al. 2010, Wong and Muthukumar 2010). For the αHL pore, there are three prominent types of blockade events, as illustrated in Figure 10.2a for the translocation of sodium (polystyrene sulfonate). These are composed of two distinct blockade levels, the shallow (level 1) and the deep (level 2) levels. From these two blockade levels, three types of events are identified: (a) *event 1* composed of only level 1, with duration τ_1; (b) *event 12* composed of two sublevels *event 12a* and *event 12b* with durations τ_{12a} and τ_{12b}, respectively; and (c) *event 2* composed of a single level 2 blockade with duration τ_2. The two distributions of the dwell time and the ionic current in the blockade states are combined into an "event plot" (Meller et al. 2000). The event diagrams for the translocation of sodium (polystyrene sulfonate) corresponding

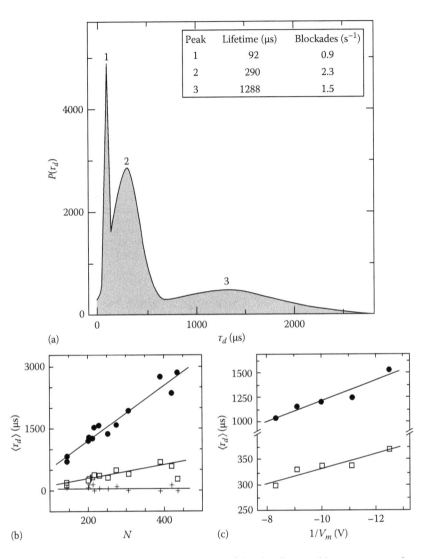

Peak	Lifetime (µs)	Blockades (s⁻¹)
1	92	0.9
2	290	2.3
3	1288	1.5

Figure 10.1 (a) Experimental histogram of the dwell time. (b) Average translocation time for the second and third peaks in (a) is proportional to N and that for the first peak is independent of N. (c) The average translocation time is inversely proportional to the applied voltage for the second and third peaks of (a). (Adapted from Kasianowicz, J.J. et al., *Proc. Natl. Acad. Sci. USA*, 93, 13770, 1996.)

to the different event types are illustrated in Figure 10.2b. The dependence of the average duration of these event types on N/V_m is given in Figure 10.2c. It is clear that the shallow blockades (level 1) are independent of N/V_m. Only the deep blockades (*events 2* and *events 12b*) correspond to successful translocation processes (Butler et al. 2006, 2007, 2008, Murphy and Muthukumar

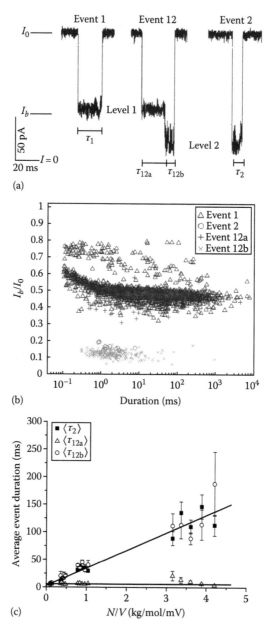

Figure 10.2 (a) The three most prominent event types of ionic current block-ades in αHL. (b) Event diagrams for these event types for the translocation of sodium (polystyrene sulfonate) (molecular weight = 16 kg/mol, pH = 7.5, 1 M KCl, and V_m = 140 mV). (c) The average dwell times for these events against N/V_m. (From Wong, C.T.A. and Muthukumar, M., *J. Chem. Phys.*, 133, 045101, 2010. With permission.)

2007, Henrickson et al. 2010, Wong and Muthukumar 2010). Taking the case of sodium (polystyrene sulfonate) (with the number of repeat units $N \simeq 87$) in 1 M KCl buffer solution with pH = 7.5 as an example, the percentage of successful translocation events among all events is given in Figure 10.3a as a function of V_m. Even at $V_m \simeq 160\,\text{mV}$ ($\approx 6k_BT$), more than 80% of the events are failed attempts to translocate. However, the extent of successful translocation can be dramatically influenced by changing the pH conditions. Instead of pH = 7.5 in the *trans* compartment (Figure 10.3a), if pH = 4.5 is

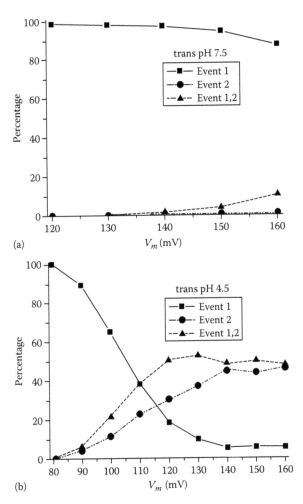

(a)

(b)

Figure 10.3 Percentage of event types at different voltage differences. The pH in *trans* compartment is 7.5 in (a) and 4.5 in (b). Other conditions are the same as in Figure 10.2b and c. (From Wong, C.T.A. and Muthukumar, M., *J. Chem. Phys.*, 133, 045101, 2010. With permission.)

maintained in the *trans* compartment, the success rate of translocation is dramatically increased, as shown in Figure 10.3b (Wong and Muthukumar 2010). A comparison of Figure 10.3a and b demonstrates the significant role played by the electrostatic environment of the pore.

10.1.2 Translocation through Solid-State Nanopores

In contrast to the protein pores of very narrow diameter allowing only single-file translocation of single-stranded DNA, protein molecules (Oukhaled et al. 2007), carbohydrates (Brun et al. 2008), and synthetic polymers (Murphy and Muthukumar 2007, Wong and Muthukumar 2010), solid-state nanopores can be made with tunable diameter to allow thicker polymers such as dsDNA and polymer complexes to pass through. Ionic current measurements in monitoring the translocation of dsDNA through solid-state nanopores have yielded considerable information and several puzzles. First, the event diagrams are more complex than those observed for ssDNA through αHL pore (Storm et al. 2005a). Second, a large capture region can develop in front of the pore (Chen et al. 2004a). Third, depending on the diameter of the pore, in spite of the fact that the persistence length of dsDNA is much larger than the pore diameters used in the nanopore experiments, the polymer translocates through the pore in quantized configurations (Li et al. 2003), such as a single file, chain with one hairpin, etc. As a puzzling result, the relative propensity of single-file (unfolded) events was found to increase with the voltage bias (Figure 10.4a). This is counterintuitive when one considers that the energy cost (Section 2.4) to bend dsDNA into a hairpin configuration should be easier to overcome at higher voltages. Also, as seen in Figure 10.4a, the percentage of unfolded translocation decreases with polymer length.

In spite of the nonlinear dependence of the percentage of single-file translocation, the net electrophoretic velocity is linear in voltage (Figure 10.4b). In addition, the velocity is independent of the molecular length, as different symbols corresponding to different DNA lengths fall on the same line in Figure 10.4b. This implies that the translocation time τ, given by the ratio of the molecular length to the constant velocity, is proportional to N. This result for dsDNA in solid-state nanopores is the same as for the case of ssDNA through the αHL pore (Equation 10.1),

$$\langle \tau \rangle \sim N. \tag{10.2}$$

Furthermore, the standard deviation σ_τ of the translocation time distribution is reported (Chen et al. 2004a) to be proportional to the average translocation time,

$$\sigma_\tau \equiv \left(\langle \tau^2 \rangle - \langle \tau \rangle^2 \right)^{1/2} \sim \langle \tau \rangle. \tag{10.3}$$

In a parallel study (Storm et al. 2005a,b), using different solid-state nanopores, but with comparable pore diameters, the translocation time for

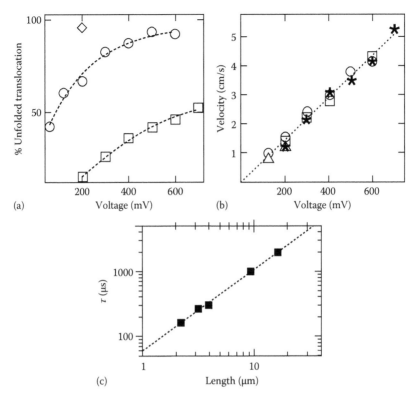

Figure 10.4 Translocation of dsDNA through alumina-coated silicon nitride nanopores (a and b) of diameter ~15 nm and silicon oxide nanopores (c) of diameter ~10 nm. (a) The percentage of unfolded events increases with voltage bias and decreases with DNA length (diamond: 3 kbp, circles: 10 kbp, squares: 48.5 kbp). (From Chen, P. et al., *Nano Lett.*, 4, 2293, 2004a. With permission.) (b) Electrophoretic velocity through the nanopore versus voltage bias. Different symbols are the same as in (a); stars represent 48.5 kbp DNA in a longer pore. (From Chen, P. et al., *Nano Lett.*, 4, 2293, 2004a. With permission.) (c) The translocation time is proportional to $N^{1.26}$. (From Storm, A.J. et al., *Phys. Rev. E*, 71, 051903, 2005a. With permission.)

single-file translocation of dsDNA was found to depend on the chain length, according to

$$\langle \tau \rangle \sim N^{1.26}, \tag{10.4}$$

as shown in Figure 10.4c. This report, based on silicon oxide nanopores of diameter of about 10 nm, is in contradiction with the results of Figure 10.4b, based on alumina-coated silicon nitride nanopores of diameter of about 15 nm. It is apparent from these sets of data that the translocation time is sensitive to the chemical details of the nanopore. Furthermore, with silicon nitride nanopores

with smaller diameters in the range 2.7–5 nm, two different kinds of translocation events have been suggested by relying heavily on data analysis (Wanunu
et al. 2008). These translocation times are reported to depend on N with even
stronger exponent than in Equations 10.2 and 10.4, namely 1.4 and 2.28. In
addition, the translocation times decrease with an increase in the pore diameter, fall off exponentially with an increase in voltage, and follow the Arrhenius
law of increase with temperature as $\exp(\Delta F / k_B T)$ (with ΔF being the free
energy of the barrier for translocation, in the range 10–$20 \, k_B T$). It is clear from
these observations that the various factors discussed in the preceding chapters,
such as the pore–polymer interaction, chain confinement, entropic barriers,
electrostatic nature of the pore, and the electroosmotic flow, contribute to the
translocation time. The different asymptotic limits corresponding to different
dominant factors and the crossover behaviors between these limits are yet to be
established.

10.1.3 Translocation through Channels

When wide channels are used for polymer translocation, instead of cylindrical nanopores, the confinement penalty is in only one direction (Section 5.4).
Nevertheless, the general physical conclusions based on the entropic barrier
for polymer translocation are expected to be applicable to wide channels as
well as nanopores. However, dramatic differences in the chain length dependence of the translocation time have been observed between wider channels
and nanopores. A nanofluidic channel device, consisting of periodically placed
entropic traps, has been used to monitor the electrophoretic mobility of dsDNA
(Han and Craighead 2000). The experimental setup is sketched in Figure 10.5a,
as a cross-sectional schematic diagram. The device consists of a periodic array
of alternating strips of deep wells of thickness $t_d = 1.5$–$3 \, \mu m$ and shallow wells
of thickness $t_s = 75$–$100 \, nm$, with a period of $4 \, \mu m$. These wells are filled with
a buffer solution containing fluorescently labeled DNA molecules in kilobase-
pairs length scale. The lateral dimension of the device is macroscopic so that
the DNA molecules are unrestricted laterally. Under a voltage difference, the
DNA molecules move electrophoretically toward the anode. The time taken by
the molecules with different molecular lengths to reach a particular location
from the starting point is monitored by measuring the fluorescence intensity
with time. Since the device is a periodic array, this time is proportional to
the average translocation time associated with the mobility from one entropic
trap to the next trap by crossing the intervening entropic barrier. These measurements showed that longer DNA molecules move faster in this device as
shown in Figure 10.5b (Han and Craighead 2000). An empirical fitting of the
experimental data of Figure 10.5b to a power law gives

$$\langle \tau \rangle \sim \frac{1}{N^{0.42}}. \tag{10.5}$$

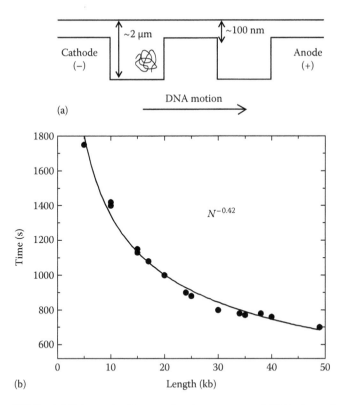

(a)

(b)

Figure 10.5 (a) Schematic diagram for the transport of DNA through a periodic array of entropic traps. (b) The average translocation time decreases with DNA length. (Adapted from Han, J. and Craighead, H.G., *Science*, 288, 1026, 2000.)

This result observed in the nanofluidic channel device, where longer molecules move faster, is counterintuitive and is the opposite of the results observed in nanopores where the longer chains take longer times to translocate.

10.2 INSIGHTS FROM SIMULATIONS

We have introduced in Section 1.4 the idea of entropic barriers for polymer translocation through nanopores, primarily based on intuitive arguments (Figure 1.8). The relevance and the extent of contributions from entropic barriers are generally hard to discern based only on experimental data, as contributions from many control variables need to be separately assessed in interpreting the data. In this context, computer simulations, particularly with the use of toy models, have helped to identify the key molecular mechanisms of polymer translocation.

Consider, for example, the escape (Muthukumar and Baumgaertner 1987, Muthukumar 1991) of a self-avoiding-walk chain ($\nu \simeq 3/5$) of N segments from a cubic cavity of inner side $2R_1$ through the gates at the centers of the walls of the cavity as shown in Figure 10.6a. Each of the six gates is taken to be of length λ, with a square cross-section of side $2R_3$. The chain conformations, dynamics of segments, and the center-of-mass diffusion of a chain were monitored by considering a periodic three-dimensional array of the cavity (Figure 10.6b), using the Monte Carlo simulation method (Muthukumar and Baumgaertner 1987). The simulations showed that the diffusion coefficient D of the center-of-mass of the chain in the asymptotic limit of long enough times followed the relation,

$$D \sim \frac{1}{N} e^{-\Delta F / k_B T}, \tag{10.6}$$

where ΔF arises entirely from restrictions on the chain conformations. The prefactor term is due to the chain connectivity and is consistent with the Rouse result (Equation 7.34), as hydrodynamic interactions among the segments are absent in the model. The value of the entropic barrier ΔF depends on R_1, R_3, and λ (Muthukumar and Baumgaertner 1987). Since the model system is deliberately chosen to be athermal, the only factors that determine the

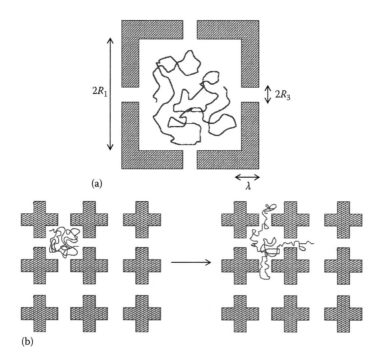

Figure 10.6 (a) Two-dimensional projection of a cubic cavity. (b) Polymer escape from entropic traps through gates.

polymer dynamics are the chain connectivity and the entropic changes associated with the conformational changes of the polymer during the translocation process.

Further insight into the relevance of the free energy landscape (Figure 1.8) for polymer translocation through a single pore was obtained by simulating the following model (Muthukumar 2001). A flexible polyelectrolyte chain was first equilibrated inside a closed sphere at a prescribed ionic strength by using the screened Debye–Hückel interaction among the polymer segments and repulsive interaction between the polymer and the cavity wall, and the Monte Carlo simulation method. Then, a single hole, just big enough to allow only one monomer at a time, was made on the surface of the sphere at the start of a clock. The expulsion of the chain from the sphere into the outside world was then followed as a function of time. As expected, the chain was trying to exit as soon as one of the two chain ends approached the hole, by ejecting a few of the segments at the chain end. Remarkably, the chain then went back inside the sphere instead of proceeding with the ejection. Once the chain went back into the sphere, the process started all over again. After rattling inside the sphere for a while, one of the two ends approached the hole again, and some monomers were put outside and then the whole chain went back in again. After about 300 such attempts, the chain put out enough monomers in the outside world and completely got out of the sphere. This sequence of the events is given in Figure 10.7, where t represents time in arbitrary units.

Such a sequence of events is typical of the nucleation and growth mechanism encountered in the kinetic evolution of a metastable state into an equilibrium state separated by a free-energy barrier. The results of Figure 10.7 are the manifestation of this nucleation and growth mechanism for the translocation of one polymer chain negotiating the entropic barrier of Figure 1.8. In addition to demonstrating the relevance of the central idea of the entropic barriers, the simulation showed that the theoretical technology that has been in place for eight decades to describe the kinetics of first-order phase transformations could be implemented for the polymer translocation through nanopores. We shall describe the theory of nucleation and growth for translocation in the following sections.

In addition to the above examples, modeling of translocation with computer simulations has become an important tool, particularly because the experimental data are only the cumulative results from many contributing factors. On the other hand, effects from selected factors can be individually monitored in computer simulations. A full review of these new developments is beyond the scope of the present discussion. Nevertheless, we briefly mention a few key points below.

Most of the simulations on polymer translocation involve uncharged flexible chains, focusing on the dependence of the average translocation time on the chain length. The exponent α in the relation,

$$\langle \tau \rangle \sim N^\alpha, \tag{10.7}$$

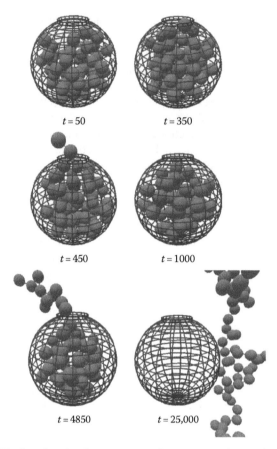

$t = 50$ $t = 350$

$t = 450$ $t = 1000$

$t = 4850$ $t = 25,000$

Figure 10.7 Simulated polymer escape demonstrates the analogy with nucleation and growth. t is time in arbitrary units. (From Muthukumar, M., *Phys. Rev. Lett.*, 86, 3188, 2001. With permission.)

has been keenly debated in the literature (Chuang et al. 2002, Kantor and Kardar 2004, Wolterink et al. 2006, Dubbeldam et al. 2007a,b, 2009, Panja et al. 2007, 2009, Luo et al. 2006, 2009, Vocks et al. 2008, Gauthier and Slater 2008a,b, Bhattacharya et al. 2009, Bhattacharya and Binder 2010, de Han and Slater 2010). However, such huge efforts in the computer modeling of the translocation of an uncharged polymer molecule through a single nanopore are not matched by experimental investigations. In all of the experiments so far exploring the chain length dependence of the translocation time, the polymer is a charged molecule. As discussed in Section 7.4, the mobility and related dynamics of polyelectrolytes are qualitatively different from those of uncharged polymers. In view of the electrically charged nature of the polymers in the present experimental context of the translocation phenomenon, we do not dwell on the ramifications of the ongoing debates on uncharged polymers presented in the above references.

For the experimentally relevant situations of translocating polyelectrolytes through nanopores, there have been a few computer simulations (Muthukumar 2001, Kong and Muthukumar 2002, Aksimentiev and Schulten 2004, 2005, Aksimentiev et al. 2004, Chen et al. 2004b, Mathe et al. 2005, Kong and Muthukumar 2005, Heng et al. 2006, Muthukumar and Kong 2006, Gracheva et al. 2006, Matysiak et al. 2006, Forrey and Muthukumar 2007, Wells et al. 2007, Luan and Aksimentiev 2008, Izmitli et al. 2008, Hernandez-Ortiz et al. 2009). We shall return to some of the simulation results in Section 10.4 when we compare the experimental results with theory and simulations.

The computer simulations on translocating polyelectrolytes have provided vivid details of the molecular processes that are responsible for the observed ionic current traces in experiments. An example is the monitoring of translocation times and the simultaneous ionic current traces, as shown in Figure 8.10. Another example is the details of the coupling between the polymer conformations and the hydrodynamic flow field during translocation (Figure 10.8) (Hernandez-Ortiz et al. 2009). As discussed in Section 7.5 and Section 9.6, the polymer chain can undergo the coil-stretch transition in the presence of strong flow fields. When the fluid flow is continuous through a narrow pore, strong velocity gradients develop near the pore entrance as discussed in Section 9.6. Using the technique of forward flux sampling method, computer simulations have been carried out for a rectangular reservoir of width $17.22\,\mu m$ feeding into a rectangular pore of width $8.6\,\mu m$ and length $8.6\,\mu m$ (Figure 10.8). Accompanying the fluid flow, the conformation and trajectory of one DNA molecule of contour length $420\,\mu m$ were monitored in the simulation. As seen from the three snapshots of the polymer in Figure 10.8, the polymer undergoes the

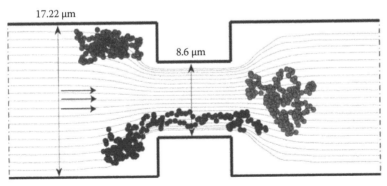

Figure 10.8 Polymer undergoes coil-stretch transition inside a pore with varying velocity gradients. Three snapshots of $420\,\mu m$ DNA translocating through a rectangular pore. (From Hernandez-Ortiz, J.P. et al., *J. Chem. Phys.*, 131, 044904, 2009. With permission.)

coil-stretch transition in locations with steeper velocity gradients. Such details from simulations provide insight into the critical concepts required to build an understanding of the vast amount of data from experiments and simulations.

10.3 THEORY OF TRANSLOCATION KINETICS

As we have discussed in Chapter 5, the typical free energy profiles for a translocating polymer as a function of the extent of translocation are given in Figures 5.3, 5.7, and 5.8. We now derive general formulas for the time-dependent probability of realizing a particular state of translocation, the average translocation time, the distribution function of the translocation time, and the probability of successful translocation, for any given free energy profile suited to a particular experimental system. A sketch of the generic free energy profile is given in Figure 10.9a. Here, F_m is the free energy of a chain of N monomers with m monomers having already translocated into the receiver compartment. In the simplest situation where the pore is a hole (Figures 5.3 and 5.8), $N - m$ monomers are in the donor compartment. When the pore is of finite length, then the free energy profile can be nonmonotonic as in Figure 5.7. Several examples of the profile F_m are given in Chapter 5. Independent of the actual shape of the free energy profile, we simply take F_m as some prescribed function of m that depends on the specificity of the pore and the polymer.

Let us imagine many translocation events, each with a particular trajectory of $m(t)$. In different realizations of the translocation process, the number of

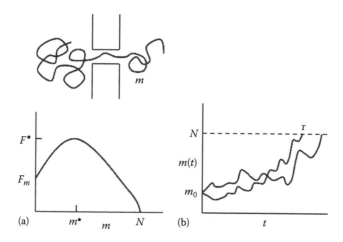

Figure 10.9 (a) Sketch of a generic free energy profile for translocation as a function of the number of monomers that are already translocated into the receiver compartment. (b) Sketch of two trajectories for the time evolution of the number of monomers translocated into the receiver compartment. The translocation time is the first passage time.

monomers that have been translocated into the receiver compartment is different at a given time. A couple of trajectories are illustrated in Figure 10.9b, where the initial value of m at $t = 0$ is m_0. The time evolution of $m(t)$ is a random process due to the diffusion of the monomer along the pore, with a biased drift arising from the driving force for translocation. The translocation time τ for one translocation event is the first passage time for one trajectory, representing the time taken by the whole chain to enter the receiver compartment for the first time. Once pulled into the receiver compartment, we assume that the chain does not reenter the pore. We follow the Fokker–Planck method described in Chapter 6 to describe the stochastic process of translocation (Sung and Park 1996, Muthukumar 1999, Lubensky and Nelson 1999).

Let the probability of finding m monomers in the receiver compartment at time t, among all translocation events, be $P(m, t)$. We assume that the value of m can change by only one monomer in one unit of time, either by translocating one more monomer into the receiver compartment or by pushing back one monomer into the donor compartment. Therefore, the kinetics of translocation in the immediate neighborhood of the value m consists of four elementary steps shown in Figure 10.10. A state with the value m can arise either from the state with value $m - 1$ by translocating one monomer or from the state with value $m + 1$ by pushing the chain back into the donor compartment by one segment. Similarly, the state with the value m can become either the state with $m + 1$ or the state with $m - 1$. Let k_m be the rate at which one monomer translocates forward to increase the number of monomers in the receiver compartment from m to $m + 1$. Similarly, let k'_m represent the rate at which one monomer goes backward such that the number of monomers in the receiver compartment decreases from m to $m - 1$. The rate of change of the probability of finding m monomers in the receiver compartment at time t is dictated by the four elementary steps in Figure 10.10 as given by

$$\frac{\partial P(m, t)}{\partial t} = k_{m-1}P(m - 1, t) - k'_m P(m, t) - k_m P(m, t) + k'_{m+1}P(m + 1, t).$$
(10.8)

This can be rewritten as

$$\frac{\partial P(m, t)}{\partial t} = -(J_m(t) - J_{m-1}(t)),$$
(10.9)

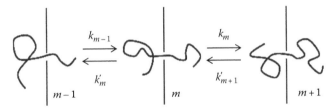

Figure 10.10 Sketch of the four elementary steps in polymer translocation.

where J_m is defined as the net rate at which the state m becomes the state $m+1$, that is, the net rate at which the $(m+1)$th monomer translocates,

$$J_m = k_m P(m, t) - k'_{m+1} P(m+1, t). \qquad (10.10)$$

We make an important assumption that the various states of the chain are in local equilibrium during the step of translocating each monomer. This allows an expression for the reverse rate constant k' in terms of the forward rate constant k, through the definition of the equilibrium constant (Hammes 1978). With the assumption of local equilibrium, the ratio of the forward rate constant to the reverse rate constant is the equilibrium constant given by the exponential of the negative of the change in the free energy in units of $k_B T$. For the second equilibrium given in Figure 10.10, the relation is

$$\frac{k_m}{k'_{m+1}} = e^{-(F_{m+1} - F_m)/k_B T}. \qquad (10.11)$$

Substitution of Equation 10.11 into Equation 10.10 gives

$$J_m = k_m P(m, t) - k_m e^{(F_{m+1} - F_m)/k_B T} P(m+1, t). \qquad (10.12)$$

For convenience, we take m as a continuous variable so that

$$F_{m+1} = F_m + \frac{\partial F_m}{\partial m} + \cdots, \qquad (10.13)$$

$$P_{m+1} = P_m + \frac{\partial P_m}{\partial m} + \cdots, \qquad (10.14)$$

and

$$J_m = J_{m-1} + \frac{\partial J_m}{\partial m} + \cdots. \qquad (10.15)$$

Expanding the exponential in Equation 10.12 and combining with Equations 10.13 and 10.14, we get

$$J_m = k_m P(m, t) - k_m \left[1 + \frac{1}{k_B T} (F_{m+1} - F_m) + \cdots \right] P(m+1, t) \qquad (10.16)$$

$$= -k_m [P(m+1, t) - P(m, t)] - \frac{k_m}{k_B T} (F_{m+1} - F_m) P(m+1, t) + \cdots \qquad (10.17)$$

$$= -k_m \frac{\partial P(m, t)}{\partial m} - \frac{k_m}{k_B T} \frac{\partial F_m}{\partial m} P(m, t) + \cdots \qquad (10.18)$$

Substituting Equation 10.15 into Equation 10.9, we get the continuity equation in the space of the translocation variable m as

$$\frac{\partial P(m, t)}{\partial t} = -\frac{\partial J_m(t)}{\partial m}, \qquad (10.19)$$

where $J_m(t)$ is the flux given by Equation 10.18, to the leading order, as

$$J_m = -k_m \frac{\partial P(m, t)}{\partial m} - \frac{k_m}{k_B T} \frac{\partial F_m}{\partial m} P(m, t). \qquad (10.20)$$

By combining the above two equations, we get

$$\frac{\partial P(m, t)}{\partial t} = -\frac{\partial}{\partial m}[A(m)P(m, t)] + \frac{\partial^2}{\partial m^2}[B(m)P(m, t)], \qquad (10.21)$$

where

$$A(m) = -k_m \frac{\partial}{\partial m}\left(\frac{F_m}{k_B T}\right) + \frac{\partial k_m}{\partial m}, \qquad (10.22)$$

and

$$B(m) = k_m. \qquad (10.23)$$

Equation 10.21 is exactly the same as the Fokker–Planck equation (Equations 6.56, 6.57, and 6.84) discussed in Sections 6.3 through 6.7. The coordinate x in Chapter 6 is replaced by the variable m in the treatment of polymer translocation. The first term on the right-hand side of Equation 10.21 represents the "drift" contribution to the translocation kinetics arising from the negative gradient in free energy during translocation. The second term on the right-hand side of Equation 10.21 denotes the "diffusion" contribution due to local friction of the monomer inside the pore. As a result, k_m may be identified as the diffusion coefficient of the mth monomer inside the pore.

If all monomers of the chain are the same, as in a homopolymer, with uniform diffusion coefficient, k_m is independent of m and the Fokker–Planck equation for the translocation kinetics becomes

$$\frac{\partial P(m, t)}{\partial t} = k_0 \left\{ \frac{\partial}{\partial m}\left[\frac{1}{k_B T}\frac{\partial F_m}{\partial m}P(m, t)\right] + \frac{\partial^2 P(m, t)}{\partial m^2} \right\}, \qquad (10.24)$$

where $k_m \equiv k_0$, independent of m. This equation is exactly the same as Equation 6.65, with the variable m replacing x and k_0 replacing D. In the translocation process, the variable m is the number of monomers that have translocated at time t (in the one-dimensional domain $0 < m < N$), whereas the variable x in Chapter 6 is the position of a particle in the one-dimensional coordinate system.

Analogous to the derivation in Section 8.4, the flux equation (Equation 10.20) can be rearranged as (Equation 8.37)

$$J_m = -k_m e^{-F_m/k_B T} \left[\frac{\partial}{\partial m} (e^{F_m/k_B T} P(m, t)) \right]. \qquad (10.25)$$

In the steady state, the flux J_m is a constant $\bar{J}$. Integrating Equation 10.25 between the limits of $m = 0$ and $m = N$, the constant steady-state flux of the translocating monomers is given by

$$\bar{J} = \frac{e^{F_0/k_B T} P(0, t) - e^{F_N/k_B T} P(N, t)}{\int_0^N dm \, \frac{1}{k_m} e^{F_m/k_B T}}. \qquad (10.26)$$

This is the same as Equation 8.45. For the translocation process, the probability $P(N)$ is zero as the translocating polymer molecules are continuously being depleted from the pore. Also, let us take $P(m = 0) \simeq 1$ during the early stages of the translocation of a chain. With these boundary conditions, the above equation simplifies to

$$\bar{J} = \frac{1}{\int_0^N dm \, \frac{1}{k_m} e^{\Delta F_m/k_B T}}. \qquad (10.27)$$

with

$$\Delta F_m \equiv (F_m - F_0). \qquad (10.28)$$

The mapping of the time-evolution equation for the translocation kinetics to the Fokker–Planck equation allows immediate deduction of the various properties of polymer translocation, directly from the equations presented in Chapter 6. The inputs in obtaining the results are the free energy landscapes derived in Chapter 5 and the diffusion constants k_m. We give below the key results for polymer translocation by copying the general solutions presented in Chapter 6. We shall take the diffusion coefficient of the monomer k_m to be uniform (k_0) in the following sections.

10.3.1 Nucleation Time

Near the free energy maximum around $m^\star$ in Figure 10.9a, the free energy difference $F_m - F_0$ can be expanded as a Taylor series,

$$\frac{\Delta F_m}{k_B T} = \frac{\Delta F_{m^\star}}{k_B T} + \frac{1}{2} \left[\frac{\partial^2 (\Delta F_m/k_B T)}{\partial m^2} \right]_{m^\star} (m - m^\star)^2 + \cdots \qquad (10.29)$$

The linear term is absent because the first derivative is zero at the maximum. The term inside the square brackets is negative around the free energy maximum. Defining

$$\frac{|\Delta F''_{m^\star}|}{k_B T} \equiv - \left(\frac{\partial^2 \Delta F_m/k_B T}{\partial m^2} \right)_{m^\star}, \qquad (10.30)$$

the steady-state flux is given by

$$\bar{J} = e^{-\frac{\Delta F_{m^\star}}{k_B T}} \left[\int_0^N dm \frac{1}{k_m} e^{-\frac{1}{2} \frac{|\Delta F''_{m^\star}|}{k_B T} (m-m^\star)^2} \right]^{-1}. \tag{10.31}$$

If the free energy barrier is sharply peaked (as shown in Section 8.5), the saddle-point approximation can be used to approximate the integral inside the square brackets to get

$$\int_0^N dm \frac{1}{k_m} e^{-|\Delta F''_{m^\star}|(m-m^\star)^2/(2k_B T)} \simeq \frac{1}{k_{m^\star}} \left(\frac{2\pi k_B T}{|\Delta F''_{m^\star}|} \right)^{1/2}. \tag{10.32}$$

The steady-state flux follows from Equation 10.31 as

$$\bar{J}_{nucleation} = k_{m^\star} \left(\frac{|\Delta F''_{m^\star}|}{2\pi k_B T} \right)^{1/2} e^{-\Delta F_{m^\star}/k_B T}. \tag{10.33}$$

We shall call this rate the nucleation rate for the polymer translocation process. Once the barrier in Figure 10.9a is crossed, the translocation process is generally a downhill process. The role of the entropic barrier in controlling the translocation is captured by Equation 10.33. It is in the familiar Arrhenius form or equivalently the Kramers-type form. The nucleation time is the average time taken by the polymer chain to put sufficient number of monomers on the receiver side, after crossing the nucleation barrier, for further progress of the translocation event. We define the nucleation time as the reciprocal of the nucleation rate,

$$\tau_{nucleation} \equiv \frac{1}{\bar{J}_{nucleation}} \sim e^{\Delta F_{m^\star}/k_B T}. \tag{10.34}$$

Any particular form of the free energy barrier height, as described in Chapter 5, can be used in evaluating the nucleation rate. As an example, the nucleation time for a flexible polyelectrolyte chain of N segments to escape from a confining spherical cavity to the outside world is obtained by combining Equations 5.31, 10.33, and 10.34 as

$$\tau_{nucleation} \sim e^{1/N^\beta}, \tag{10.35}$$

with the apparent exponent $\beta \simeq 0.2$. A few other examples will be discussed in Section 10.4.

10.3.2 Translocation Time

Using the equations for a general stochastic process with the drift and diffusion factors (A and B in Equation 6.84), derived in Section 6.6, explicit expressions for the distribution function of the translocation time, the average translocation time, and the probability of successful translocation can be readily obtained. The specificity of the free energy profile F_m and the monomer diffusion coefficients k_m for a polymer-pore system enters through Equations 10.22 and 10.23, and the resulting factors $A(m)$ and $B(m)$ are to be used in the formulas of Section 6.6. As noted already, the translocation time is the average first passage time $\tau_1 = \langle \tau \rangle$ for the stochastic translocation process.

Returning to the basic free energy profile of Equation 5.6, let us consider the simple example where the gain in the electrochemical potential per monomer in the receiver compartment is dominant over the entropic part from chain conformations. Under this condition, Equation 5.6 gives

$$\frac{F_m}{k_B T} = -m \frac{\Delta \mu}{k_B T}, \tag{10.36}$$

where $\Delta \mu \equiv \mu_1 - \mu_2$, with μ_1 and μ_2 being the chemical potentials of the polymer per monomer in the donor and the receiver compartments, respectively. Further, let us assume that the diffusion coefficients of the various monomers are the same ($k_m = k_0$). Combining Equations 10.36, 10.22, and 10.23, we get,

$$A(m) = k_0 \frac{\Delta \mu}{k_B T}, \tag{10.37}$$

$$B(m) = k_m = k_0. \tag{10.38}$$

In the present simple model, both the translocation force arising from the free energy profile and the diffusion constant of the translocating monomer are constants. Substituting the above equations in Equation 10.21, we get the familiar drift–diffusion equation,

$$\frac{\partial P(m, t)}{\partial t} = -k_0 \frac{\Delta \mu}{k_B T} \frac{\partial P(m, t)}{\partial m} + k_0 \frac{\partial^2 P(m, t)}{\partial m^2}. \tag{10.39}$$

This is exactly the same equation as Equation 6.112, where the drift velocity is $k_0 \Delta \mu / k_B T$ and the diffusion coefficient is k_0,

$$v \equiv \frac{k_0 \Delta \mu}{k_B T} \tag{10.40}$$

$$D \equiv k_0. \tag{10.41}$$

With this mapping between the translocation process and one-dimensional drift–diffusion of a particle, all results in Section 6.7 can be simply copied for the various boundary conditions. The other changes in the variables are $x_0 \to m_0$ and $L \to N$. Therefore, the results of Section 6.7 for the position of a particle undergoing a one-dimensional biased random walk can be mapped to the corresponding results for the number of monomers of a polymer chain translocated into the receiver compartment, with the following substitutions,

$$x_0 \to m_0, \quad v \to \frac{k_0 \Delta \mu}{k_B T}, \quad D \to k_0, \quad L \to N. \qquad (10.42)$$

The hard labor of going through the general formulas in Section 6.7 now bears fruit by enabling the use of the same formulas for translocation kinetics as well by simply substituting the relevant variables. This is demonstrated in the following sections.

Consider the stochastic translocation process given by Equation 10.39. Let m_0 be the number of monomers, which have been nucleated in the receiver compartment, to begin with. Starting from this initial condition, the probability distribution function of the first passage time τ is given by Equation 6.86. Let us now consider the key results for the boundary conditions $BC1$ and $BC2$ (Table 6.1). The results for the radiation boundary condition $BC3$ can be obtained similarly by looking up the results in Section 6.7.4.

It must be mentioned that the model of translocation as a ratchet is a subset of the one-dimensional drift–diffusion process (Peskin et al. 1993). In the ratchet model, every monomer being translocated gets reflected back at every site along the pore in the direction of translocation and undergoes drift–diffusion between successive sites along the pore. The final results of this model are readily derived from the same drift–diffusion equation (Equation 10.39).

10.3.2.1 Two Absorbing Barriers

Using the boundary condition $BC1$, where both boundaries at $m = 0$ and $m = N$ are absorbing, the probability distribution of the first passage time is given by Equation 6.116,

$$g(m_0, \tau) = \frac{2}{N} \sum_{p=1}^{\infty} \beta_p \left[1 - \cos(\beta_p N) e^{\left(\frac{\Delta \mu N}{2 k_B T} \right)} \right]$$

$$\times \sin(\beta_p m_0) e^{\left[-\frac{\Delta \mu m_0}{2 k_B T} - \left(\beta_p^2 + \left(\frac{\Delta \mu}{2 k_B T} \right)^2 \right) k_0 \tau \right]}, \qquad (10.43)$$

where $\beta_p = \pi p / N$.

Some representative results from Equation 10.43 are already given in Figure 6.8, where Equation 10.21 and the mapping of the symbols in Equation 10.42 are to be used. Two graphs in Figure 6.8 are replotted in Figure 10.11. The dependence of the probability distribution function of the first passage time, $N^2 g(m_0, \tau)/2\pi$, on $k_0 \tau / N^2$ is given in Figure 10.11a and b, for the initial positions at $m_0 = N/5$ and $m_0 = N/2$ at $N\Delta\mu/k_B T = 7$. It is seen from Figure 10.11a that there are two peaks in the distribution when the initial number of monomers in the receiver compartment is small. Now, the chain has a finite chance to get back to the donor compartment, even against the drift, in addition to translocating completely into the receiver compartment. The higher

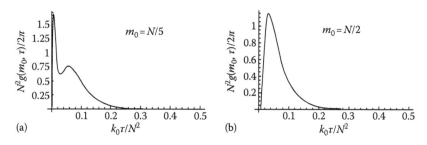

Figure 10.11 Probability distribution function $g(m_0, \tau)N^2/2\pi$ of the translocation time τ, for the absorbing conditions at $m = 0$ and $m = N$. The initial value of m is (a) $m_0 = N/5$ and (b) $m_0 = N/2$.

peak at shorter translocation times represents exit through $m = 0$ boundary, and the other peak represents the exit through $m = N$ boundary. For $m_0 = N/2$, there is only one peak (Figure 10.11b), and now, the chance to retract to the donor compartment is essentially zero. The probability distribution function is asymmetric in shape; it rises sharply, reaches a maximum, and then decreases roughly algebraically. As discussed in Section 6.7.2 (Figure 6.8d), the distribution function for the first passage time is compressed to shorter times at higher driving forces.

The mean first passage time is obtained from Equations 6.99 through 6.101 as

$$\langle \tau \rangle = \tau_1(m_0) = \frac{k_B T}{k_0 \Delta \mu} \frac{\left[N(1 - e^{-m_0 \Delta \mu/k_B T}) - m_0(1 - e^{-N\Delta \mu/k_B T}) \right]}{(1 - e^{-N\Delta \mu/k_B T})}. \tag{10.44}$$

The limits of this equation for $\Delta \mu/k_B T \to 0$ and $\Delta \mu/k_B T \to \infty$ are

$$\langle \tau \rangle = \tau_1(m_0) = \begin{cases} \frac{m_0(N-m_0)}{2k_0}, & N\Delta \mu/k_B T \to 0 \\ \frac{k_B T(N-m_0)}{k_0 \Delta \mu}, & N\Delta \mu/k_B T \to \infty. \end{cases} \tag{10.45}$$

In the drift-dominated regime, the average translocation time is proportional to the chain length (for $N \gg m_0$),

$$\langle \tau \rangle = \frac{k_B T}{k_0} \frac{N}{\Delta \mu}. \tag{10.46}$$

In the diffusion-limited regime, the mean first passage time is a maximum if the starting point is at the middle ($m_0 = N/2$) and is zero at either boundary (due to the imposed $BC1$), as presented in Figure 6.9a.

The probability of successful translocation by exiting through the boundary at $m = N$, without ever reaching the other boundary at $m = 0$, follows from

Equation 6.106 as

$$\pi_+(m_0) = \frac{(1 - e^{-m_0 \Delta\mu/k_B T})}{(1 - e^{-N\Delta\mu/k_B T})}. \tag{10.47}$$

The conditional probability to exit through the other boundary is $\pi_-(m_0) = 1 - \pi_+(m_0)$. The dependence of $\pi_+(m_0)$ on the strength of the drift ($N\Delta\mu/k_B T$) is exactly the same as in Figure 6.10, with the changes in the variables (Equation 10.42). Similarly, the mean translocation time $\tau_+(m_0)$ through the boundary at $m = N$ and the mean translocation time $\tau_-(m_0)$ through the boundary at $m = 0$ are given by Equations 6.122 and 6.123, with the substitutions given in Equation 10.42. The role of the driving force $\Delta\mu$ on the dependence of τ_+ and τ_- on the initial number of nucleated monomers in the translocation process can be read from Figure 6.11 with the use of Equation 10.42.

10.3.2.2 One Absorbing Barrier and One Reflecting Barrier

Let the boundary at $m = 0$ be reflecting and the boundary at $m = N$ be absorbing, namely the boundary condition $BC2$ in Table 6.1. The probability distribution function for the first passage time is given by Equation 6.124 as

$$g(m_0, \tau)$$

$$= 2 \sum_{p=1}^{\infty} \frac{\beta_p \left(\beta_p^2 + \left(\frac{\Delta\mu}{2k_B T}\right)^2\right) e^{\frac{\Delta\mu}{2k_B T}(N - m_0) - \left(\beta_p^2 + \left(\frac{\Delta\mu}{2k_B T}\right)^2\right) k_0 \tau} \sin(\beta_p(N - m_0))}{\left[N\left(\beta_p^2 + \left(\frac{\Delta\mu}{2k_B T}\right)^2\right) + \frac{\Delta\mu}{2k_B T}\right]}, \tag{10.48}$$

with

$$\beta_p \cot(\beta_p N) = -\frac{\Delta\mu}{2k_B T}. \tag{10.49}$$

The mean first passage time for the boundary condition $BC2$ follows from Equations 6.99, 6.100, and 6.102 as

$$\tau_1(m_0) = \int_{m_0}^{N} dy e^{-\int_0^y dy' \frac{A(y')}{B(y')}} \int_0^y dz \frac{e^{\int_0^z dz' \frac{A(z')}{B(z')}}}{B(z)}. \tag{10.50}$$

Substituting Equations 10.22 and 10.23 in Equation 10.50, with $k_m = k_0$, we get

$$\tau_1(m_0) = \frac{1}{k_0} \int_{m_0}^{N} dy e^{F_y/k_B T} \int_0^y dz e^{-F_z/k_B T}. \tag{10.51}$$

The mean first passage time for the process of Equation 10.36 is obtained by performing the two integrations in Equation 10.51 as

$$\tau_1(m_0) = \frac{k_B T}{k_0 \Delta\mu} \left[N - m_0 + \frac{k_B T}{\Delta\mu} (e^{-N\Delta\mu/k_B T} - e^{-m_0\Delta\mu/k_B T}) \right]. \tag{10.52}$$

The asymptotic limits of this result, for the weak and strong strengths of the drift versus diffusion, follow as

$$\langle \tau \rangle = \tau_1(m_0) = \begin{cases} \frac{N^2 - m_0^2}{2k_0}, & N\Delta\mu/k_B T \to 0 \\ \frac{k_B T(N - m_0)}{k_0 \Delta\mu}, & N\Delta\mu/k_B T \to \infty. \end{cases} \tag{10.53}$$

Therefore, the two limits in the diffusion- and drift-dominated regimes for the average translocation time, for $N \gg m_0$, are given by

$$k_0\langle \tau \rangle \sim \begin{cases} \frac{N^2}{2} & \text{(diffusion)}, \quad N\Delta\mu/k_B T \to 0 \\ \frac{k_B T}{\Delta\mu}N & \text{(drift)}, \quad N\Delta\mu/k_B T \to \infty. \end{cases} \tag{10.54}$$

The crossover formula for the chain length dependence of the average translocation time, (Equation 10.52) is plotted in Figure 10.12 for several values of the driving force $\Delta\mu/k_B T = 0, 0.1$, and 0.5. In this figure, the initial number of monomers m_0 is taken as zero. Similar plots to those in Figure 10.12 can be readily drawn for any choice of m_0 in Equation 10.52. In the absence of the drift, the average translocation time (in units of $1/k_0$) is $N^2/2$, and the slope in the double logarithmic plot of Figure 10.12 is 2. In the presence of drift, the asymptotic slope is 1, that is, $\langle \tau \rangle \sim N$. The drift-dominated regime can be achieved for lower chain lengths, if the driving force $\Delta\mu/k_B T$ is higher. Naturally, in the crossover region, an effective exponent for the dependence of $\langle \tau \rangle$ on N is in between the drift-dominated value of 1 and the diffusion-dominated value of 2. In obtaining Equations 10.52 through 10.54, the monomer diffusion coefficient k_0 is assumed to depend on the local friction of a monomer against the pore. If one uses the hypothesis that k_0 is proportional to the chain length, then the average translocation time is proportional to N^3 in the diffusion-dominated

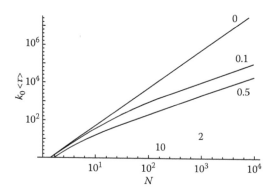

Figure 10.12 Double logarithmic plot of the average translocation time against chain length for $\Delta\mu/k_B T = 0, 0.1$, and 0.5.

regime and to N^2 in the drift-dominated regime (Sung and Park 1996). The experimental evidences so far support the assumption that the local k_0 is independent of N.

The above analytical results are derived only for the simple model of the inside of a pore, with the drift force being a constant independent of the translocation coordinate m. We consider next several examples of more complex situations.

10.3.3 Effect of Pore–Polymer Interactions

Consider the translocation processes depicted in Figure 10.13. In Figure 10.13a, a confined chain inside a spherical cavity escapes into the outside world through a tiny hole, which allows only one monomer at a time. In Figure 10.13b, two spherical cavities exchange the polymer through single-file translocation. In Figure 10.13c, the translocation of the polymer from a donor spherical cavity to a receiver spherical cavity occurs through a nanopore. When a pore is embedded in a membrane, there can be membrane potentials across the pore, as discussed in Chapter 8. We shall consider the translocation of a polyelectrolyte chain through such a pore, as depicted in Figure 10.13d. In addition to the voltage gradient across the pore, there can be an electrochemical potential gradient between the donor and receiver compartments. We have already

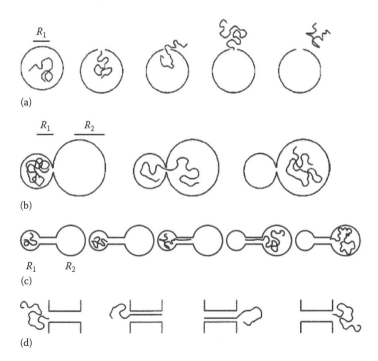

Figure 10.13 (a–d) Some examples of translocation processes.

derived, in Chapter 5, the free energy profiles for translocation corresponding to the situations in Figure 10.13a through c. We shall now present the consequences of the free energy profiles on the average translocation time for these examples.

10.3.3.1 Escape from a Spherical Cavity

Let the radius of the spherical cavity be R_1 and the interaction energy between one segment and the hole-like pore be ϵ_p. The free energy profile for the translocation of the polymer in this geometry is presented in Figure 5.8f. In constructing this profile, the Gaussian chain statistics, along with the ground state approximation, were used, for illustrative purposes. As discussed in Section 5.2.1, the primary contribution to the free energy barrier arises from the search by the chain end for the hole. Therefore, the overall rate of translocation events is dominated by the nucleation barrier discussed in Section 10.3.1. Once a monomer is placed outside the hole, the average translocation time taken by the chain to reach the state of Figure 5.8d with $N - 1$ segments in the outside world is obtained from Equation 10.51 as

$$\langle \tau \rangle = \frac{1}{k_0} \int_1^{N-1} dy \int_0^y dz e^{(F_y - F_z)/k_B T}. \tag{10.55}$$

Here, F_y and F_z are given by Equation 5.19. Substituting the combination of Equations 5.13, 5.17, through 5.19 into Equation 10.55, the computed result is presented in Figure 10.14. In Figure 10.14, the average translocation time

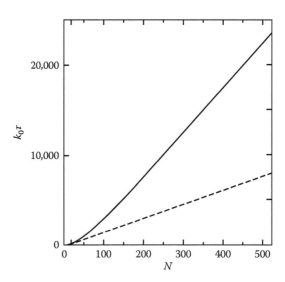

Figure 10.14 Chain length dependence of the average translocation time for Figures 10.13a (dashed) and b (solid). (From Muthukumar, M., *J. Chem. Phys.*, 118, 5174, 2003. With permission.)

(in units of $1/k_0$) is plotted against the chain length for $R_1 = 5\ell$ (with ℓ being the Kuhn length). It should be noted that in calculating the translocation time, the contribution from the conformational entropic barrier is included, although the main barrier associated with the search for the hole is not included. It turns out that the conformational entropic part plays only a minor role. For large values of N, the translocation time is linear with N. In addition, the interaction energy ϵ_p between a segment and the hole does not affect the average translocation time as only a difference in free energy appears in Equation 10.55.

10.3.3.2 Translocation between Two Spherical Cavities

For the situation depicted in Figure 10.13b, where a polymer molecule undergoes translocation from one spherical cavity of radius R_1 to another spherical cavity of radius R_2 through a hole, the free energy profile is given by Equation 5.20. The result for a Gaussian chain is illustrated in Figure 5.9d for $N = 100, R_1 = 5\ell$, and $R_2 = 5\ell$ and 6ℓ. The net driving force arises solely from chain entropy, which is a function of R_1, R_2, and the radius of gyration R_g, as given in Equation 5.21. Substituting Equation 5.20 into Equation 10.55, the average translocation time is calculated with the result given in Figure 10.14 for $R_1 = 5\ell$ and $R_2 = 6\ell$. The confinement of the translocated polymer into a finite region leads to an increase in the average translocation time, as seen from a comparison between the two curves in Figure 10.14.

Analogous to the result of Equation 10.52, the average translocation time is given by (Muthukumar 2003)

$$k_0 \langle \tau \rangle = \frac{k_B T}{\Delta \mu} \left[N - \frac{k_B T}{\Delta \mu} \left(1 - e^{-\frac{N \Delta \mu}{k_B T}} \right) \right], \tag{10.56}$$

where the driving force for a Gaussian chain is

$$\frac{\Delta \mu}{k_B T} = \frac{\pi^2 \ell^2}{6} \left(\frac{1}{R_1^2} - \frac{1}{R_2^2} \right). \tag{10.57}$$

The driving force arises only through the chain entropy and the radii of the confining spheres. The nature of the driving force is modified when intrachain excluded volume effect is present (Kong and Muthukumar 2004). The asymptotic results in the diffusion-limited and the drift-limited regimes follow from Equation 10.56 as the same as given in Equation 10.54. For values of N not small, the scaling exponent for the N-dependence of the average translocation time is 1,

$$\langle \tau \rangle \sim N. \tag{10.58}$$

10.3.3.3 Two Spherical Cavities Connected by a Pore

Let a narrow pore, allowing only single-file translocation, of length $M\ell$ be present between two spherical cavities of radii R_1 and R_2 (Figure 10.13c). Let ϵ_p be the interaction energy between the polymer and the pore per polymer segment. The free energies of the various states in Figure 10.13c are discussed in Section 5.2.1 and are given by Equations 5.22 and 5.23. As an example, the free energy profile for $N = 100, M = 10, \epsilon_p/k_BT = -0.1, R_1 = 5\ell$, and $R_2 = 6\ell$ is given in Figure 5.10. In this figure, the ordinate η_t is the translocation order parameter denoting the fraction of monomers depleted from the donor compartment. η_t is $0, 0.1$, and 1.0 for the second, third, and fourth states of Figure 10.13c, respectively. During the last stage of expulsion of the chain from the filled pore, $\eta_t > 1.0$. In Figure 5.10, the pore is assumed to be attractive for the polymer, by choosing a negative value for ϵ_p. As a comparison between Figures 5.9d and 5.10 reveals, the presence of an interacting pore of finite length can significantly modify the free energy profile for polymer translocation. An adsorbing polymer inside the pore leads to lowering of free energy as the polymer is sucked into the pore. However, this adsorption results in a barrier during the late stage of translocation where the polymer needs to be pulled out against the polymer–pore attraction. These two steps of suction and expulsion of the polymer are separated by the entropic barrier associated with the transfer of monomers outside the pore.

The average times associated with the three steps of filling the pore, transferring the rest of the monomers, and depleting the pore are calculated by using the same equation as Equation 10.55. The average time $\langle \tau \rangle_1$ taken by the chain to fill the pore of length $M\ell$ with $N - M$ segments inside the donor sphere of radius R_1 is

$$\langle \tau \rangle_1 = \frac{1}{k_0} \int_0^M dy \int_0^y dz e^{(F_y - F_z)/k_BT}, \tag{10.59}$$

where F_y and F_z are F_s given by

$$\frac{F_s}{k_BT} = s \frac{\epsilon_p}{k_BT} - \ln P_1(N - s, R_1), \tag{10.60}$$

with P_1 defined in Equation 5.13. With the ground state approximation for P_1, the result is

$$k_0 \langle \tau \rangle_1 = \frac{k_BT}{\Delta\mu_1} \left[M - \frac{k_BT}{\Delta\mu_1} \left(1 - e^{-\frac{M\Delta\mu_1}{k_BT}} \right) \right], \tag{10.61}$$

where

$$\frac{\Delta\mu_1}{k_BT} = \frac{\pi^2\ell^2}{6R_1^2} - \frac{\epsilon_p}{k_BT}. \tag{10.62}$$

The average duration $\langle \tau \rangle_2$ of the second step of translocating $N - M$ segments from the donor to the receiver compartment is the same as given by Equation 10.52 with $m_0 = 0$ and N replaced by $N - M$. Analogous to the derivation of Equation 10.61, the average duration $\langle \tau \rangle_3$ of the last step of chain expulsion from the pore is (Muthukumar 2003)

$$k_0 \langle \tau \rangle_3 = \frac{k_BT}{\Delta\mu_3} \left[M - \frac{k_BT}{\Delta\mu_3} \left(1 - e^{-\frac{M\Delta\mu_3}{k_BT}} \right) \right], \tag{10.63}$$

with

$$\frac{\Delta\mu_3}{k_BT} = \frac{\epsilon_p}{k_BT} - \frac{\pi^2\ell^2}{6R_2^2}. \tag{10.64}$$

The sum of $\langle\tau\rangle_1, \langle\tau\rangle_2$, and $\langle\tau\rangle_3$ is approximately the mean translocation time for the process in Figure 5.7b through e. The different contributions to the average translocation time are given in Figure 10.15a, for $N = 100, R_1 = 5\ell, R_2 = 6\ell$, and $\epsilon_p = -0.1k_BT$, as functions of the pore length $M\ell$. The dominant contributions are from the entropic transfer of segments and the peeling of the chain from the pore. As the pore length increases, the time for filling the pore increases, as expected. If the pore is too long, the number of segments to be transferred through the entropic barrier is small, and as a result, $\langle\tau\rangle_2$ decreases with M. On the other hand, if the polymer-attractive pore is longer, the time required to deplete the chain from the pore gets longer, as seen in Figure 10.15a. The opposite trends of $\langle\tau\rangle_2$ and $\langle\tau\rangle_3$ result in an optimum pore length M_0 at which the average translocation time is a minimum for a given set of values of R_1, R_2, ϵ_p, and N. Also, there exists a pore length M_c, beyond which the translocation time is longer than the value for the case of the pore replaced by a hole. For $M < M_c$, the translocation process is dominated by the entropic barrier mechanism. For $M > M_c$, the interaction between the pore and the polymer dominates. These two regimes are depicted in Figure 10.15b. If the pore repels the segments ($\epsilon_p > 0$), the roles of the first and third steps are reversed and the uphill free energy associated with the pore-filling dominates $\langle\tau\rangle$, resulting in the same nonmonotonic behavior of Figure 10.15b.

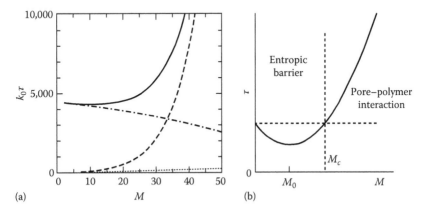

Figure 10.15 (a) Contributions from different stages of Figure 10.13c to the overall translocation time, $\langle\tau_1\rangle$ (dotted line), $\langle\tau_2\rangle$ (dot-dashed line), $\langle\tau_3\rangle$ (long dashed line), and $\langle\tau\rangle$ (solid line). (b) Entropic barrier and pore–polymer interaction regimes for the pore length dependence of the translocation time. (From Muthukumar, M., J. Chem. Phys., 118, 5174, 2003. With permission.)

10.3.3.4 Translocation with a Membrane Potential

Consider the passage of a polyelectrolyte molecule through a single pore embedded into a planar membrane with thickness $M\ell$ and membrane potential V_m (Figure 10.13d). For the state sketched in Figure 10.16a, where s monomers are inside the pore and $N - s$ monomers are in the donor compartment, the free energy is the sum of the entropy associated with the tail conformation of the chain and the electrostatic energy associated with the monomers inside the pore. The potential energy of a segment with an effective charge $e|z_p|\alpha$ (with α being the degree of ionization, as discussed in Chapter 4) at a distance $s'\ell$ from the pore entrance is $-e\alpha|z_p|V_m|s'/M$. Here we assume that the voltage gradient across the pore is linear and favors the translocation from the donor compartment. The electrostatic energy for all segments filling the pore up to the distance $s\ell$ in Figure 10.16a is

$$\frac{\Delta F_{el}}{k_B T} = \int_0^s ds' \frac{-e\alpha|z_p|V_m|s'}{k_B TM} = -\frac{e\alpha|z_p|V_m|s^2}{2k_B TM}. \tag{10.65}$$

As a result, the pore–polymer interaction parameter ϵ_p becomes s-dependent (Equation 5.24).

The free energy profile for the present situation can be composed as follows. To be general, let us assume that there is an additional electrochemical potential

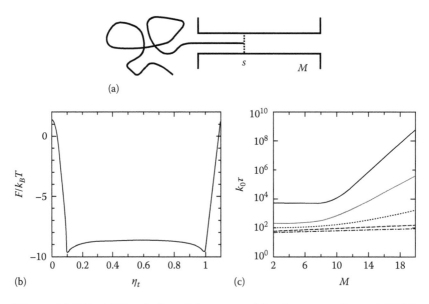

(a)

(b)

(c)

Figure 10.16 (a) Sketch of partial insertion of the chain in an electric field. (b) A deep well can occur in the presence of a membrane potential. (c) Dependence of the average translocation time on the pore length for $\Delta\mu/k_B T = 0$ (solid), 0.5 (dotted), 1.0 (dashed), 1.5 (long-dashed), and 2.0 (dot-dashed). (From Muthukumar, M., J. Chem. Phys., 118, 5174, 2003. With permission.)

difference $\Delta\mu$ driving the polymer from the donor to the receiver compartment. Also, as derived in Equation 5.5, the free energy associated with m segments of a flexible polyelectrolyte chain in the receiver compartment and $N-m$ segments in the donor compartment is

$$\frac{F_m}{k_BT} = (1-\gamma')\ln[m(N-m)] - \frac{m\Delta\mu}{k_BT}, \tag{10.66}$$

where $\gamma' \simeq 0.69$ if the salt concentration in the solution is high enough (Equation 5.3). In writing the above equation, we have assumed the critical exponent γ' to be the same in both the donor and receiver compartments. By combining Equations 10.65 and 10.66, the free energy of a chain with s segments inside the pore and $N-s$ segments in the donor compartment is given by

$$\frac{F_s}{k_BT} = s\left(\frac{\epsilon_p}{k_BT} - \frac{e\alpha|z_pV_m|s}{2k_BTM}\right) + (1-\gamma')\ln(N-s). \tag{10.67}$$

For a polymer conformation with m segments in the receiver compartment and $N-M-m$ segments in the donor compartment, F_m is

$$\frac{F_m}{k_BT} = M\left(\frac{\epsilon_p}{k_BT} - \frac{e\alpha|z_pV_m|}{2k_BT}\right) + (1-\gamma')\ln(N-M-m) - \frac{m\Delta\mu}{k_BT}. \tag{10.68}$$

During the last stage of chain release from the pore, the free energy of a chain with $N-M-p$ segments in the receiver compartment and $M-p$ segments inside the pore is given by

$$\frac{F_p}{k_BT} = (M-p)\left(\frac{\epsilon_p}{k_BT} - \frac{e\alpha|z_pV_m|(M+p)}{2k_BTM}\right) + (1-\gamma')\ln(N-M+p)$$

$$- \frac{(N-M+p)\Delta\mu}{k_BT}. \tag{10.69}$$

The free energy profile for polyelectrolyte translocation through a membrane of membrane potential V_m, based on Equations 10.67 through 10.69, is given in Figure 10.16b. The values of the parameters are $\Delta\mu = 0$, $e\alpha|z_pV_m|/k_BT = 2.0$, $\epsilon_p = -0.1k_BT$, $M = 10$, and $N = 100$. The abscissa is the translocation order parameter η_t, as defined above. It is evident from Figure 10.16b that a voltage drop across the pore and the attractive interaction between pore and the polymer result in a basin in the free energy profile for translocation. The uphill part at the later stages of translocation is due to the peeling of the chain from the attractive pore. If the additional driving force $\Delta\mu$ that stabilizes the segments in the receiver compartment is present, the profile in Figure 10.16b is distorted toward a ramp-like profile (Muthukumar 2003).

Substitution of Equations 10.67 through 10.69 into Equation 10.55 yields the average translocation time. The result is given in Figure 10.16c as a function of the pore length $M\ell$. In this figure, $\gamma' = 0.69$, $N = 100$, $e\alpha|z_p V_m|/k_B T = 2.0$, and $\epsilon_p = -0.1 k_B T$. The average translocation time is essentially independent of M for low values of M, and it increases linearly with M. Presence of an additional driving force decreases the value of the average translocation time, as illustrated with $\Delta\mu = 0, 0.5, 1.0, 1.5$, and 2.0, in Figure 10.16c.

The above specific examples are only for illustrative purposes. These examples show how to implement the technology of Fokker–Planck formalism and how to apply the derived equations for a given situation. Variations in the shape of the pore, entropic barrier, pore–polymer interaction, electrical forces inside and outside the pore, hydrodynamic flows, electroosmotic flows, pressure gradients, polymer sequence, boundary conditions, etc., can be readily addressed by performing calculations analogous to those presented above.

10.4 COMPARISON BETWEEN EXPERIMENTAL DATA AND THEORY

In general, almost every detail of the polymer–pore system appears to contribute to the observed experimental facts on polymer translocation. On the other hand, theoretical efforts are mainly focused on identifying only a few major concepts in order to understand the experimental data. Just as in experiments, theoretical treatments show that the translocation phenomenon is a multivariable problem and that a particular data set might be in crossover regimes. In this sense, the limiting laws derived for the various asymptotic regimes might not be observed in one polymer–pore system when only a narrow range of experimental variables is explored. Nevertheless, there have been several systematic measurements in order to understand the physics of polymer translocation. Some of these measurements were presented in Section 10.1. In particular, the dependencies of the translocation time on chain length and applied voltage have been investigated over wide ranges. We shall consider again the examples given in Section 10.1 and compare the experimental data with theoretical predictions.

10.4.1 Translocation through α-Hemolysin Pore

One of the robust features of the theory of polymer translocation, with the mapping to the drift–diffusion Fokker–Planck equation, is that the average translocation time $\langle\tau\rangle$ is proportional to the chain length N and inversely proportional to the driving force $\Delta\mu$,

$$\langle\tau\rangle \sim \frac{N}{\Delta\mu}, \tag{10.70}$$

in the drift-dominated regime. This is clearly seen in experiments, as evident from Figures 10.1b, c, and 10.2c, where $\Delta\mu$ is the applied voltage.

The shape of the distribution function for the translocation time given by the drift–diffusion equation (Figure 10.11b) is similar to experimentally observed histograms, as long as the multiple peaks of Figure 10.1a are deconvoluted. When the reflecting boundary condition is used at the pore entry, theory yields only one peak in the distribution function of the translocation time. On the other hand, experiments show that there are two peaks, after the first peak corresponding to faster collisions between the polymer and pore is deleted. This feature is particular to the αHL protein pore and ssDNA, and is also based on the way the data are analyzed.

There have been several explanations for the occurrence of the two peaks of interest in Figure 10.1a. The shape of the αHL pore is such that there is a large vestibule region in front of the β-barrel pore. The volume of the vestibule is so large that it can trap hundreds of monomers and allow them to explore their conformations that would be forbidden inside the β-barrel. As a result, the vestibule acts as an entropic trap. Trapping the polymer inside the vestibule before the eventual translocation through the β-barrel leads to a delay in the translocation time. There is also a finite probability with which the chain end can go directly to the β-barrel, which will result in a shorter translocation time. The histograms obtained from the Langevin dynamics simulations (Kong and Muthukumar 2002, Muthukumar and Kong 2006) of a coarse-grained polymer translocating through αHL pore have shown that there are two peaks. By analyzing the actual trajectories of the various events constituting the histogram, it is possible to identify the molecular origin of these two peaks. Indeed, the longer average time corresponds to the trajectories where the vestibule acts as an entropic trap, and the shorter average time corresponds to essentially the direct threading through the β-barrel.

Another contributing factor for the occurrence of the two peaks is the orientation of the polymer backbone. Although the translocating DNA molecules are identical, the molecule can enter the β-barrel either with its $3'$-end or the $5'$-end. Since the bases are oriented at an angle with the chain backbone, the friction of the monomer inside the pore can be different for different polymer orientation. Molecular dynamics simulations, and measurements with polymers consisting of blocks of DNA with different orientations, have clearly established that the chains with different orientations have different average translocation times (Mathe et al. 2005, Butler et al. 2006).

The significant role played by the vestibule in the αHL pore is clearly seen in the prominent occurrence of the mid-level blockade of ionic current (Figure 10.2b). In fact, a substantial number of the blockade events are likely to be failed attempts of translocation, as is evident in Figure 10.3a. In the original data analysis in composing Figure 10.1a, all blockade events were combined together instead of the procedure followed in the later investigations (Butler et al. 2006, 2007, 2008, Henrickson et al. 2010, Wong and Muthukumar 2010). A combination of better data analysis and use of different protein pore such as the MspA (Butler et al. 2008) or a solid-state nanopore for ssDNA

(Kowalczyk et al. 2010) is likely to shed more light into the origin of multiple peaks in the histogram of translocation times.

As shown in Figure 10.3a and b, a change in pH of the solution or the presence of salt concentration gradient can lead to significant modification in the signatures of translocation. In general, the contributions from the electrostatic environment created by the pore and the polymer entropy are combined together in generating the free energy barrier for polymer translocation. The relative weights of these enthalpic and entropic contributions are yet to be fully established.

10.4.2 Translocation through Solid-State Nanopores

We have seen in Section 10.1.2 several unexpected aspects about the translocation phenomenon in solid-state nanopores. These may be catalogued as the existence of a large capture region in front of some pores, unexpected higher propensity of unfolded conformations of dsDNA at higher voltages, and somewhat nonuniversal chain length dependence of the translocation time.

One possible explanation of the existence of a large capture region in front of the pore is the electroosmotic flow, as discussed in the previous chapter. There is no analytic theory yet to predict the relative weight of hairpins versus unfolded conformations for translocating a semiflexible chain into a nanopore with an applied voltage. Langevin dynamics simulations have led to some insight on this issue. As inferred from the experimentally measured ionic traces as dsDNA molecules pass through solid-state nanopores, simulations (Forrey and Muthukumar 2007) show that the coarse-grained model chain of dsDNA penetrates into the pore either as a single file or as a hairpin. Also, once the chain enters the pore, it may retract by failing to cross the nucleation barrier or proceed successfully with the translocation. Independent of whether the polymer succeeded or not, there would be a blockade in ionic current. The conclusion that the fraction of single-file translocation increases with applied voltage, as shown in Figure 10.4a, is based only on counting the number of times of occurrence of the ionic current blockades corresponding to the single-file blockade. This is done without any consideration of whether the event is a successful translocation or not. In simulations, where the trajectories are monitored, it is easy to identify the fraction of events that are successful. When all events of single-file encounters between the polymer and the pore are included in the statistics, simulations with electric field present only inside the pore showed the same trend as Figure 10.4a, as given in Figure 10.17a for different chain lengths. The fraction of single-file events increases with applied potential difference across the pore and decreases with chain length. On the other hand, if the failed attempts are excluded from the statistics, the fraction of single-file translocation decreases with the applied voltage as shown in Figure 10.17b. This is in accordance with expectations based on the ease with which a stiff chain can bend at higher voltages.

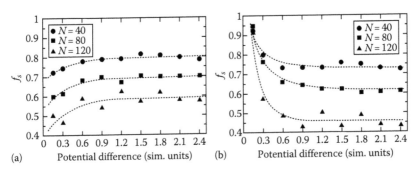

Figure 10.17 (a) Fraction of single-file events increases with voltage bias, when both unsuccessful and successful translocation events are analyzed. (b) Fraction of single-file events decreases with voltage bias, when only successful translocation events are analyzed. (From Forrey, C. and Muthukumar, M., *J. Chem. Phys.*, 127, 015102, 2007. With permission.)

The different values of the exponent for the N-dependence of the single-file translocation time described in Section 10.1.2 are difficult to rationalize at present. The above-mentioned simulations (Forrey and Muthukumar 2007) showed that the average translocation time is proportional to N and inversely proportional to the applied voltage. This conclusion is in agreement with the results of Equation 10.2. However, another simulation (Izmitli et al. 2008) showed that the value of the exponent is ≈ 1.3, in closer agreement with Equation 10.4. The experimentally observed higher values (Wanunu et al. 2008) of the exponent are yet to be understood. The crossover behaviors arising from the pore–polymer interaction, electroosmotic flow, and pore geometry need to be addressed.

Similar to the derivation of the average translocation time, higher moments of the distribution function for the first passage time can be derived from Equation 6.85. For the simple model treated in Section 10.3.2.2, the average translocation time (with the initial value $m_0 = 0$ in Equation 10.52) is

$$\langle \tau \rangle = \frac{k_B T}{k_0 \Delta \mu} \left[N - \frac{k_B T}{\Delta \mu} \left(1 - e^{-\frac{N \Delta \mu}{k_B T}} \right) \right], \tag{10.71}$$

and the variance is

$$\langle \tau^2 \rangle - \langle \tau \rangle^2 = \frac{(k_B T)^4}{k_0^2 (\Delta \mu)^4} \left[-5 + 2\frac{N \Delta \mu}{k_B T} + 4e^{-N \Delta \mu / k_B T} \left(1 + \frac{N \Delta \mu}{k_B T} \right) \right.$$
$$\left. + e^{-2N \Delta \mu / k_B T} \right]. \tag{10.72}$$

Therefore, the standard deviation is related to the average translocation time as

$$\sigma_\tau = \begin{cases} \sqrt{\dfrac{2\langle\tau\rangle}{k_0}\dfrac{k_BT}{\Delta\mu}}, & N\Delta\mu/k_BT \gg 1 \\ \sqrt{2/3}\langle\tau\rangle, & N\Delta\mu/k_BT \ll 1. \end{cases} \tag{10.73}$$

Therefore, it is possible to realize the experimentally observed behavior of Equation 10.3 if the diffusion dominates over the drift in the simplest model illustrated here. However, if the drift does not dominate, a higher value of the exponent than 1 is expected for the N-dependence of the average translocation time. More systematic analysis of the data on the first and second moments of the histogram of the translocation time, along with comparison with Equations 10.72 and 10.73, and analogous equations pertinent to the experimental systems are needed to clarify these issues.

10.4.3 Translocation through Channels

The translocation time in the periodic array of entropic traps decreases with chain length, as given by Equation 10.5 (Figure 10.5a and b). This trend is the opposite of an increase in the translocation time with chain length in all translocation experiments involving nanopores. It would be undesirable to invoke a new set of concepts to describe translocation in channels that is different from the concepts for nanopores. Therefore, we seek to apply the entropic barrier model to the channels as well.

Let us consider the model of nucleation and threading in the context of the experiment described in Figure 10.5a. For each period of the setup, which consists of a deep region followed by a shallow region, a DNA molecule first travels through the deep region with time τ_1 (Figure 10.18). It then stops

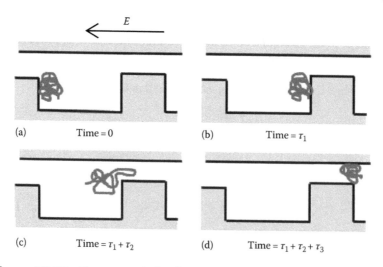

Figure 10.18 Three steps, (a–b), (b–c), and (c–d) contributing to translocation in one period of the array. (From Wong, C.T.A. and Muthukumar, M., *Biophys. J.*, 95, 3619, 2008b. With permission.)

momentarily at the interface between the deep and shallow regions and takes τ_2 to enter the shallow region. The momentary stagnation of the chain near the entry into the shallow region is due to the entropic barrier. Finally, the molecule takes τ_3 to pass through the shallow region. The DNA molecule thus takes the time $\tau_1 + \tau_2 + \tau_3$ to travel through one period. Because of the height difference, the electric field in the deep region is much lower than that in the shallow region, making τ_3 negligible compared to $\tau_1 + \tau_2$. Also, the confinement effect is unimportant in the deep region so that the electrophoretic behavior of the DNA molecule is the same as that in a free solution. As we have seen in Section 7.4, the electrophoretic mobility of the polyelectrolyte chain in the deep region is independent of the chain length. Therefore, the observed chain length dependence originates essentially from τ_2, the time taken by the DNA molecule to enter the shallow region from the deep region. The above description is consistent with fluorescence microscopy observations (Han and Craighead 2000).

The entropic barrier associated with the passage of the chain forward at the interface between the deep and shallow regions is considered as follows. When a polymer chain tries to enter the shallow region, part of the chain partitions into the shallow region and the rest of the chain resides in the deep region, as shown in Figure 10.19. We consider two modes of translocation, namely linear and hairpin translocation. In the linear translocation (Figure 10.19a), the chain enters the shallow region of height $D_p\ell$ with a chain end, forming n blobs in a linear series, with m segments in the shallow region and a tail of $N - m$ segments in the deep region. In a hairpin translocation, the chain enters the shallow region as a hairpin, with m segments in the shallow region and two tails of j and $N - m - j$ segments in the deep region. In the shallow region, there are two linear chains of blobs, each with $n/2$ blobs.

The free energy of the whole chain is the sum of free energies of polymer segments inside the shallow region and those inside the deep region. The free

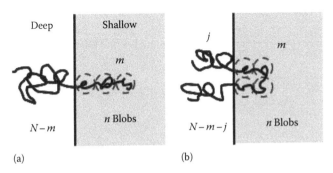

Figure 10.19 Two modes of nucleation into the shallow region: (a) linear mode and (b) hairpin mode. (From Wong, C.T.A. and Muthukumar, M., *Biophys. J.*, 95, 3619, 2008b. With permission.)

energy of segments inside the shallow region has two competing contributions: (a) energy gain due to the electric field and (b) loss in conformational entropy due to confinement. As a result, an entropic barrier emerges as sketched in Figure 10.9a. The free energy of confinement of m segments inside the shallow region of height $D_p\ell$ is derived in Section 5.4 as

$$\frac{F_c(m)}{k_B T} = \frac{m}{D_p^{1/\nu}}. \tag{10.74}$$

Here, D_p is the height in units of the Kuhn length ℓ. The electrostatic energy gain for a linear chain of m segments in the shallow region with a constant electric field follows from Equation 10.65 as (Wong and Muthukumar 2008b)

$$\frac{F_{el,l}(m)}{k_B T} = -\frac{|ez_p E|\ell m^2}{2k_B T D_p^{1/\nu - 1}}, \tag{10.75}$$

where $|ez_p|$ is the effective charge of a segment. Similarly, the electrostatic energy gain for two chains of m segments in the shallow region is (Wong and Muthukumar 2008b)

$$\frac{F_{el,h}(m)}{k_B T} = -\frac{|ez_p E|\ell m^2}{4k_B T D_p^{1/\nu - 1}}. \tag{10.76}$$

As derived in Section 5.1, the free energy associated with a tail of j monomers in the deep region is

$$\frac{F_t(j)}{k_B T} = (1 - \gamma')\ln(j). \tag{10.77}$$

Here, γ' is the effective exponent given in Equation 5.3. We assume that there are no confinement effects on the chain in the deep region.

Combining Equations 10.74 through 10.77, the free energies of linear and hairpin conformations are

$$\frac{F_l(m)}{k_B T} = -\frac{|ez_p E|\ell m^2}{2k_B T D_p^{1/\nu - 1}} + \frac{m}{D_p^{1/\nu}} + (1 - \gamma')\ln(N - m), \tag{10.78}$$

$$\frac{F_h(m,j)}{k_B T} = -\frac{|ez_p E|\ell m^2}{4k_B T D_p^{1/\nu - 1}} + \frac{m}{D_p^{1/\nu}} + (1 - \gamma')\ln(j) + (1 - \gamma')\ln(N - m - j), \tag{10.79}$$

respectively. The relative probability $P_l(m)$ of realizing a state given in Figure 10.19a with m segments in the shallow region is

$$P_l(m) = \exp\left(-\frac{F_l(m)}{k_B T}\right), \tag{10.80}$$

and the relative probability $P_h(m, i)$ of realizing a hairpin state with m segments in the shallow region and two tails of lengths j and $N-m-j$ in the deep region is

$$P_h(m, i) = \exp\left(-\frac{F_h(m, j)}{k_B T}\right). \tag{10.81}$$

Translocation into the shallow region is initially unfavorable due to the confinement free energy $\sim m/D_p^{1/\nu}$. After a critical number of inserted segments $m^\star$, the translocation process becomes favorable and the free energy profile is dominated by the electrostatic energy $\sim |ez_p E|\ell m^2/D_p^{1/\nu-1}$. The translocation time is largely determined by the nucleation barrier $F(m^\star)$ relative to the state in the deep region. The critical values of $m^\star$ for the linear and hairpin modes of translocation are obtained by maximizing Equations 10.78 and 10.79 with respect to m. For long chain lengths $N \gg D_p^{(1-\nu)/2\nu}/(|ez_p E|\ell/k_B T)^{1/2}$, the critical values of the number of segments for nucleation to occur, for the linear and hairpin modes, are (Wong and Muthukumar 2008b)

$$m_l^\star \simeq \frac{k_B T}{|ez_p E|\ell D_p}, \tag{10.82}$$

$$m_h^\star \simeq \frac{2k_B T}{|ez_p E|\ell D_p}. \tag{10.83}$$

Substitution of the above expressions in Equations 10.78 and 10.79 gives the free energy barriers for the linear and hairpin modes of translocation. Using the free energy barriers in Equations 10.80 and 10.81 yields the probabilities of linear and hairpin modes of nucleation. For the linear mode of nucleation, the result follows from Equations 10.78, 10.80, and 10.82 as

$$P_l^\star \simeq N^{\gamma'-1} \exp\left(-\frac{k_B T}{2|ez_p E|\ell D_p^{1/\nu+1}}\right), \tag{10.84}$$

for large N such that $N \gg m_l^\star$. Similarly, Equations 10.79, 10.81, and 10.83 give the probability of nucleation as a hairpin with two tails of length j and $(N - j - m_h^\star) \simeq (N - j)$ as

$$P_h^\star(j) \simeq (N-j)^{\gamma'-1}j^{\gamma'-1} \exp\left(-\frac{k_B T}{|ez_p E|\ell D_p^{1/\nu+1}}\right). \tag{10.85}$$

The total probability of hairpin mode of nucleation is obtained by integrating over all allowed values of j between 0 and $N - m_h^\star$ as

$$P_h^\star \simeq AN^{2\gamma'-1} \exp\left(-\frac{k_B T}{|ez_p E|\ell D_p^{1/\nu+1}}\right), \tag{10.86}$$

where the numerical factor $A = \Gamma(\gamma')^2/\Gamma(2\gamma')$ and $\Gamma(x)$ is the Gamma function (Abramowitz and Stegun 1965). For $\gamma' = 0.69, A = 1.94$.

As discussed in Section 10.3.1, the nucleation time is proportional to the reciprocal of the probability of nucleation (Equations 10.33 and 10.34). Therefore, it follows from Equations 10.84 and 10.86 that the nucleation times τ_l and τ_h for linear and hairpin modes of translocation are given by

$$\tau_l \sim N^{1-\gamma'} \exp\left(\frac{k_B T}{2|ez_p E|\ell D_p^{1/\nu+1}}\right), \tag{10.87}$$

$$\tau_h \sim N^{1-2\gamma'} \exp\left(\frac{k_B T}{|ez_p E|\ell D_p^{1/\nu+1}}\right). \tag{10.88}$$

The value of γ' for the conditions used in the experiments (Han and Craighead 2000) is $\simeq 0.69$. Taking this value for γ', and $\nu \simeq 3/5$, the nucleation times for the linear and hairpin modes of translocation follow as

$$\tau_l \sim N^{0.31} \exp\left(\frac{k_B T}{2|ez_p E|\ell D_p^{8/3}}\right), \tag{10.89}$$

$$\tau_h \sim N^{-0.38} \exp\left(\frac{k_B T}{|ez_p E|\ell D_p^{8/3}}\right). \tag{10.90}$$

It has been shown (Wong and Muthukumar 2008b) that the probability of the linear mode of nucleation decreases with chain length and the strength of the electric field, and in general, there is a crossover from dominance by linear mode at shorter chain lengths to dominance by hairpin mode at longer chain lengths. In the first regime, the average nucleation time increases with chain length, whereas it decreases with chain length for longer chains according to

$$\tau_h \sim \frac{1}{N^{0.38}}. \tag{10.91}$$

This theoretical result is close to the experimental result given by Equation 10.5. Therefore, for the translocation of DNA between a deep channel and a shallow channel, the nucleation rate across the entropic barrier appears to dictate the translocation time. It should be noted that when the translocation time with the nucleation process becomes very small for longer chains, the other stages of DNA mobility in the deep and shallow regions become rate-determining. Additional experimental validation of the theoretical predictions in terms of the height of the shallow region and the strength of the electric field would enable the general applicability of the entropic barrier theory for polyelectrolyte separation with nanofabricated devices.

10.5 SUMMARY

Once a polymer molecule is brought near the vicinity of a pore, its subsequent translocation through the pore is initiated by first crossing an entropic barrier and then threading through the pore. We have mapped the crossing of the barrier as a classical nucleation process and the threading kinetics as a drift–diffusion stochastic process. Depending on the experimental conditions, either the nucleation or the threading process might dominate the kinetics of translocation. In some situations, both processes need to be considered simultaneously.

We have presented a theoretical framework for treating the nucleation process, and explicit formulas are derived for the nucleation rate in terms of the free energy barrier. The results derived in Chapter 5 for the free energy barriers corresponding to different experimental conditions provide the input in predicting the nucleation rate. When the nucleation step is rate-determining, the translocation time is the reciprocal of the nucleation rate. We have given an example for such a scenario by considering the translocation of DNA through a nanofabricated periodic channel array.

The kinetics of threading, where a certain number of monomers $m(t)$ are in the receiver compartment at time t, is derived as a drift–diffusion equation with the local free energy gradient with respect to m providing the drift force and the local friction of the monomer against the pore providing the diffusion. We have mapped this equation into the drift–diffusion equation for the position of a biased random walker, developed in Chapter 6. This allows the implementation of the Fokker–Planck formalism to compute the average translocation time as the average first passage time, for any given free energy profile. We have applied the equations derived in Chapter 6 to several examples of threading kinetics. These exercises may be used as examples to implement the various formulas for translocation kinetics to new experimental systems. In general, once sufficient number of monomers are nucleated into the pore, the average translocation time $\langle \tau \rangle$ is directly proportional to the chain length N and inversely proportional to the driving force $\Delta\mu$,

$$\langle \tau \rangle \sim \frac{N}{\Delta\mu}, \tag{10.92}$$

in the drift-dominated regime. This result is prevalent in most of the experimental conditions, and deviations from this law can be addressed with the derived crossover formulas in terms of various experimental variables.

A few examples of experimental data from αHL protein pore, solid-state nanopores, and nanofluidic channels are also discussed in the context of the nucleation and threading model developed in this chapter. The theoretical predictions and experimental data are in agreement on all substantial points.

11

FURTHER ISSUES

The real world of polymer translocation is far more complex than the rudimentary physical models developed in the preceding chapters. The complexity of the phenomenon arises due to particular combinations of entropic contributions from polymer conformations and energetic contributions from polymer–pore interactions that are specific to the particular translocation systems. The spirit of the description of translocation developed so far is along the direction of identifying the most general universal model, which forms the basic skeleton of the process, with the entropy and energy contributions being parametrized. The chemical details, and the physicochemical processes in the physical vicinity of the translocation event, must be addressed in interpreting these parameters for a specified system. We shall briefly mention below a few examples of how chemical details are amplified in the observed features of polymer translocation. While the general premise of translocation processes is the sequence of capture/recognition, nucleation, and threading, these chemical details must be accounted for in quantitative descriptions. Complementary to the challenges in addressing the diversity in chemical details, there are also challenges in understanding the nonequilibrium polymer conformations particularly when the threading is much faster than the equilibration times for polymer conformations. We illustrate these issues briefly here to underline the ongoing investigations.

11.1 NONEQUILIBRIUM CONFORMATIONS DURING THREADING

In the entropic barrier model accompanied by the drift–diffusion formalism used to compute the molecular weight dependence of the translocation time, we have assumed that the chain is in quasi-equilibrium during translocation. The friction of every monomer at the pore is so great that the translocation time is much longer than the relaxation times (Equations 7.47 and 7.113) of the polymer tails in the donor and receiver compartments. The construction of the free energy landscape for the polymer translocation, as described in the previous chapters, hinges upon this assumption. For all polymer chains used so far in the single-file translocation experiments through the α-hemolysin protein pore, this assumption is valid. This can be seen by comparing the experimentally observed translocation time with the Zimm time τ_{Zimm} as given by Equation 7.47. Under these conditions, the robust result is that the average translocation time is simply linearly proportional to the chain length (Equation 10.70),

$$\langle \tau \rangle \sim N. \tag{11.1}$$

This result is clearly seen in all single-file experiments involving the α-hemolysin protein pore (Figures 10.1 and 10.2) (Kasianowicz et al. 1996, Wong and Muthukumar 2010).

The basic assumption behind the above law must be violated for very large molecular weights, where τ_{Zimm} can be longer than the translocation time. If the chain does not have sufficient time to equilibrate in the process of being threaded through the pore, it is in a nonequilibrium conformation. It is now difficult to construct the appropriate free energy landscape for the translocation process. Pulling a chain at one of its monomers will generate a tensile force that would propagate to the remote parts of the chain, which in turn would influence the segmental orientations and local monomer density. The memory effect that is generated by this mechanism could be long-lived for very large chains in comparison with the translocation time (Krasilnikov et al. 2006). Basically, the free energy landscape becomes time-dependent and it is difficult to derive general laws. The challenge in describing these nonequilibrium situations becomes even more daunting when the polymer is electrically charged and the counterions are also contributing to the local chain dynamics.

An indication of the richness of the memory effects in the nonequilibrium behavior of translocating polymers has been seen in the simulations of uncharged single chains through nanopores and theoretical analyses. Although these simulations and theoretical arguments are not directly pertinent to the current experimental situations on polyelectrolytes, they provide insights into the role of nonequilibrium effects on translocation. The main consequence of the memory effect is that the average translocation time of an uncharged polymer in the presence of a pulling force depends on the chain length with a higher value of the exponent than unity (Kantor and Kardar 2004, Sakaue 2007, Dubbeldam et al. 2007a, Vocks et al. 2008, Gauthier and Slater 2008b, Kolomeisky 2008)

$$\langle \tau \rangle \sim N^{\alpha}, \quad \alpha > 1. \tag{11.2}$$

Different values for the exponent α have been proposed by different authors. Assuming that the average time is the ratio of the distance (comparable to the average radius of gyration) the polymer would travel to the average velocity of the center of mass of the chain (which is inversely proportional to the chain length under a constant force), α was originally proposed (Kantor and Kardar 2004) to be

$$\alpha = 1 + \nu, \tag{11.3}$$

where ν is the size exponent discussed in Section 2.1. For good solutions, $\alpha \simeq 1.6$. By considering the local taut of the chain at the pore as a propagating defect, it was shown that (Sakaue 2007)

$$\alpha = \frac{1 + 3\nu}{2} \simeq 1.4, \tag{11.4}$$

with hydrodynamic interaction and

$$\alpha = \frac{1 + \nu + 2\nu^2}{1 + \nu} \simeq 1.45, \tag{11.5}$$

without hydrodynamic interaction. By accounting for the nonequilibrium conformations with a partial Fokker–Planck description, α was derived to be (Dubbeldam et al. 2007a)

$$\alpha = 2\nu + 1 - \gamma_1 \simeq 1.5, \tag{11.6}$$

where γ_1 is a critical index (Dubbeldam et al. 2007a). A different treatment (Vocks et al. 2008) of the memory effect of polymer conformations under translocation led to

$$\alpha = \frac{1 + 2\nu}{1 + \nu} \simeq 1.38. \tag{11.7}$$

The above theoretical arguments suggest that the value of the exponent α is in the range of 1.38–1.6. There has been a substantial investment in computer simulations by various researchers to identify the value of the exponent α for an uncharged polymer. According to one of the recent accounts (Bhattacharya et al. 2009, Bhattacharya and Binder 2010, Bhattacharya 2010), the parts of the polymer chain in the donor and receiver compartments are clearly seen to be in nonequilibrium states. The value of the exponent α seen in these simulations ($\alpha \approx 1.2$–1.36) is not in conformity with the above theoretical predictions. Also, α depends on the details of the pore, with its value decreasing below the lower bounds set by some of the above theoretical arguments, as the pore diameter is decreased. It is likely that the conclusions on this ideal model system will become more firm in the immediate future. However, these simulations are not yet directly relevant to the experimental systems. The transfer of conclusions on the dynamics of uncharged polymers to those of polyelectrolyte chains is impossible as we have repeatedly seen in the preceding chapters. In efforts to gain insight into the nonequilibrium conformations of the polyelectrolyte molecules undergoing translocation, the polyelectrolyte chains need to be simulated as a collective system of the polymer, pore, counterions, electrolyte ions, and the hydrodynamics of the solvent. Such simulation results should then be compared with experimental data in order to fully understand the hierarchy of timescales in polymer translocation.

11.2 AMPLIFICATION OF CHEMICAL DETAILS

The chemical details of the pore and the polymer play significant role in determining the numerical prefactor in the relation of Equation 11.1 and the width of the distribution of the histogram of the translocation time. Quantitative aspects of these measures require a thorough understanding of the local interactions among the polymer and the pore. Such fine features are yet to be quantified in a predictive manner. We shall illustrate the effects by considering the protein pores and the translocating polymers separately.

11.2.1 Chemical Decoration of the Pore

As a specific example, consider the internal surface of the α-hemolysin protein pore that is exposed to the polymer undergoing translocation. This surface is decorated with hydrophobic, hydrophilic, and ionic residues, with substantial spatial correlations of their locations (Figure 11.1). The charge distribution on the pore wall offers a corrugated energy landscape for the polymer to pass through. The 1.5 nm constriction at the beginning of the β-barrel is composed of seven lysines (Lys147, in each of the heptamers of the α-hemolysin pore assembly) and seven glutamic acids (Glu111). At pH 7.5, the Lys147 residues are positively charged and the Glu111 residues are negatively charged, making the overall charge of the constriction zero. At the other end of the β-barrel, there is a highly charged ring composed of 14 aspartic acids (Asp127 and Asp128) and 7 lysines (Lys131). At pH 7.5, the net charge of the ring is -7 electron charges. This net negative charge at the end of the pore is likely to create an electrostatic repulsion for penetration of a negatively charged polymer. This feature might be responsible for the relatively large fraction (more than 90%) of failures in the translocation, as shown in Figure 10.3a.

When the pH in the chamber adjacent to the β-barrel is reduced, the net charges on the amino acid residues are modified. By assuming the standard ionic equilibria for the amino acid residues and their pK_a values, the estimated net charges on the constriction and the ring of the β-barrel are given in Figure 11.1b as a function of the pH in the receiver compartment. As the net charge on the ring becomes less negative at lower pH values, the electrostatic repulsion for the threading of a negatively charged polymer is reduced. This is then expected to improve the efficiency of successful threading of the polymer. Indeed, as seen in Figure 10.3b, the success rate is higher than 90% at pH 4.5, in comparison with $\sim$5% at pH 7.5 (Wong and Muthukumar 2010). An alternate route to design chemical details of the pore is to generate different mutations of the amino acid sequences of the protein molecules constituting the protein pore (Maglia et al. 2008).

Another example of protein pores that exhibits significant chemical effects on the translocation characteristics is the MspA pore (Figure 11.2) (Butler et al. 2008). The vestibule of this pore is more repulsive and the pore is shorter in comparison with the α-hemolysin pore. In contrast to the case of α-hemolysin pore, the DNA does not enter the MspA pore. This has been attributed to the high density of negative charge in the internal surface of the MspA pore. Replacement of three negative charges in the wild-type MspA by three positive charges into the mutant M1MspA (D90N/D91N/D93N

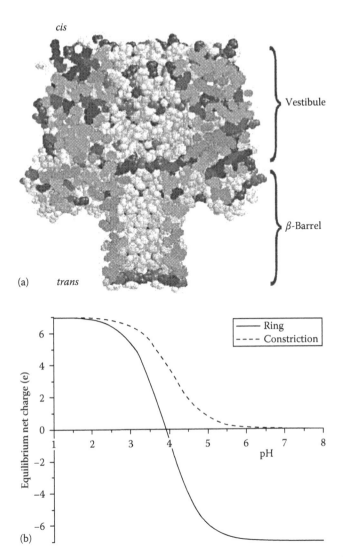

Figure 11.1 (a) The space-filling model of α-hemolysin protein pore. Different shades represent the hydrophobic, hydrophilic, and ionic amino acid residues. (b) The electrostatic nature of the pore can be tuned with pH. (From Wong, C.T.A. and Muthukumar, M., *J. Chem. Phys.*, 133, 045101, 2010. With permission.)

Figure 11.2) and further modification into the mutant M2MspA (where additional three negative charges were made into positive charges, D90N/D91N/D93N/D118R/D134R/E139K Figure 11.2) have led to DNA translocation events for ssDNA with 50 nucleotides. These pores exhibit much faster translocation in comparison with the α-hemolysin pore. Furthermore, a comparison between the data for the M1MspA and M2MspA pores clearly has shown that

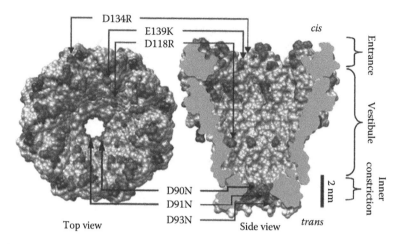

Figure 11.2 Structure of the MspA pore and its mutants. The vestibule is more repulsive for a negatively charged polymer and the pore is shorter in comparison with α-hemolysin pore. (Adapted from Butler, T.Z. et al., *Proc. Natl. Acad. Sci. USA*, 105, 20647, 2008.)

an additional change of just three negative charges into positive charges on the pore makes a huge difference in the characteristics of translocation. Specifically, the ionic current blockade rates for M2MspA were ~20 times higher than for M1MspA, M2MspA required much lower threshold voltage for translocation compared to M1MspA, and the partial blockades in M2MspA were about 100 times longer than for M1MspA. All of these observations demonstrate the crucial role played by the electrical charges on the internal surface of the protein pores in affecting the polymer translocation behavior. In addition, the geometry of the pore is also expected to affect the polymer significantly.

More research is required to design the most appropriate charge patterns to be embedded on the internal surface of the nanopores in order to achieve precisely controllable translocation times. Addressing this issue is expected to be at the forefront of nanopore design in the immediate future in the context of polymer translocation.

11.2.2 Secondary Structures of the Translocating Polymer

The ability of certain polymer sequences to spontaneously form secondary structures influences their migration through nanopores. As examples, let us consider the translocation events of single-stranded polynucleotides through α-hemolysin protein pore in the electrophysiology experiments. The event plots of translocation (every data point in the plot referring to the dwell time and the ionic current of a particular blockade event) for polycytidylic acid (poly C) and polydeoxycytidylic acid (poly dC) are compared in Figure 11.3a,

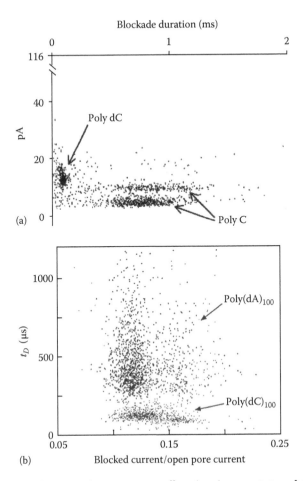

Figure 11.3 The secondary structures affect the characteristics of the translocation events. (a) Comparison between poly dC and poly C, under identical experimental conditions. (From Akeson, M. et al., *Biophys. J.*, 77, 3227, 1999. With permission.) (b) Comparison between poly dA and poly dC. (Adapted from Meller, A. et al., *Proc. Natl. Acad. Sci. USA*, 97, 1079, 2000.)

under equivalent experimental conditions (Akeson et al. 1999). The number of nucleotides in poly C is 130 ± 20 and that in poly dC is 100. The event plots for poly dC and polydeoxyadenylic acid (poly dA) are also quite distinct from each other, as seen in Figure 11.3b (Meller et al. 2000). In Figure 11.3b, t_D is the dwell time. It has also been observed that the translocation of poly C blocks the ionic current more than that of poly A, despite the fact that the physical size of the pyrimidines in poly C are smaller than the purines in poly A (Akeson et al. 1999). Also, the dwell times for poly C are shorter, by a factor of about 4, than those for poly A. Furthermore, the dwell times for poly C are longer by a factor of 7.5 than those for poly dC. To rationalize these observations, it has been suggested (Akeson et al. 1999, Lin et al. 2010) that poly C, poly A, and poly dC are able to assume helical conformations to different extents. The mapping between the secondary structures of the polymer and the translocation event plots remains to be fully understood. Even the simpler situation of a polymer with heterogeneous sequences needs to be fully understood (Muthukumar 2002b, Mohan et al. 2008, Kowalczyk et al. 2010).

Another active experimental situation is to electrophoretically translocate a single-stranded polynucleotide, which is covalently attached to a block of hybridized double-stranded polynucleotide, as a single file. If the single-stranded piece is pulled with a large force, the double-stranded part can melt. The physical consequences of the sequence-dependent melting of the hybridized portion on the translocation characteristics are not fully understood. If the pulling force is weak, the threading of the single-stranded block would stop as soon as the double-stranded block is at the mouth of the pore (which allows only single-file translocation). This kind of hybridization-based strategies has been experimentally investigated by several research groups in the context of sequencing technology (Vercoutere et al. 2001, Howorka et al. 2001, Butler et al. 2008, Cockroft et al. 2008, Branton et al. 2008, Singer et al. 2010). The data from these experiments turn out to be very rich. A predictive understanding of these data is yet to evolve (Kotsev and Kolomeisky 2006, 2007). The situation is similar regarding the translocation of protein molecules, which are capable of readily assuming secondary structures (Movileanu 2008, Finkelstein 2009).

11.3 BIOLOGICAL EXAMPLES

In biological contexts, the phenomenon of polymer translocation is ubiquitous, with a very large number of examples. We mention below only a few to emphasize the generality of the basic physical process discussed in the previous chapters and the emergence of diversity due to specific chemical and biological functions.

11.3.1 Mitochondrial Transport

The transport of proteins from one domain into another through a hydrophobic barrier is carried out by a protein complex, generically called translocon, which constitutes the pore (Gold et al. 2007, Yuan et al. 2010). There have been experiments on protein translocation, conducted in bacteria and archaea, where the translocons are embedded in the plasma membrane. The analogous process in eukaryotes, occurring in the endoplasmic reticulum, has also been investigated experimentally (Rapoport 2007). In general, the translocon provides a route for the protein to pass through the hydrophobic barrier separating the outside and inside of the system. The basic aspects are maintenance of the protein molecule in an unfolded state by binding with other protein molecules, targeting the translocon by recognizing a signal sequence on the protein, and the eventual translocation induced by a combination of electrochemical potential gradients and energy sources.

One of the well-studied examples of polymer translocation in biology is the protein transport into mitochondria (Glick and Schatz 1991, Beasley et al. 1992, Pfanner et al. 1992). A simplified cartoon of this process is illustrated in Figure 11.4, and the original references must be consulted for the elaborated details of the transport machinery. The protein transport occurs as a single-file translocation process through pores embedded in the outer and inner membranes of the mitochondria. As a necessary step, the protein molecule carrying an additional 20–80-residue-long signal peptide gets recognized first by a receptor protein on the outer membrane. The recognition appears to be mediated by the electrostatic interaction, as the signal peptide is usually endowed with about five positive charges at one end of the sequence. Once recognition and binding have occurred, there are two distinct steps by which the translocation proceeds. In the first step, a certain number of residues are inserted into the pore. This is achieved with the help of an electrochemical potential gradient across the pore. In the second step, the protein is threaded through the pore completely and ends up in the inside of the mitochondrion. This step is accomplished by ATP hydrolysis. In realistic conditions, proteins in the hsp70 family bind to the translocating protein both in the cytosol and the mitochondrion. During the threading step, the cytosolic hsp70 proteins bound to the translocating polymer unbind by requiring ATP hydrolysis and feed the polymer for threading. At the same time, the polymer is pulled by sequential binding of

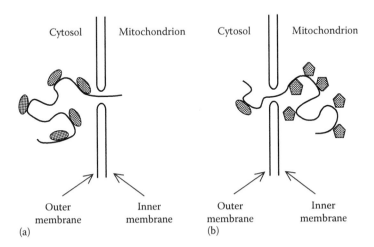

Figure 11.4 Cartoon of protein import into mitochondria. (a) Protein insertion by an electrochemical potential gradient. (b) Threading driven by unbinding of cytosolic hsp70 proteins, accompanying ATP hydrolysis, and binding of mitochondrial hsp70 proteins.

mitochondrial hsp70 proteins. Experiments have revealed that the insertion and threading are two distinct steps. As an example, if the experiment is carried out at 5°C, instead of 20°C, only the insertion step can occur without full translocation. All of these observations are consistent with the basic model of polymer translocation based on the capture, nucleation, and subsequent threading. The capture is facilitated by chemical signals; the nucleation is accomplished by electrochemical potential gradient and ATP hydrolysis, and the threading is carried out by the force arising from the unbinding and binding of proteins on the translocating polymer, which in turn is controlled by ATP hydrolysis.

Although the basic elements of this biological process seem to be the same as in the simpler single-molecule electrophysiology experiments, the actual quantitative measures of the driving force and the free energy barriers are yet to be established for import of mitochondrial proteins.

11.3.2 Bacterial Conjugation

Another example of polymer translocation is bacterial conjugation, where the genetic information is transferred between bacterial cells through a narrow passage made by direct cell-to-cell contact or a bridge-like long pore made by pili. Once the conjugation process is initiated by a signal, one of the strands of circular F-plasmid, 100 kb long, is nicked by an enzyme (relaxase). The nicked strand is unwound from the plasmid and translocated into the recipient cell. The single strands are then replicated in both the donor and recipient cells. Only qualitative description is known in terms of the rate of translocation, and the energetics associated with the driving force is yet to be established (Russi et al. 2008).

11.3.3 Transport through Nuclear Pore

The exchange of macromolecules between the cell nucleus and the cytoplasm occurs through the nuclear pore complexes (NPCs) situated across the nuclear envelope (Peters 2006). Unlike the mitochondrial pores, which allow only single-file translocation, the NPC is very large allowing the passage of folded proteins and polymer complexes as particles. The NPC is made of more than 30 different nucleoporins (Nups) in multiple copies. It is massive (120 MD) with a diameter of about 120 nm. The overall span in the direction normal to the nuclear envelope is about 90 nm. There is a narrow cylindrical pore of about 9 nm diameter, which has the capacity to transport molecules of diameter 40 nm (Pante and Kann 1998). The NPC regulates the bidirectional transport of various molecules. Many protein molecules necessary for the functioning of the nucleus are imported through NPC. Some nuclear proteins and ribosome subunits, as well as RNAs and mRNAs, are exported from the nucleus to the cytosol through NPC. The chemical nature of the Nups and the heterogeneity in the porous structure of the NPC control the selectivity of the bidirectional traffic of small and large macromolecules through NPC. In the import process of proteins into the nucleus, selectivity is executed by the presence of nuclear localization signals (NLS), which are 4–8 residue sequence rich in positive charge. Some cytosolic proteins bind to the NLS and somehow the complex is deposited at the NPC. The formation of such complexes is accompanied by conformational changes of the protein. Active translocation through NPC is then carried out by ATP hydrolysis. Overall, there is a hierarchy of control arising from the macromolecule being transported, the binding proteins and the pore itself (Terry et al. 2007). There are also substantial evidences for the flexible fibrils, which are located near the nuclear and cytoplasmic peripheries of the NPC, providing an entropic barrier for the navigation of polymer translocation (Lim et al. 2006, 2007).

The mechanism of the export of proteins is similar to the import mechanism. Proteins with nuclear export sequence (NES) are recognized by some binding proteins and then actively transported to the cytosol. The export of RNA is also signal-mediated with different set of receptor proteins in the nucleus that recognize RNA molecules or the proteins bound to the RNA. The details of the transport through NPC are rather extensive. There is promise in the direction of organizing these details into a working model, where these details can be reliably reduced into a few parameters. In a recent study (Nielsen et al. 2006), a simple attempt has been made to use the Fokker–Planck formalism (Equation 6.84) to model the nuclear transport. The chemical details of the heterogeneous porous structure of NPC and the spatial variations of the free energy landscape and local diffusion coefficient need to be accounted for in connecting with experimental data.

11.3.4 Genome Packing in Bacteriophages

The process of packaging single double-stranded DNA molecules (viral genomes) into the rigid confining capsids of bacteriophages is an example of translocation against a resistive free energy barrier. Genome packing in bacteriophages is one of the well-studied experimental systems involving dsDNA (Cerritelli et al. 1997, Smith et al. 2001). Linear dimensions of capsids are typically tens of nanometers, whereas the length of the genome to be packaged is generally three to four orders of magnitude longer (Knipe and Howley 2001). The persistence length of the genome is $\sim$50 nm in physiological conditions and is comparable to the linear dimensions of a typical viral capsid. The bending required of the polymer as it is wound tightly inside the capsid leads to a buildup of energy that is large in comparison with the thermal energy. Furthermore, the presence of the negatively charged phosphate linkages along the dsDNA backbone leads to large repulsive electrostatic energy inside the densely packed capsid.

Thus, the process of viral genome packaging is a conflict of scales, wherein a long molecule must be compressed within a length scale on which it resists bending and to a monomer density at which it must also overcome strong repulsive forces from electrostatic and excluded volume interactions. Such a conflict leads to buildup of large pressures inside the capsid. Several x-ray diffraction and cryo-transmission electron microscopy studies (Cerritelli et al. 1997) have determined the three-dimensional structure of the packaged genome. The time-dependent buildup of force, as the genome is packaged inside the $\phi29$ bacteriophage by a motor protein, has been investigated (Smith et al. 2001) by the single-molecule optical tweezers technique. From these two kinds of measurements, details of the kinetics of genome packaging and of the final structure of the genome inside the capsid are beginning to emerge. It must be remarked that there is a qualitative difference between the packaging of bacteriophages and the assembly of RNA-viruses. In the case of bacteriophages involving the semiflexible dsDNA, the packaging is an uphill process requiring a motor protein and an energy supply such as ATP-hydrolysis. On the other hand, in viruses with the flexible ssRNA chains, the packaging is a spontaneous process of co-assembly of the RNA and the capsid proteins. In the latter case, electrostatic interactions between the polynucleotide and the proteins dominate the virus assembly (Belyi and Muthukumar 2006), whereas in the bacteriophages, the DNA stiffness and excluded volume interactions dominate the active packaging process.

Complementary to the experimental investigations, there have been theoretical analysis (Purohit et al. 2003) and computer simulations of the genome packaging into phages (Kindt et al. 2001, Forrey and Muthukumar 2006, Petrov et al. 2007, Angelescu and Linse 2008). These simulations have qualitatively reproduced the experimental results on force profiles, x-ray diffraction, and cryo-transmission electron microscopy. A typical trajectory of genome packaging for a T7 bacteriophage is given in Figure 11.5. Analysis of the detailed forces reveals that the genome-packaging process is fundamentally different from the previously popular inverse spool model and that it is dominated by entropy associated with polymer dynamics (Forrey and Muthukumar 2006). A hysteresis in the force–density occurs between the packaging and the subsequent ejection of the genome, depending on the rate of packaging. A satisfactory theoretical formulation, by taking into account the stochastic nature

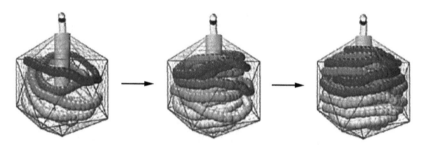

Figure 11.5 Modeling of packaging of dsDNA into T7 bacteriophage. (From Forrey, C. and Muthukumar, M., *Biophys. J.*, 91, 25, 2006. With permission.)

of the translocation process against an uphill free energy barrier, is yet to be developed to fully describe the phenomenology of genome packaging.

11.4 SUMMARY

The basic steps of recognition, insertion, and threading of the polymer are common in all translocation processes. While the general phenomenon is relatively easy to visualize in terms of the physical mechanism of capture-nucleation-threading, proper description of the various chemical forces in the particular translocation systems remains as a challenge. Although absolute values of the various characteristics of the translocation behavior in complex systems are difficult to predict, the relative effects, due to variations in a few control variables, may be predicted based on the physical models presented in the preceding chapters. We have attempted in these chapters to organize the apparently diverse sets of experimental data on a variety of translocation systems into a small universal set of conceptual ideas.

REFERENCES

Abramowitz, M. and Stegun, I.A., 1965. *Handbook of Mathematical Functions*, New York: Dover.

Aidley, D.J. and Stanfield, P.R., 1996. *Ion Channels*, Cambridge, U.K.: Cambridge University Press.

Akeson, M., Branton, D., Kasianowicz, J.J., Brandin, E., and Deamer, D.W., 1999. Microsecond time-scale discrimination among polycytidylic acid, polyadenylic acid, and polyuridylic acid as homopolymers or as segments within single RNA molecules, *Biophys. J.*, 77, 3227–3233.

Aksimentiev, A., Heng, J.B., Timp, G., and Schulten, K., 2004. Microscopic kinetics of DNA translocation through synthetic nanopores, *Biophys. J.*, 87, 2086–2097.

Aksimentiev, A. and Schulten, K., 2004. Extending molecular modeling methodology to study insertion of membrane nanopores, *Proc. Natl. Acad. Sci. USA*, 101, 4337–4338.

Aksimentiev, A. and Schulten, K., 2005. Imaging α-hemolysin with molecular dynamics: Ionic conductance, osmotic permeability, and the electrostatic potential map, *Biophys. J.*, 88, 3745–3761.

Alberts, B., Johnson, A., Lewis, J., Raff, M., Roberts, K., and Walter, P., 2008. *Molecular Biology of the Cell*, New York: Garland Science.

Alfrey, T., Berg, P.W., and Morawetz, H., 1951. The counterion distribution in solutions of rod-shaped polyelectrolytes, *J. Polym. Sci.*, 7, 543–547.

Ambjornsson, T., Apell, S.P., Konkoli, Z., DiMarzio, E.A., and Kasianowicz, J.J., 2002. Charged polymer membrane translocation, *J. Chem. Phys.*, 117, 4063–4073.

Ammenti, A., Cecconi, F., Marconi, U.M.B., and Vulpiani, A., 2009. A statistical model for translocation of structured polypeptide chains through nanopores, *J. Phys. Chem. B*, 113, 10348–10356.

Angelescu, D.G. and Linse, P., 2008. Viruses as supramolecular self-assemblies: Modelling of capsid formation and genome packaging, *Soft Matter*, 4, 1981–1990.

Arfken, G.B. and Weber, H.J., 2001. *Mathematical Methods for Physicists*, San Diego, CA: Academic Press.

Arseniev, A.S., Lomize, A.L., Barsukov, I.L., and Bystrov, V.F., 1986. Gramicidin a transmembrane ion-channel. Three-dimensional structure reconstruction based on nmr spectroscopy and energy refinement, *Biol. Membr.*, 3, 1077–1104.

Atkins, E. and Taylor, M., 2004. Elongational flow studies on DNA in aqueous solution and stress-induced scission of the double helix, *Biopolymers*, 32, 911–923.

Babcock, H., Teixeira, R., Hur, J., Shaqfeh, E., and Chu, S., 2003. Visualization of molecular fluctuations near the critical point of the coil-stretch transition in polymer elongation, *Macromolecules*, 36, 4544–4548.

Banavar, J.R., Hoang, T.X., and Maritan, A., 2005. Proteins and polymers, *J. Chem. Phys.*, 122, 234910.

Barthel, J.M.G., Krienke, H., and Kunz, W., 1998. *Physical Chemistry of Electrolyte Solutions*, New York: Springer.

Beasley, E., Washter, C., and Schatz, G., 1992. Putting energy into mitochondrial protein import, *Curr. Opin. Cell Biol.*, 4, 646–651.

Beer, M., Schmidt, M., and Muthukumar, M., 1997. The electrostatic expansion of linear polyelectrolytes: Effects of gegenions, co-ions, and hydrophobicity, *Macromolecules*, 30, 8375–8385.

Belyi, V.A. and Muthukumar, M., 2006. Electrostatic origin of the genome packing in viruses, *Proc. Natl. Acad. Sci. USA*, 103, 17174–17178.

Berezhkovskii, A.M., Hummer, G., and Bezrukov, S.M., 2006. Identity of distributions of direct uphill and downhill translocation times for particles traversing membrane channels, *Phys. Rev. Lett.*, 97, 020601.

Berezhkovskii, A.M., Pustovoit, M.A., and Bezrukov, S.M., 2002. Channel-facilitated membrane transport: Transit probability and interaction with the channel, *J. Chem. Phys.*, 116, 9952–9956.

Berg, H.C., 1983. *Random Walks in Biology*, Princeton, NJ: Princeton University Press.

Berne, B.J. and Pecora, R., 1976. *Dynamic Light Scattering*, New York: John Wiley.

Besteman, K., Eijk, K.V., and Lemay, S.G., 2007. Charge inversion accompanies DNA condensation by multivalent ions, *Nat. Phys.*, 3, 641–644.

Bezrukov, S.M., 2000. Ion channels as molecular Coulter counters to probe metabolite transport, *J. Membrane Biol.*, 174, 1–13.

Bezrukov, S.M., Berezhkovskii, A.M., Pustovoit, M.A., and Szabo, A., 2000. Particle number fluctuations in a membrane channel, *J. Chem. Phys.*, 113, 8206–8211.

Bhattacharya, A., 2010. How local factors affect the exponents of forced polymer translocation through a nano-pore, *Phys. Proc.*, 3, 1411–1416.

Bhattacharya, A. and Binder, K., 2010. Out-of-equilibrium characteristics of a forced translocating chain through a nanopore, *Phys. Rev. E*, 81, 041804.

Bhattacharya, A., Morrison, W.H., Luo, K., Ala-Nissila, T., Ying, S.C., Milchev, A., and Binder, K., 2009. Scaling exponents of forced polymer translocation through a nanopore, *Eur. Phys. J. E*, 29, 423–429.

Branton, D., Deamer, D.W., Marziali, A., Bayley, H., Benner, S.A., Butler, T., Ventra, M.D., Garaj, S., Hibbs, A., Huang, X., Jovanovich, S.B., Krstic, P.S., Lindsay, S., Ling, X.S., Mastrangelo, C.H., Meller, A., Oliver, J.S., Pershin, Y.V., Ramsey, J.M., Riehn, R., Soni, G.V., Tabard-Cossa, V., Wanunu,

M., Wiggin, M., and Schloss, J.A., 2008. The potential and challenges of nanopore sequencing, *Nat. Biotechnol.*, 26, 1146–1153.

Brilliantov, N.V., Kuznetsov, D.V., and Klein, R., 1998. Chain collapse and counterion condensation in dilute polyelectrolyte solutions, *Phys. Rev. Lett.*, 81, 1433–1436.

Brown, R., 1828. A brief account of microscopical observations, made in the months of June, July, and August, 1827 on the particles contained in the pollen of plants; and on the general existence of active molecules in organic and inorganic bodies, *Edin. New Phil. J.*, 5, 358–371.

Brun, L., Pastoriza-Gallego, M., Oulhaled, G., Mathe, J., Bacri, L., Auvray, L., and Pelta, J., 2008. Dynamics of polyelectrolyte transport through a protein channel as a function of applied voltage, *Phys. Rev. Lett.*, 100, 158302.

Butler, T.Z., Gundlach, J.H., and Troll, M.A., 2006. Determination of RNA orientation during translocation through a biological nanopore, *Biophys. J.*, 90, 190–199.

Butler, T.Z., Gundlach, J.H., and Troll, M.A., 2007. Ionic current blockades from DNA and RNA molecules in the α-hemolysin nanopore, *Biophys. J.*, 93, 3229–3240.

Butler, T.Z., Pavlenok, M., Derrington, M., Niederweis, M., and Gundlach, J.H., 2008. Single-molecule DNA detection with an engineered MSPA protein nanopore, *Proc. Natl. Acad. Sci. USA*, 105, 20647–20652.

Cacciuto, A. and Luijten, E., 2006. Self-avoiding flexible polymers under spherical confinement, *Nano. Lett.*, 6, 901–905.

Cantor, C.R. and Schimmel, P.R., 1980. *Biophysical Chemistry Part III*, New York: W.H. Freeman and Company.

Cardenas, A.E., Coalson, R.D., and Kurnikova, M.G., 2000. Three-dimensional Poisson–Nernst–Planck theory studies: Influence of membrane electrostatics on gramicidin a channel conductance, *Biophys. J.*, 79, 80–93.

Carri, G.A. and Muthukumar, M., 1999. Attractive interactions and phase transitions in solutions of similarly charged rod-like polyelectrolytes, *J. Chem. Phys.*, 111, 1765–1777.

Carslaw, H.S. and Jaeger, J.C., 1986. *Conduction of Heat in Solids*, Oxford: Clarendon Press.

Casassa, E.F., 1967. Equilibrium distribution of flexible polymer chains between a macroscopic solution phase and small voids, *Polym. Lett. Part B*, 5, 773–778.

Casassa, E.F., 1972. Gel chromatography of once-broken rod molecules, *J. Polym. Sci. Phys. Ed.*, 10, 381–384.

Casassa, E.F. and Tagami, Y., 1969. An equilibrium theory for exclusion chromatography of branched and linear polymer chains, *Macromolecules*, 2, 14–26.

Cerritelli, M.E., Cheng, N.Q., Rosenberg, A.H., McPherson, C.E., Booy, F.P., and Steven, A.C., 1997. Encapsidated conformation of bacteriophage t7 DNA, *Cell*, 91, 271–280.

Chandrasekhar, S., 1943. Stochastic processes in physics and astronomy, *Rev. Mod. Phys.*, 15, 1–89.

Chen, D., Lear, J., and Eisenberg, R.S., 1997. Permeation through an open channel: Poisson–Nernst–Planck theory of a synthetic ion channel, *Biophys. J.*, 72, 97–116.

Chen, P., Gu, J., Brandin, E., Kim, Y.R., Wang, Q., and Branton, D., 2004a. Probing single DNA molecule transport using fabricated nanopores, *Nano Lett.*, 4, 2293–2298.

Chen, Y.L., Graham, M.D., de Pablo, J.J., Randall, G.C., Gupta, M., and Doyle, P.S., 2004b. Conformation and dynamics of single DNA molecules in parallel-plate slit microchannels, *Phys. Rev. E*, 70, 060901.

Chuang, J., Kantor, Y., and Kardar, M., 2002. Anomalous dynamics of translocation, *Phys. Rev. E*, 65, 011802.

Cockroft, S.L., Chu, J., Amorin, M., and Ghadiri, M.R., 2008. A single-molecule nanopore device detects DNA polymerase activity with single-nucleotide resolution, *J. Am. Chem. Soc.*, 130, 818–820.

Cooper, K., Jacobsson, E., and Wolynes, P., 1985. The theory of ion transport through membrane channels, *Prog. Biophys. Molec. Biol.*, 46, 51–96.

Corry, B., Kuyucak, S., and Chung, S.H., 2000. Tests of continuum theories as models of ion channels. II. Poisson–Nernst–Planck theory versus Brownian dynamics, *Biophys. J.*, 78, 2364–2381.

Cotton, J.P., Decker, D., Benoit, H., Farnoux, B., Higgins, J., Jannink, G., Ober, R., Picot, C., and des Cloizeaux, J., 1974. Conformation of polymer chain in the bulk, *Macromolecules*, 7, 863–872.

Cox, D.R. and Miller, H.D., 1968. *The Theory of Stochastic Processes*, New York: John Wiley.

Dautzenberg, H., Jaeger, W., Kotz, J. Phillip, B. Seidel, C., and Stscherbina, D., 1994. *Polyelectrolytes*, New York: Hanser Publishers.

de Gennes, P.G., 1971. Reptation of a polymer chain in the presence of fixed obstacles, *J. Chem. Phys.*, 55, 572–579.

de Gennes, P.G., 1974. Coil-stretch transition of dilute flexible polymers under ultrahigh velocity gradients, *J. Chem. Phys.*, 60, 5030–5042.

de Gennes, P.G., 1979. *Scaling Concepts in Polymer Physics*, Ithaca, NY: Cornell University Press.

de Gennes, P.G., 1999. Flexible polymers in nanopores, *Adv. Polym. Sci.*, 138, 91–105.

de Gennes, P.G., Pincus, P., and Brochard, F., 1976. Remarks on polyelectrolyte conformation, *J. Phys. France*, 37, 1461–1473.

de Han, H.W. and Slater, G.W., 2010. Mapping the variation of the translocation α scaling exponent with nanopore width, *Phys. Rev. E*, 81, 051802.

de la Cruz, M.O., Belloni, L., Delsanti, M., Dalbiez, J.P., Spalla, O., and Drifford, M., 1995. Precipitation of highly charged polyelectrolyte solutions in the presence of multivalent salts, *J. Chem. Phys.*, 103, 5781–5791.

des Cloizeaux, J. and Jannink, G., 1990. *Polymers in Solution*, Oxford: Clarendon Press.

DiMarzio, E.A. and Mandell, A.J., 1997. Phase transition behavior of a linear macromolecule threading a membrane, *J. Chem. Phys.*, 107, 5510–5514.

Doi, M. and Edwards, S.F., 1986. *The Theory of Polymer Dynamics*, Oxford: Clarendon Press.

Drifford, M. and Dalbiez, J.P., 1985. Effect of salt on sodium polystyrene sulfonate measured by light scattering, *Biopolymers*, 24, 1501–1514.

Dubbeldam, J.L.A., Milchev, A., Rostiashvili, V.G., and Vilgis, T.A., 2007a. Driven polymer translocation through a nanopore: A manifestation of anomalous diffusion, *Europhys. Lett.*, 79, 18002.

Dubbeldam, J.L.A., Milchev, A., Rostiashvili, V.G., and Vilgis, T.A., 2007b. Polymer translocation through a nanopore: A showcase of anomalous diffusion, *Phys. Rev. E*, 76, 010801(R).

Dubbeldam, J.L.A., Milchev, A., Rostiashvili, V.G., and Vilgis, T.A., 2009. Comment on 'anomalous dynamics of unbiased polymer translocation through a nanopore' and other recent papers by D. Panja, G. Barkema and R. Ball, *J. Phys.: Condens. Matter*, 21, 098001.

Edwards, S.F., 1966. The theory of polymer solutions at intermediate concentration, *Proc. Phys. Soc.*, 88, 265–280.

Eisenberg, R.S., 1996. Computing the field in proteins and channels, *J. Membr. Biol.*, 150, 1–25.

Eisenriegler, E., 1993. *Polymers Near Surfaces*, Singapore: World Scientific.

Essafi, W., Lafuma, F., and Williams, C.E., 1995. Effect of solvent quality on the behavior of highly charged polyelectrolytes, *J. Phys. II France*, 5, 1269–1275.

Evans, D.F. and Wennerstrom, H., 1999. *The Colloidal Domain*, New York: Wiley-VCH.

Finkelstein, A., 2009. Proton-coupled protein transport through the anthrax toxin channel, *Phil. Trans. R. Soc.*, B364, 209–215.

Fitzkee, N.Z. and Rose, G.D., 2004. Reassessing random-coil statistics in unfolded proteins, *Proc. Natl. Acad. Sci. USA*, 101, 12497–12502.

Flory, P.J., 1953. *Principles of Polymer Chemistry*, Ithaca, NY: Cornell University Press.

Flory, P.J., 1969. *Statistical Mechanics of Chain Molecules*, New York: Interscience Publishers.

Forrey, C. and Muthukumar, M., 2006. Langevin dynamics simulations of genome packing in bacteriophage, *Biophys. J.*, 91, 25–41.

Forrey, C. and Muthukumar, M., 2007. Langevin dynamics simulations of ds-DNA translocation through synthetic nanopores, *J. Chem. Phys.*, 127, 015102.

Forster, S. and Schmidt, M., 1995. Polyelectrolytes in solution, *Adv. Polym. Sci.*, 120, 51–133.

Forster, S., Schmidt, M., and Antonietti, M., 1990. Static and dynamic light scattering by aqueous polyelectrolyte solutions: Effect of molecular weight, charge density and added salt, *Polymer*, 31, 781–792.

Frohlich, H., 1958. *Theory of Dielectrics*, Oxford: Clarendon Press.

Fujita, H., 1990. *Polymer Solutions*, Amsterdam, the Netherlands: Elsevier.

Gardiner, C.W., 1985. *Handbook of Stochastic Methods*, Berlin, Germany: Springer.

Gauthier, M.G. and Slater, G.W., 2008a. A Monte Carlo algorithm to study polymer translocation through nanopores: II. Scaling laws, *J. Chem. Phys.*, 128, 205103.

Gauthier, M.G. and Slater, G.W., 2008b. A Monte Carlo algorithm to study polymer translocation through nanopores: I. Theory and numerical approach, *J. Chem. Phys.*, 128, 065103.

Ghosh, K., Carri, G.A., and Muthukumar, M., 2001. Configurational properties of a single semiflexible polyelectrolyte, *J. Chem. Phys.*, 115, 4367–4375.

Glick, B. and Schatz, G., 1991. Import of proteins into mitochondria, *Annu. Rev. Genet.*, 25, 21–44.

Gold, V.A., Duong, F., and Collinson, I., 2007. Structure and function of the bacterial Sec translocon, *Mol. Membr. Biol.*, 24, 387–394.

Gong, H., Hocky, G., and Freed, K.F., 2008. Influence of nonlinear electrostatics on transfer energies between liquid phases: Charge burial is far less expensive than Born model, *Proc. Natl. Acad. Sci. USA*, 105, 11146–11151.

Gracheva, M.E., Aksimentiev, A., and Leburton, J.P., 2006. Electric signatures of single-stranded DNA with single base mutations in a nanopore capacitor, *Nanotechnology*, 17, 3160–3165.

Griffiths, D.J., 1999. *Introduction to Electrodynamics*, Upper Saddle River, NJ: Prentice Hall.

Grosberg, A.Y. and Khokhlov, A.R., 1994. *Statistical Physics of Macromolecules*, New York: AIP Press.

Gross, R.J. and Osterle, J.F., 1968. Membrane transport characteristics of ultrafine capillaries, *J. Chem. Phys.*, 49, 228–234.

Hammes, G.G., 1978. *Principles of Chemical Kinetics*, New York: Academic Press.

Han, J. and Craighead, H.G., 2000. Separation of long DNA molecules in a microfabricated entropic trap array, *Science*, 288, 1026–1029.

Heng, J.B., Aksimentiev, A., Ho, C., Marks, P., Grinkova, Y.V., Sligar, S., Schulten, K., and Timp, G., 2006. The electromechanics of DNA in a synthetic nanopore, *Biophys. J.*, 90, 1098–1106.

Henrickson, S.E., DiMarzio, E.A., Wang, Q., Stanford, V.M., and Kasianowicz, J.J., 2010. Probing single nanometer-scale pores with polymeric molecular rulers, *J. Chem. Phys.*, 132, 135101.

Henrickson, S.E., Misakian, M., Robertson, B., and Kasianowicz, J.J., 2000. Driven DNA transport into an asymmetric nanometer-scale pore, *Phys. Rev. Lett.*, 85, 3057–3060.

Hernandez-Ortiz, J.P., Chopra, M., Geier, S., and de Pablo, J.J., 2009. Hydrodynamic effects on the translocation rate of a polymer through a pore, *J. Chem. Phys.*, 131, 044904.

Hiemenz, P.C. and Rajagopalan, R., 1997. *Principles of Colloid and Surface Chemistry*, New York: Marcel Dekker.

Hiergeist, C. and Lipowsky, R., 1996. Elastic properties of polymer-decorated membranes, *J. Phys. II France*, 6, 1465–1481.

Hille, B., 2001. *Ion Channels of Excitable Membranes*, Sunderland, MA: Sinauer Associates, Inc.

Hoagland, D., Arvanitidou, E., and Welch, C., 1999. Capillary electrophoresis measurements of the free solution mobility for several model polyelectrolyte systems, *Macromolecules*, 32, 6180–6190.

Holm, C., Joanny, J.F., Netz, R.R., Reineker, P., Seidel, C., Vilgis, T.A., and Winkler, R.G., 2004a. Polyelectrolyte theory, *Adv. Polym. Sci.*, 166, 67–111.

Holm, C., Rehahn, M., Oppermann, W., and Ballauff, M., 2004b. Stiff-chain polyelectrolytes, *Adv. Polym. Sci.*, 166, 1–27.

Howorka, S., Movileanu, L., Braha, O., and Bayley, H., 2001. Kinetics of duplex formation for individual DNA strands within a single protein nanopore, *Natl. Acad. Sci. USA*, 98, 12996–13001.

Huber, K., 1993. Calcium-induced shrinking of polyacrylate chains in aqueous solution, *J. Phys. Chem.*, 97, 9825–9830.

Ikeda, Y., Beer, M., Schmidt, M., and Huber, K., 1998. Ca^{2+} and Cu^{2+} induced conformational changes of sodium polymethacrylate in dilute aqueous solution, *Macromolecules*, 31, 728–733.

Israelachvili, J., 1992. *Intermolecular and Surface Forces*, London: Academic Press.

Izmitli, A., Schwartz, D.C., Graham, M.D., and de Pablo, J.J., 2008. The effect of hydrodynamic interactions on the dynamics of DNA translocation through pores, *J. Chem. Phys.*, 128, 085102.

Jackson, M.B., 2006. *Molecular and Cellular Biophysics*, Cambridge, U.K.: Cambridge University Press.

Jendrejack, R.M., de Pablo, J.J., and Graham, M.D., 2002. Stochastic simulations of DNA in flow: Dynamics and the effects of hydrodynamic interactions, *J. Chem. Phys.*, 116, 7752–7759.

Kaji, K., Urakawa, H., Kanaya, T., and Kitamaru, R., 1988. Phase diagram of polyelectrolyte solutions, *J. Phys. France*, 49, 993–1000.

Kandel, E.R., Schwartz, J.H., and Jessell, T.M., 1991. *Principles of Neural Science*, Norwalk, CT: Appleton and Lange.

Kantor, Y. and Kardar, M., 2004. Anomalous dynamics of forced translocation, *Phys. Rev. E*, 69, 021806.

Kasianowicz, J.J., Brandin, E., Branton, D., and Deamer, D.W., 1996. Characterization of individual polynucleotide molecules using a membrane channel, *Proc. Natl. Acad. Sci. USA*, 93, 13770–13773.

Kato, T., Miyaso, K., Noda, I., Fujimoto, T., and Nagasawa, M., 1970. Thermodynamic and hydrodynamic properties of linear polymer solutions. i. Light scattering of monodisperse poly(α-methylstyrene), *Macromolecules*, 3, 777–786.

Khokhlov, A.R., 1980. On the collapse of weakly charged polyelectrolytes, *J. Phys. A*, 13, 979–987.

Khokhlov, A.R. and Kramarenko, E.Y., 1994. Polyelectrolyte/ionomer behavior in polymer gel collapse, *Macromol. Theory Simul.*, 3, 45–59.

Kindt, J., Tzlil, S., Ben-Shaul, A., and Gelbart, W.M., 2001. DNA packaging and ejection forces in bacteriophage, *Proc. Natl. Acad. Sci. USA*, 98, 13671–13674.

Kirkwood, J. and Riseman, J., 1948. The intrinsic viscosities and diffusion constants of flexible macromolecules in solution, *J. Chem. Phys.*, 16, 565–573.

Knipe, D.M. and Howley, P.M., 2001. *Fundamental Virology*, Philadelphia, PA: Lippincott Williams and Wilkins.

Kohn, J.E., Millett, I.S., Jacob, J., Zagrovic, B., Dillon, T.M., Cingel, N., Dothager, R.S., Seifert, S., Thiyagarajan, P., Sosnick, T.R., Hasan, M.Z., Pande, V.S., Ruczinski, I., Doniach, S., and Plaxco, K.W., 2004. Random-coil behavior and the dimensions of chemically unfolded proteins, *Proc. Natl. Acad. Sci. USA*, 101, 12491–12496.

Kolomeisky, A.B., 2007. Channel-facilitated molecular transport across membranes: Attraction, repulsion, and asymmetry, *Phys. Rev. Lett.*, 98, 048105.

Kolomeisky, A.B., 2008. How polymers translocate through pores: Memory is important, *Biophys. J.*, 94, 1547–1548.

Kolomeisky, A.B. and Kotsev, S., 2008. Effect of interactions on molecular fluxes and fluctuations in the transport across membrane channels, *J. Chem. Phys.*, 128, 085101.

Kong, C.Y. and Muthukumar, M., 2002. Modeling of polynucleotide translocation through protein pores and nanotubes, *Electrophoresis*, 23, 2697–2703.

Kong, C.Y. and Muthukumar, M., 2004. Polymer translocation through a nanopore. ii. Excluded volume effect, *J. Chem. Phys.*, 120, 3460–3466.

Kong, C.Y. and Muthukumar, M., 2005. Simulations of stochastic sensing of proteins, *J. Am. Chem. Soc.*, 127, 18252–18261.

Kotsev, S. and Kolomeisky, A.B., 2006. Effect of orientation in translocation of polymers through nanopores, *J. Chem. Phys.*, 125, 084906.

Kotsev, S. and Kolomeisky, A.B., 2007. Translocation of polymers with folded configurations across nanopores, *J. Chem. Phys.*, 127, 185103.

Kowalczyk, S.W., Tuijtel, M.W., Donkers, S.P., and Dekker, C., 2010. Unraveling single-stranded DNA in a solid-state nanopore, *Nano Lett.*, 10, 1414–1420.

Krasilnikov, O.V., Rodrigues, C.G., and Bezrukov, S.M., 2006. Single polymer molecules in a protein nanopore in the limit of a strong polymer-pore attraction, *Phys. Rev. Lett.*, 97, 018301.

Kumar, R., Kundagrami, A., and Muthukumar, M., 2009. Counterion adsorption on flexible polyelectrolytes: Comparison of theories, *Macromolecules*, 42, 1370–1379.

Kumar, R. and Muthukumar, M., 2008. Confinement free energy of flexible polyelectrolytes in spherical cavities, *J. Chem. Phys.*, 128, 184902.

Kumar, R. and Muthukumar, M., 2009. Origin of translocation barriers for polyelectrolyte chains, *J. Chem. Phys.*, 131, 194903.

Kundagrami, A. and Muthukumar, M., 2008. Theory of competitive counterion adsorption on flexible polyelectrolytes: Divalent salts, *J. Chem. Phys.*, 128, 244901.

Kundagrami, A. and Muthukumar, M., 2010. Effective charge and coil-globule transition of a polyelectrolyte chain, *Macromolecules*, 43, 2574–2581.

Kurnikova, M.G., Coalson, R.D., Graf, P., and Nitzan, A., 1999. A lattice relaxation algorithm for three-dimensional Poisson–Nernst–Planck theory with application to ion transport through the Gramicidin A channel, *Biophys. J.*, 76, 642–656.

Lamm, G. and Pack, G.R., 1997. Calculation of dielectric constants near polyelectrolytes in solution, *J. Phys. Chem. B*, 101, 959–965.

Landau, L.D. and Lifshitz, E.M., 1959. *Fluid Mechanics*, Oxford: Pergamon Press.

Larson, R.G., 2005. The rheology of dilute solutions of flexible polymers: Progress and problems, *J. Rheol.*, 49, 1–70.

Lauger, P., 1973. Ion transport through pores: A rate-theory analysis, *Biochim. Biophys. Acta*, 311, 423–441.

Lee, C.L. and Muthukumar, M., 2009. Phase behavior of polyelectrolyte solutions with salt, *J. Chem. Phys.*, 130, 024904.

Levich, V., 1962. *Physicochemical Hydrodynamics*, Englewood Cliffs, NJ: Prentice-Hall, Inc.

Li, J., Gershow, M., Stein, D., Brandin, E., and Golovchenko, J., 2003. DNA molecules and configurations in a solid-state nanopore microscope, *Nat. Mater.*, 2, 611–615.

Lim, R., Huang, N., Koeser, J., Deng, J., Lau, K., Schwarz-Herion, K., Fahrenkrog, B., and Aebi, U., 2006. Flexible phenylalanine-glycine nucleoporins as entropic barriers to nucleocytoplasmic transport, *Proc. Natl. Acad. Sci. USA*, 103, 9512–9517.

Lim, R.Y.H., Fahrenkrog, B., Koeser, J., Schwarz-Herion, K., Deng, J., and Aebi, U., 2007. Nanomechanical basis of selective gating by the nuclear pore complex, *Science*, 318, 640–643.

Lin, J., Kolomeisky, A., and Meller, A., 2010. Helix-coil kinetics of individual polyadenylic acid molecules in a protein channel, *Phys. Rev. Lett.*, 104, 158101.

Liu, S., Ashok, B., and Muthukumar, M., 2004. Brownian dynamics simulations of bead-rod-chain in simple shear flow and elongational flow, *Polymer*, 45, 1383–1389.

Liu, S., Ghosh, K., and Muthukumar, M., 2003. Polyelectrolyte solutions with added salt: A simulation study, *J. Chem. Phys.*, 119, 1813–1823.

Liu, S. and Muthukumar, M., 2002. Langevin dynamics simulation of counterion distribution around isolated flexible polyelectrolyte chains, *J. Chem. Phys.*, 116, 9975–9982.

Lodish, H., Berk, A., Kaiser, C.A., Krieger, M., Scott, M.P., Bretscher, A., Ploegh, H., and Matsudaira, P., 2007. *Molecular Cell Biology*, New York: W.H. Freeman and Company.

Loh, P., Deen, R., Vollmer, D., Fischer, K., Schmidt, M., Kundagrami, A., and Muthukumar, M., 2008. Collapse of linear polyelectrolyte chains in a poor solvent: When does a collapsing polyelectrolyte collect its counterions?, *Macromolecules*, 41, 9352–9358.

Long, D., Viovy, J.L., and Ajdari, A., 1996. Simultaneous action of electric fields and nonelectric forces on a polyelectrolyte: Motion and deformation, *Phys. Rev. Lett.*, 76, 3858–3861.

Luan, B. and Aksimentiev, A., 2008. Electro-osmotic screening of the DNA charge in a nanopore, *Phys. Rev. E*, 78, 021912.

Lubensky, D.K. and Nelson, D.R., 1999. Driven polymer translocation through a narrow pore, *Biophys. J.*, 77, 1824–1838.

Luo, K., Huopaniemi, I., Ala-Nissila, T., and Ying, S.C., 2006. Polymer translocation through a nanopore under an applied electric field, *J. Chem. Phys.*, 124, 114704.

Luo, K., Metzler, R., Ala-Nissila, T., and Ying, S.C., 2009. Polymer translocation out of confined environments, *Phys. Rev. E*, 80, 021907.

Lynden-Bell, R.M. and Rasaiah, J.C., 1996. Mobility and solvation of ions in channels, *J. Chem. Phys.*, 105, 9266–9280.

Maglia, G., Restrepo, M.R., Mikhailova, E., and Bayley, H., 2008. Enhanced translocation of single DNA molecules through α-hemolysin nanopores by manipulation of internal charge, *Proc. Natl. Acad. Sci. USA*, 105, 19720–19725.

Manning, G.S., 1969. Limiting laws and counterion condensation in polyelectrolyte solutions. I: Colligative properties. II: Self-diffusion of the small ions, *J. Chem. Phys.*, 51, 924–938.

Manning, G.S., 1978. The molecular theory of polyelectrolyte solutions with application to the electrostatic properties of polynucleotides, *Quarter. Rev. Biophys.*, 11, 179–246.

Mathe, J., Aksimentiev, A., Nelson, D.R., Schulten, K., and Meller, A., 2005. Orientation discrimination of single-stranded DNA inside the α-hemolysin membrane channel, *Proc. Natl. Acad. Sci. USA*, 102, 12377–12382.

Matsumoto, T., Nishioka, N., and Fujita, H., 1972. Excluded-volume effects in dilute polymer solutions. iv. Polyisobutylene, *J. Polym. Sci. A-2*, 10, 23–42.

Matysiak, S., Montesi, A., Kolomeisky, A.B., and Clementi, C., 2006. Dynamics of polymer translocation through nanopores: Theory meets experiment, *Phys. Rev. Lett.*, 96, 118103.

Mazo, R.M., 2002. *Brownian Motion*, Oxford: Clarendon Press.

McQuarrie, D.A., 1976. *Statistical Mechanics*, New York: Harper and Row.

Mehler, E.L. and Eichele, G., 1984. Electrostatic effects in water-accessible regions of proteins, *Biochemistry*, 23, 3887–3891.

Meller, A. and Branton, D., 2002. Single molecule measurements of DNA transport through a nanopore, *Electrophoresis*, 23, 2583–2591.

Meller, A., Nivon, L., Brandin, E., Golovchenko, J., and Branton, D., 2000. Rapid nanopore discrimination between single polynucleotide molecules, *Proc. Natl. Acad. Sci. USA*, 97, 1079–1084.

Menasveta, M. and Hoagland, D., 1992. Molecular weight dependence of the critical strain rate for flexible polymer solutions in elongational flow, *Macromolecules*, 25, 7060–7062.

Miyaki, Y., Einaga, Y., and Fujita, H., 1978. Excluded-volume effects in dilute polymer solutions. 7. Very high molecular weight polystyrene in benzene and cyclohexane, *Macromolecules*, 11, 1180–1186.

Miyaki, Y., Einaga, Y., Hirosye, T., and Fujita, H., 1977. Solution properties of poly(d-β-hydroxybutyrate). 2. Light scattering and viscosity in trifluoroethanol and behavior of highly expanded polymer coils, *Macromolecules*, 10, 1356–1364.

Mohan, A., Kolomeisky, A.B., and Pasquali, M., 2008. Effect of charge distribution on the translocation of an inhomodeneously charged polymer through nanopore, *J. Chem. Phys.*, 128, 125104.

Morrison, F.A., 2001. *Understanding Rheology*, New York: Oxford University Press.

Movileanu, L., 2008. Squeezing a single polypeptide through a nanopore, *Soft Matter*, 4, 925–931.

Moy, G., Corry, B., Kuyucak, S., and Chung, S.H., 2000. Tests of continuum theories as models of ion channels. I. Poisson-Boltzmann theory versus Brownian dynamics, *Biophys. J.*, 78, 2349–2363.

Murphy, R.J. and Muthukumar, M., 2007. Threading synthetic polyelectrolytes through protein pores, *J. Chem. Phys.*, 126, 051101.

Muthukumar, M., 1987. Adsorption of a polyelectrolyte chain to a charged surface, *J. Chem. Phys.*, 86, 7230–7235.

Muthukumar, M., 1991. Entropic barrier model for polymer diffusion in concentrated polymer solutions and random media, *J. Non-Cryst. Solids*, 131–133, 654–666.

Muthukumar, M., 1996a. Double screening in polyelectrolyte solutions: Limiting laws and crossover formulas, *J. Chem. Phys.*, 105, 5183–5199.

Muthukumar, M., 1996b. Theory of electrophoretic mobility of a polyelectrolyte in semidilute solutions of neutral polymers, *Electrophoresis*, 17, 1167–1172.

Muthukumar, M., 1997. Dynamics of polyelectrolyte solutions, *J. Chem. Phys.*, 107, 2619–2635.

Muthukumar, M., 1999. Polymer translocation through a hole, *J. Chem. Phys.*, 111, 10371–10374.

Muthukumar, M., 2001. Translocation of a confined polymer through a hole, *Phys. Rev. Lett.*, 86, 3188–3191.

Muthukumar, M., 2002a. Phase diagram of polyelectrolyte solutions: Weak polymer effect, *Macromolecules*, 35, 9142–9145.

Muthukumar, M., 2002b. Theory of sequence effects on DNA translocation through proteins and nanopores, *Electrophoresis*, 23, 1417–1420.

Muthukumar, M., 2003. Polymer escape through a nanopore, *J. Chem. Phys.*, 118, 5174–5184.

Muthukumar, M., 2004. Theory of counterion condensation on flexible polyelectrolytes: Adsorption mechanism, *J. Chem. Phys.*, 120, 9343–9350.

Muthukumar, M., 2007. Mechanism of DNA transport through pores, *Annu. Rev. Biophys. Biomol. Struct.*, 36, 435–450.

Muthukumar, M., 2010. Theory of capture rate in polymer translocation, *J. Chem. Phys.*, 132, 195101.

Muthukumar, M. and Baumgaertner, A., 1987. Effects of entropic barriers on polymer dynamics, *Macromolecules*, 22, 1937–1941.

Muthukumar, M. and Edwards, S.F., 1982a. Extrapolation formulas for polymer solution properties, *J. Chem. Phys.*, 76, 2720–2730.

Muthukumar, M. and Edwards, S.F., 1982b. Screening concepts in polymer solution dynamics, *Polymer*, 23, 345–348.

Muthukumar, M. and Edwards, S.F., 1983. Screening of hydrodynamic interaction in a solution of rodlike macromolecules, *Macromolecules*, 16, 1475–1478.

Muthukumar, M., Hua, J., and Kundagrami, A., 2010. Charge regularization in phase separating polyelectrolyte solutions, *J. Chem. Phys.*, 132, 084901.

Muthukumar, M. and Kong, C.Y., 2006. Simulation of polymer translocation through protein channels, *Proc. Natl. Acad. Sci. USA*, 103, 5273–5278.

Muthukumar, M. and Nickel, B.G., 1984. Perturbation theory for a polymer chain with excluded volume interaction, *J. Chem. Phys.*, 80, 5839–5850.

Muthukumar, M. and Nickel, B.G., 1987. Expansion of a polymer chain with excluded volume interaction, *J. Chem. Phys.*, 86, 460–476.

Nguyen, T.T., Rouzina, I., and Shklovskii, B.I., 2000. Reentrant condensation of DNA induced by multivalent counterions, *J. Chem. Phys.*, 112, 2562–2568.

Nielsen, B., Jeppesen, C., and Ipsen, J.H., 2006. Managing free-energy barriers in nuclear pore transport, *J. Biol. Phys.*, 32, 465–472.

Nierlich, M., Williams, C., Boue, F., Cotton, J.P., Daoud, M., Farnoux, B., Jannink, G., Picot, C., Moan, M., Wolff, C., Rinaudo, M., and de Gennes, P.G., 1979. Small angle neutron scattering by semi-dilute solutions of polyelectrolyte, *J. Phys. France*, 40, 701–704.

Nishida, K., Kaji, K., and Kanaya, T., 2001. High concentration crossovers of polyelectrolyte solutions, *J. Chem. Phys.*, 114, 8671–8677.

Nkodo, A., Garnier, J., Tinland, B., Ren, H., Desruisseaux, C., McCormick, L., Drouin, G., and Slater, G., 2001. Diffusion coefficient of DNA molecules during free solution electrophoresis, *Electrophoresis*, 22, 2424–2432.

Nonner, W., Chen, D.P., and Eisenberg, B., 1999. Progress and prospects in permeation, *J. Gen. Physiol.*, 113, 773–782.

Nykypanchuk, D., Strey, H.H., and Hoagland, D.A., 2002. Brownian motion of DNA confined within a two-dimensional array, *Science*, 297, 987–990.

Odell, J. and Keller, A., 1986. Flow-induced chain fracture of isolated linear macromolecules in solution, *J. Polym Sci., Polym. Phys.*, 24, 1889–1916.

Odijk, T., 1977. Polyelectrolytes near the rod limit, *J. Polym Sci., Polym. Phys.*, 15, 477–483.

Odijk, T., 1983. On the statistics and dynamics of confined or entangled stiff polymers, *Macromolecules*, 16, 1340–1344.

Odijk, T., 1986. Theory of lyotropic polymer liquid crystals, *Macromolecules*, 19, 2313–2329.

Olivera, B.M., Baine, P., and Davidson, N., 1964. Electrophoresis of the nucleic acids, *Biopolymers*, 2, 245–257.

Onsager, L., 1949. The effects of shape on the interaction of colloidal particles, *Ann. N.Y. Acad. Sci.*, 51, 627–659.

Ou, Z. and Muthukumar, M., 2005. Langevin dynamics of semiflexible polyelectrolytes: Rod- toroid-globule-coil structures and counterion distribution, *J. Chem. Phys.*, 123, 074905.

Oukhaled, G., Mathe, J., Biance, A.L., Bacri, L., Betton, J.M., Lairez, D., Pelta, J., and Auvray, L., 2007. Unfolding of proteins and long transient conformations detected by single nanopore recording, *Phys. Rev. Lett.*, 98, 158101.

Ozisik, M.N., 1993. *Heat Conduction*, New York: Wiley-Interscience.

Panja, D., Barkema, G.T., and Ball, R.C., 2007. Anomalous dynamics of unbiased polymer translocation through a nanopore, *J. Phys.: Condens. Matter*, 19, 432202.

Panja, D., Barkema, G.T., and Ball, R.C., 2009. Reply to the comment on 'anomalous dynamics of unbiased polymer translocation through a nanopore' and other recent papers by D. Panja, G. Barkema and R. Ball, *J. Phys.: Condens. Matter*, 21, 098002.

Pante, N. and Kann, M., 1998. Nuclear pore complex is able to transport macromolecules with diameters of 39 nm, *Mol. Biol. Cell*, 13, 425–434.

Park, P.J. and Sung, W., 1998a. Polymer release out of a spherical vesicle through a pore, *Phys. Rev. E*, 57, 730–734.

Park, P.J. and Sung, W., 1998b. Polymer translocation induced by adsorption, *J. Chem. Phys.*, 108, 3013–3018.

Pearson, K., 1905a. The problem of the random walk, *Nature*, 72, 294.

Pearson, K., 1905b. The problem of the random walk, *Nature*, 72, 342.

Perkins, T., Smith, D., and Chu, S., 1997. Single polymer dynamics in an elongational flow, *Science*, 276, 2016–2021.

Peskin, C.S., Odell, G.M., and Oster, G.F., 1993. Cellular motions and thermal fluctuations: The Brownian ratchet, *Biophys. J.*, 65, 316–324.

Petera, D. and Muthukumar, M., 1999. Brownian dynamics simulation of bead-rod chains under shear with hydrodynamic interaction, *J. Chem. Phys.*, 111, 7614–7623.

Peters, R., 2006. Introduction to nucleocytoplasmic transport: Molecules and mechanisms, *Methods Mol. Biol.*, 322, 235–258.

Petrov, A.S., Lim-Hing, K., and Harvey, S.C., 2007. Packaging of DNA by bacteriophage Epsilon 15: Structure, forces, and thermodynamics, *Structure*, 15, 807–812.

Pfanner, N., Rassow, J., van der Klei, I., and Neupert, W., 1992. A dynamical model of the mitochondrial protein import machinery, *Cell*, 68, 999–1002.

Pincus, P., 1991. Colloid stabilization with grafted polyelectrolytes, *Macromolecules*, 24, 2912–2919.

Prabhu, V.M., Amis, E.J., Bossov, D.P., and Rossov, N., 2004. Counterion associative behavior with flexible polyelectrolytes, *J. Chem. Phys.*, 121, 4424–4429.

Prabhu, V.M., Muthukumar, M., Wignall, G.D., and Melnichenko, Y.B., 2003. Polyelectrolyte chain dimensions and concentration fluctuations near phase boundaries, *J. Chem. Phys.*, 119, 4085–4098.

Probstein, R.F., 1989. *Physicochemical Hydrodynamics*, Boston, MA: Butterworths.

Purohit, P.K., Kondev, J., and Phillips, R., 2003. Mechanics of DNA packaging in viruses, *Proc. Natl. Acad. Sci. USA*, 100, 3173–3178.

Rapoport, T.A., 2007. Protein translocation across the eukaryotic endoplasmic reticulum and bacterial plasma membranes, *Nature*, 450, 663–669.

Rayleigh, 1880. On the resultant of a large number of vibrations of the same pitch and of arbitrary phase, *Phil. Mag.*, 10, 73–78.

Rayleigh, 1899. On James Bernoulli's theorem in probabilities, *Phil. Mag.*, 47, 246–251.

Redner, S., 2001. *A Guide to First-Passage Processes*, Cambridge, U.K.: Cambridge University Press.

Rice, S.A. and Nagasawa, M., 1961. *Polyelectrolyte Solutions*, New York: Academic Press.

Riseman, J. and Kirkwood, J.G., 1950. The intrinsic viscosity, translational and rotatory diffusion constants of rod-like macromolecules in solution, *J. Chem. Phys.*, 18, 512–516.

Risken, H., 1989. *The Fokker–Planck Equation*, Berlin, Germany: Springer.

Rouse, P.E., 1953. A theory of the linear viscoelastic properties of dilute solutions of coiling polymers, *J. Chem. Phys.*, 21, 1272–1280.

Roux, B. and Karplus, M., 1994. Molecular dynamics simulations of the gramicidin channel, *Annu. Rev. Biophys. Biomol. Struct.*, 23, 731–761.

Russi, S., Boer, R., and Coll, M., 2008. Molecular machinery for DNA translocation in bacterial conjugation, in *Plasmids: Current Research and Future Trends* (ed. G. Lipps), Norwich, U.K.: Caiser Academic Press.

Sakaue, T., 2007. Nonequilibrium dynamics of polymer translocation and straightening, *Phys. Rev. E*, 76, 021803.

Sakaue, T. and Raphael, E., 2006. Polymer chains in confined spaces and flow-injection problems: Some remarks, *Macromolecules*, 39, 2621–2628.

Schiessel, H., 1999. Counterion condensation on flexible polyelectrolytes: Dependence on ionic strength and chain concentration, *Macromolecules*, 32, 5673–5680.

Sedlak, M. and Amis, E.J., 1992a. Concentration and molecular weight regime diagram of salt-free polyelectrolyte solutions as studied by light scattering, *J. Chem. Phys.*, 96, 826–834.

Sedlak, M. and Amis, E.J., 1992b. Dynamics of moderately concentrated salt-free polyelectrolyte solutions: Molecular weight dependence, *J. Chem. Phys.*, 96, 817–825.

Severin, M., 1993. Thermal maximum in the size of short polyelectrolyte chains: A Monte Carlo study, *J. Chem. Phys.*, 99, 628–633.

Singer, A., Wanunu, M., Morrison, W., Kuhn, H., Frank-Kamenetskii, M., and Meller, A., 2010. Nanopore based sequence specific detection of duplex DNA for genomic profiling, *Nano Lett.*, 10, 738–742.

Skolnick, J. and Fixman, M., 1977. Electrostatic persistence length of a wormlike polyelectrolyte, *Macromolecules*, 10, 944–948.

Slagowski, E., Tsai, B., and McIntyre, D., 1976. The dimensions of polystyrene near and below the theta temperature, *Macromolecules*, 9, 687–688.

Smith, D., Babcock, H., and Chu, S., 1999. Single-polymer dynamics in steady shear flow, *Science*, 283, 1724–1727.

Smith, D., Tan, S.J., Smith, S.B., Grimes, S., Anderson, D.L., and Bustamante, C., 2001. The bacteriophage phi 29 portal motor can package DNA against a large internal force, *Nature*, 413, 748–752.

Sommerfeld, A., 1949. *Partial Differential Equations in Physics*, New York: Academic Press.

Song, L.Z., Hobaugh, M.R., Shustak, C., Cheley, S., Bayley, H., and Gouaux, J.E., 1996. Structure of staphylococcal α-hemolysin, a heptameric transmembrane pore, *Science*, 274, 1859–1866.

Stein, D., van der Heyden, F.H.J., Koopmans, W.J.A., and Dekker, C., 2006. Pressure-driven transport of confined DNA polymers in fluidic channels, *Proc. Natl. Acad. Sci. USA*, 103, 15853–15858.

Stellwagen, E., Lu, Y., and Stellwagen, N., 2003. Unified description of electrophoresis and diffusion for DNA and other polyions, *Biochemistry*, 42, 11745–11750.

Stevens, M.J. and Kremer, K., 1995. The nature of flexible linear polyelectrolytes in salt free solution: A molecular dynamics study, *J. Chem. Phys.*, 103, 1669–1690.

Storm, A.J., Chen, J.H., Zandbergen, H.W., and Dekker, C., 2005a. Translocation of double-stranded DNA through a silicon oxide nanopore, *Phys. Rev. E*, 71, 051903.

Storm, A.J., Storm, C., Chen, J., Zandbergen, H., Joanny, J.F., and Dekker, C., 2005b. Fast DNA translocation through a solid-state nanopore, *Nano Lett.*, 5, 1193–1197.

Sung, W. and Park, P.J., 1996. Polymer translocation through a pore in a membrane, *Phys. Rev. Lett.*, 77, 783–788.

Syganow, A. and von Kitzing, E., 1999. (In)validity of the constant field and constant current assumptions in theories of ion transport, *Biophys. J.*, 76, 768–781.

Terry, L.J., Shows, E.B., and Wente, S.R., 2007. Crossing the nuclear envelope: Hierarchical regulation of nucleocytoplasmic transport, *Science*, 318, 1412–1416.

van Kampen, N.G., 2007. *Stochastic Processes in Physics and Chemistry*, Amsterdam, the Netherlands: Elsevier.

Vercoutere, W., Winters-Hilt, S., Olsen, H., Deamer, D., Haussler, D., and Akeson, M., 2001. Rapid discrimination among individual DNA hairpin molecules at single-nucleotide resolution using an ion channel, *Nat. Biotechnol.*, 19, 248–252.

Verwey, E.J.W. and Overbeek, J.T.G., 1999. *Theory of the Stability of Lyophobic Colloids*, Mineola, NY: Dover.

Vocks, H., Panja, D., Barkema, G.T., and Ball, R.C., 2008. Driven polymer translocation through a nanopore: A manifestation of anomalous diffusion, *J. Phys.: Condens. Matter*, 20, 095224.

Volk, N., Vollmer, D., Schmidt, M., Oppermann, W., and Huber, K., 2004. Polyelectrolyte theory, *Adv. Polym. Sci.*, 166, 29–65.

Wanunu, M., Morrison, W., Rabin, Y., Grosberg, A.Y., and Meller, A., 2010. Electrostatic focusing of unlabelled DNA into nanoscale pores using a salt gradient, *Nat. Nanotechnol.*, 5, 160–165.

Wanunu, M., Sutin, J., McNally, B., Chow, A., and Meller, A., 2008. DNA translocation governed by interactions with solid-state nanopores, *Biophys. J.*, 95, 4716–4725.

Weiss, T.F., 1996. *Cellular Biophysics. Volume 1: Transport*, Cambridge, MA: The MIT Press.

Wells, D.B., Abramkina, V., and Aksimentiev, A., 2007. Exploring transmembrane transport through α-hemolysin with grid-steered molecular dynamics, *J. Chem. Phys.*, 127, 125101.

Wignall, G.D., Ballard, D.G.H., and Schelten, J., 1974. Measurements of persistence length and temperature dependence of the radius of gyration in bulk atactic polystyrene, *Eur. Polym. J.*, 10, 861–865.

Winkler, R.G., Gold, M., and Reineker, P., 1998. Collapse of polyelectrolyte macromolecules by counterion condensation and ion pair formation: A molecular dynamics simulation study, *Phys. Rev. Lett.*, 80, 3731–3734.

Wittmer, J., Johner, A., and Joanny, J.F., 1995. Precipitation of polyelectrolytes in the presence of multivalent salts, *J. Phys. II France*, 5, 635–654.

Wolterink, J.K., Barkema, G.T., and Panja, D., 2006. Passage times for unbiased polymer translocation through a nanopore, *Phys. Rev. Lett.*, 96, 208301.

Wong, C.T.A. and Muthukumar, M., 2007. Polymer capture by electroosmotic flow of oppositely charged nanopores, *J. Chem. Phys.*, 126, 164903.

Wong, C.T.A. and Muthukumar, M., 2008a. Polymer translocation through a cylindrical channel, *J. Chem. Phys.*, 128, 154903.

Wong, C.T.A. and Muthukumar, M., 2008b. Scaling theory of polymer translocation into confined regions, *Biophys. J.*, 95, 3619–3627.

Wong, C.T.A. and Muthukumar, M., 2010. Polymer translocation through α-hemolysin pore with tunable polymer-pore electrostatic interaction, *J. Chem. Phys.*, 133, 045101.

Yamakawa, H., 1971. *Modern Theory of Polymer Solutions*, New York: Harper and Row.

Yamakawa, H., 1997. *Helical Wormlike Chains in Polymer Solutions*, Berlin, Germany: Springer.

Young, H.D. and Freedman, R.A., 2000. *University Physics*, San Francisco, CA: Addison-Wesley.

Yuan, J., Zweers, J.C., van Dijl, J.M., and Dalbey, R.E., 2010. Protein transport across and into cell membranes in bacteria and archaea, *Cell Mol. Life Sci.*, 67, 179–199.

Zimm, B., 1956. Dynamics of polymer molecules in dilute solution: Viscoelasticity, flow birefringence and dielectric loss, *J. Chem. Phys.*, 24, 269–278.

INDEX

A

Amplification
  chemical decoration
    α-hemolysin protein pore,
      316–317
    MspA pore, 317–318
    secondary structures, translocating
      polymer
    electrophoretic translocation, 320
    polycytidylic vs.
      polydeoxycytidylic acid,
      318–320
Average translocation time
  α-Hemolysin pore, 272–273
  membrane potential, 302
  shallow channels, 278–279
  solid-state nanopores, 305

B

Bacterial conjugation, 322
Biased random walk
  drift–diffusion
    diffusion coefficient, 150–151
    drift and diffusion term,
      149–150
    probability distribution,
      149, 151
    step length, 150
    trajectory, continuous curve,
      147, 149
  Gaussian chain statistics, 148–149
  master equation, 148

net displacement, 147
probability, 146–147
realization, one dimension,
  146–147
standard deviation, 147
Biological contexts
  "bird's eye view," 3
  charged macromolecules, eukaryotic
    cell, 1–2
  "Coulomb soup," 1–2
  DNA threading, bacteriophage, 3
  gene swapping, 3–4
  isolated translocation process, 2
  nuclear pore complex, 2–3
Brownian motion
  Brownian particle, external force
    correlation time, 154
    diffusion constant, 155
    Einsteinian and Newtonian
      dynamics, 154
    friction coefficient, 155
    Langevin equation, 156
    mean square displacement, center
      of mass, 154
    Stokes–Einstein law, 156
    Stokes law of friction, 155
    velocity correlation functions,
      153–154
  evolution and definition, 145
  fluctuation-dissipation theorem, 153
  rigid spherical particle, 152
  white noise/Gaussian noise, 153
Brownian particle collection,
  159–160

C

Charge density, ion flow
electrolyte conductivity, 215
Poisson equation, 215
relaxation rate and time, 216
Charged interfaces
charged cylinder
charge density and total charge, 69
Debye length, 70–71
dimensionless Coulomb strength
parameter, 70
electric field, 69
electric potential, 69–70
Stern regime, 71
charged line, 72–74
cylindrical pore
charges, central axis, 68
cylindrical coordinate system,
66–67
Debye–Hückel approximation, 67
Debye length, 67–69
internal wall, 65–66
net charge density, 68
radius and length, surface
potential, 66
dielectric constants, 57
dielectric mismatch
analogous calculations, 76–77
effective dielectric constant, 76
image charge, 75–76
medium, 75
point charge, 74–75
planar interface
salt-free solutions, 61–62
salty solutions (see Salty solutions)
spherical macroions, 57–58
spherical particles
coions, 58
electric potential, 58–59
mobility of macroions, 59
Poisson–Boltzmann equation, 58
radius, 58–59
surface potential, 59
Constant field approximation, ion flow
concentration profile, 221
electric potential, 219
GHK current equation, 220–221
resting potential, 221–223

Convective diffusion, 251–252
Cylindrical pore
flexible chain
confined polymer conformation,
135, 138
free energy, 137
Gaussian chains, 138
number of blobs, 138
scaling law, 135–137
uniform diameter, 135
geometry, 115–116
scaling arguments, 135
semiflexible chain
arc length, 139
bending energy, 140–141
hairpin conformations, 135, 141
mean square end-to-end distance,
139
rod-like conformation, 140
sector, 135, 139
wormlike chain, 138–139

D

Debye–Hückel theory
charge distribution, 54–56
Debye–Hückel potential, 50
electric potential, 50
electrostatic interaction, 49
electrostatic screening length,
50–53
free energy, 56–57
ion size effect, 52–54
Debye length, see Electrostatic screening
length
Diffusion-limited polymer capture
capture rate, 249–250
diffusion coefficient, 249
net polymer flux, 248
polymer concentration, 247–248
Drift–diffusion process
absorbing boundary, 167–168
biased random walk
diffusion coefficient, 150–151
diffusion term, 149–150
drift term, 149–150
probability distribution,
149, 151

step length, 150
trajectory, continuous curve,
147, 149
constant velocity and diffusion
coefficient, 167
one absorbing barrier, 168–169
radiation boundary condition,
173–174
reflecting boundary, 172–173
two absorbing barrier, 169–172
Drift-limited polymer capture
crossover factor, 250
electrophoretic mobility, 250–251
polymer flux, 250–251

E

Edwards screening method, 42
Einsteinian relation, 145–146
Electric potential, 82–84
Electrolyte solutions
Coulomb interaction and Bjerrum
length
aqueous solutions, 48
chain conformations and
counterions, 47
definition, 47
electric field and potential, 46
electrostatic energy, 47
Debye–Hückel theory
charge distribution, 54–56
Debye–Hückel potential, 50
electric potential, 50
electrostatic interaction, 49
electrostatic screening length,
50–53
free energy, 56–57
ion size effect, 52–54
ion charges, 45
ion distribution, 46
Poisson–Boltzmann equation,
48–49
Electroosmotic flow
capture radius, 255–256
capture rate, 255
critical capture radius, 254
divergence of flux, 255
electric potential, 252–253

ion flow
cylindrical pores, 236–238
planar interfaces, 234–236
schematic diagram, 253
velocity gradient, 253–254
Electrophoretic mobility
average velocity, 198
counterions electric force, 202
dilute solutions, 200–201
drift-limited regime, 250–251
Einsteinian relation, 199–202
electric force, 199
external velocity, 202
frictional force, 199
long-range hydrodynamic
interaction, 202
net electric force, 203
salt concentrations, 204
Electrostatic blob
blob energy, 104
end-to-end distance, 105
linear size, chain, 104–105
scaling laws, 105–106
Electrostatic screening length
vs. Bjerrum length, 51
definition, 50
dependence, 52–53
electrostatic strength parameter, 51
ionic strength, 51
monovalent salts, 52
nanometer, electrolyte concentrations
and valencies, 52
number density, 51
Electrostatic swelling
flexible and semiflexible
polyelectrolytes
crossover formula, 106–107
effective charge density, 100
electrostatic blob, 104–106
high salt limit, 100–102
low salt limit, 102–104
monovalent, divalent, and trivalent
counterions, 95
swollen semiflexible coil, 89–90
Entropic barrier, 257, 260
conformational entropy role, 8–9
free energy, 7–8
genesis, 8
net driving potential, 8

Equilibrium vs. steady state, 160–162
Excluded-volume parameter, 24

F

Finite boundaries
    FPT, 162–163
    general equations
        forward and backward
            Fokker–Planck equations,
            163–164
        leading moments, 164
        mean exit time, 164–165
        probability distribution function,
        163
        transition probability, 163
    general solutions
        absorbing and reflecting condition,
        165
        conditional probability, 166
        drift and diffusion coefficient, 167
        mean FPT, 165–166
        radiation boundary condition, 165
    particle, two barriers, 162–163
First passage time (FPT), 162
Flexible and semiflexible polyelectrolytes
    chain connectivity, 82–84
    chemical charge density, 80
    concentration effects
        dilute, semidilute, and
            concentrated solutions,
            111–112
        first-order phase transition,
        112–113
        free energy, 112
        long-ranged correlations, 111
        polyelectrolyte concentrations,
        111
    Coulomb strength, 81
    counterion worm, 84–85
    Debye length, 81–82
    dielectric mismatch, 85–86
    effective charge, 79
    electroneutrality condition, 81
    electrostatic excluded volume, 87–88
    electrostatic persistence length
        intrinsic stiffness, 88–89
        Odijk–Skolnick–Fixman theory,
        89

    swollen semiflexible coil, 89–90
        total persistence length, 89
    electrostatic swelling
        crossover formula, 106–107
        effective charge density, 100
        electrostatic blob, 104–106
        high salt limit, 100–102
        low salt limit, 102–104
    experimental results
        multivalent counterions, 92
        polyelectrolyte solutions, 90
        polymer size determination,
        91–92
        precise light-scattering
            results, 91
        radius of gyration, 91–92
    hydrophobic interaction, 90
    polyelectrolyte chain, 79–80
    repeat unit, 80
    self-regularization
        adsorbed ion entropy, 108
        chain free energy, 109
        coil-to-globule transition, 110
        counterion condensation, 107
        degree of ionization, 109–110
        dissociated ion correlations, 108
        energy, adsorbed ion, 109
        free energy, 108
        ion pairs interaction, 109
        polymer chain behavior, 110–111
        polymer conformation, 107
        self-consistent procedure,
        109–110
        single chain, N monomers,
        107–108
        unadsorbed ion entropy, 108
    simulation results
        AY_2 salt type, 99
        charge fraction, 96–97
        computer simulations, 93
        Coulomb energy, 93
        Coulomb strength parameter,
        93–94
        counterion distribution, 96
        degree of ionization, 98
        divalent vs. monovalent
            counterions, 99–100
        effective charge, 98
        Langevin dynamics, 93

radius of gyration (*see* Radius of gyration)
repulsive Lennard-Jones potential, 93
salt-free solutions, 98–99
Fokker–Planck–Smoluchowski equation
  drift and diffusion terms, 157
  external force, 158
  flux, 158
  probability distribution function, 157, 159
  Smoluchowski equation, 158
  stochastic process, 157–158
Fractal dimension, polymer, 16
Free energy barriers
  cylindrical pore
    flexible chain, 135–138
    geometry, 115–116
    scaling arguments, 135
    semiflexible chain, 138–141
  free energy landscape, 115
  infinitely wide channels, 141–142
    confinement free energy, 142
    geometry, 115–116
    size exponent, 142
    slab/slit, 141
  spherical cavity
    Gaussian chain (*see* Gaussian chain)
    geometry, 115–116
    polyelectrolyte chain, 132–134
    translocation, two compartments, 121–122
    uncharged polymer, excluded volume, 128–132
  thin planar membrane
    chain length effect, 119–120
    critical number, nucleation, 119–120
    critical value and maximum free energy, 119–120
    disallowed polymer conformation, 116–117
    downhill and uphill processes, 119
    electrochemical potential, 118
    entropic barrier, 120–121
    geometry, 115–116
    parameters, 118–119
    partition sum, 117

    polymer tails, semi-infinite spaces, 116–117
    polymer translocation, transition, 116
    prime on exponent γ, 117
    semi-infinite space, 117
    tail, N segments, 117–118
Free energy, polymer capture
  barrier, 256–257
  curve, 245
  gain, 258
Freely jointed chain model, 24–25

G

Gaussian chain
  chain swelling, excluded volume effect, 36
  confinement free energy, 123
  entropic barrier, 122, 124–126
  explicit closed-form analytical expressions, 28
  finite extensibility, 31
  fixed end-to-end distance
    entropy, 29
    free energy, 30
    force to pull, 30–31
    fractal dimension and form factor, 29
  free energy, 125, 127
  GSD approximation, 123–124
  hydrodynamic radius, 28
  interactive pore, 127
  mean-square end-to-end distance and gyration radius, 28
  probability, end-to-end distance, 28
  segmental density profile, 29
  size exponent and shape factor, 29
  statistical properties, 122
  statistics, 148–149
  suction and expulsion processes, 128
  total number of conformations, 29
  total probability, 122–123
Genome packing, bacteriophages
  electrostatic interactions, 324
  force analysis, 324–325
  genome length, 323
  genome structure, 323–324

T7 bacteriophage, 324
volume interactions, 323–324
Goldman–Hodgkin–Katz (GHK)
    equations, 209, 219
    current equation
        asymptotic limits, 220
        barriers, 223–224
        graphical representation,
            220–221
        Ohm's law, 220
        permeability, 220
    voltage equation, 222
Gouy–Chapman length, 61
Gouy–Chapman theory, 60
Gramicidin A channel
    current–voltage curve,
        225–226
    electric potential vs. ion
        concentrations, 226–227
Ground-state dominance (GSD)
    approximation, 123–124

H

α-Hemolysin pore
    average translocation time, 272–273,
        302
    β-barrel pore, 303
    configuration, 226–228
    data analysis, 303–304
    dwell time, 272–273
    electric potential variations, 228
    ionic current blockades, 3 event types,
        272, 274
    ionic current trace, 228–229
    N/V_m vs. dwell times, 273–274
    PNP methodology and Langevin
        dynamics method, 228–229
    polymer orientation, 303
    pore current, 272
    shape, 303
    sodium translocation, event diagrams,
        273–274
    space-filling model, 228
    trans compartment, 275–276
    vestibule, 303
High salt limit
    degree of ionization, 102

effective excluded volume parameter,
    100–101
radius of gyration, 101–102
Hydrodynamic interaction
    incompressible fluid, 178
    monomer, 179
    Oseen tensor, 179
    shear flow, 180
    solvent molecules, 178
    streamlines, flow, 180–181
    two fluid elements, 179
    velocity fields, 178–179
    zero-frequency, 178

I

Ideal/theta temperature, 24
Ion flow
    barriers effect
        charge flux, 223–224
        effective permeability, 224
        potential barrier, 223
    cylindrical pore
        electrophoretic mobility, 211
        ion concentration, 209–210
        ion flux, 210
        ionic current, 211–212
        ion velocity, 210–211
        PNP equation, 212
    electroosmotic flow
        cylindrical pores, 236–238
        planar interfaces, 234–236
    equilibrium
        electrophoretic mobility, 212
        ion concentration, 212–213
        Nernst potential, 213
        partition coefficients, 213
    ionic current fluctuations
        equivalent electrical circuit,
            231–232
        α HL protein pore, 230
        spectral density, 233–234
    polymer molecule inside pore,
        209–210
    protein pores
        complications, 224–225
        gramicidin A channel,
            225–226

α-hemolysin pore, 226–230
PNP formalism, 225
steady state with drift and diffusion
  analogy to electrical circuits,
   217–218
  charge flux, 216–217
  constant field approximation,
   219–223
  PNP equations, 218–219
steady state without diffusion
  charge density, 215–216
  ionic current and Ohm's law,
   214–215
Ionic current
  fluctuations
   equivalent electrical circuit,
    231–232
   α HL protein pore, 230
   spectral density, 233–234
  Ohm's law
   conductance, 215
   ion conductivity, 214
   membrane potential, 214–215
   resistance of pore, 215
   steady-state flux, 214
  protein pores
   complications, 224–225
   gramicidin A channel, 225–226
   α-hemolysin pore, 226–230
   PNP formalism, 225

K

Kratky–Porod model, *see* Wormlike
  chain model
Kuhn chain model, 25–27

L

Langevin equation, 152, 156–157
Low salt limit
  electrostatic interaction, 102, 104
  Flory scaling form, 103
  optimum value, end-to-end distance,
   103
  threshold value, charge fraction,
   104

M

Macromolecules
  chain swelling, excluded volume
   effect
   closed-form expressions, 37
   critical exponent, 37
   excluded volume parameter, 34
   Fixman parameter, 36
   Flory exponent, 36, 39
   Flory–Huggins χ parameter, 34
   Flory radius, 36, 38
   free energy, 35
   Gaussian chain, 36, 38
   monomer density with radial
    distance, 37
   normalized probability
    distribution, 38
   pseudo-potential, 35
   scattering wave vector, 38
   second virial coefficient, 34
   size exponent, 37–38
  coarse-grained models
   freely jointed chain, 24–25
   Gaussian chain, 27–31
   Kuhn chain model, 25–27
   wormlike chain, 32–34
  coil-globule transition, 39–40
  concentration effects
   correlation length, 42–43
   different concentration regimes,
    41–42
   free energy, 43
   free energy density, 44
   mean field part, Flory–Huggins
    theory, 43
   monomer concentration, 41
   radius of gyration, 42–43
   screening, intrachain excluded
    volume interaction, 42
  excluded volume interaction,
   23–24
  polymer conformations
   coil-like conformation, 14
   form factor, 19–20
   free energy, constrained chain,
    20–21
   globular structure, 14
   hydrodynamic radius, 15–16

mean square end-to-end
distance, 15
monomer density distribution,
18–19
monomer–solvent interaction, 14
persistence length, 17–18
polyethylene chain, 14–15
radius of gyration, 15
rod-like structure, 14
semiflexible polymers, 14
shape factor, 16
size exponent, 16
skeletal structure, 15
total number of conformations, 20
universal behavior, 21–23
Manning condensation, 72, 74
Memory effect, 314–315
Mitochondrial transport
illustration, 321
insertion and threading, 321–322
polymer capture, 322
translocon, 320–321
MspA pore, 317–318

N

Nernst potential, 209, 213
Nuclear localization signals (NLS), 323
Nuclear pore complexes (NPCs),
322–323

P

Pearson walk, 145
Poisson–Boltzmann–Edwards equations,
133
Poisson–Nernst–Planck (PNP) equations,
212, 218–219
Polyelectrolyte dynamics
coil stretch, 204–206
electrophoretic mobility
average velocity, 198
counterions electric force, 202
dilute solutions, 200–201
Einsteinian relation, 199–202
electric force, 199
external velocity, 202

frictional force, 199
long-range hydrodynamic
interaction, 202
net electric force, 203
salt concentrations, 204
hydrodynamic interaction
incompressible fluid, 178
monomer, 179
Oseen tensor, 179
shear flow, 180
solvent molecules, 178
streamlines, flow, 180–181
two fluid elements, 179
velocity fields, 178–179
zero-frequency, 178
polyelectrolyte chain diffusion
center-of-mass, 194
counterion concentration, 196
coupled diffusion coefficient, 197
coupled equation, 196–197
coupling theory, 198
dilute and semidilute solutions,
198–199
electrophoretic mobility, 196
high salt concentrations, 198
net charge density and
permittivity, 196
salt concentration, upper branch,
194–195
salt-free dilute solutions,
194–195
uncharged polymer
center of mass, 181
correlation time, 183
diffusion coefficient, 181
entanglement regime, 193
friction coefficient, 181–182
monomer concentration, 182
radius of gyration, 182–183
reduced viscosity, 182
Rouse model (*see* Rouse model)
semidilute solutions and
hydrodynamic screening,
189–193
Zimm dynamics, 186–189
Polymer capture
capture rate
barrier height and chain
length, 261

barrier height and voltage
   difference, 262
chain length, 261–262
crossover, 260–261
entropic barrier, 260
polymer flux vs. voltage
   difference, 259–260
steady-state flux, 259–260
convective diffusion, 251–252
diffusion-limited capture
   diffusion coefficient, 249
   net flux, 248
   polymer concentration, 247–248
   rate, 249–250
drift-limited regime, 250–251
electric potential, 258
electroosmotic flow
   capture radius, 255–256
   capture rate, 255
   critical capture radius, 254
   divergence of flux, 255
   electric potential, 252–253
   schematic diagram, 253
   velocity gradient, 253–254
entropic barrier, 257
free energy
   barrier, 256–257
   curve, 245
   gain, 258
friction coefficient, 246
phenomenological chemical
      kinetics model
   chemical reaction scheme,
      266–267
   free energy barriers, 266–267
   polymer flux, 266
   probability of finding final
      state, 267
   rate of concentration change, 266
   schematic diagram, 265–266
   steady-state flux, 267–268
   success ratio of translocation
      events, 266
polymer flux, 245–246
polymer translocation process,
   244–245
probability of successful translocation
   absorbing boundary conditions,
      263–264

barrier height, 264–265
Fokker–Planck equation, 263
probability distribution, 262–263
voltage difference, 264–265
single-molecule nanopore-based
      translocation experiments
   capture rate, 241–243
   time-resolved fluorescence
      monitoring, 242–244
Polymer translocation, 1, 7
bacterial conjugation, 322
genome packing, bacteriophages
   electrostatic interactions, 324
   force analysis, 324–325
   genome length, 323
   genome structure, 323–324
   T7 bacteriophage, 324
   volume interactions, 323–324
mitochondrial transport
   illustration, 321
   insertion and threading, 321–322
   polymer capture, 322
   translocon, 320–321
nuclear pore, 322–323
polymer capture, 244–245
Pore–polymer interactions
membrane potential
   average translocation time, 302
   free energy profile, 300–301
   partial chain insertion, electric
      field, 300
narrow pore
   average time, 298
   entropic barrier, 299
   free energy, 297–298
   single-file translocation, 297
   suction and expulsion, 298
spherical cavity, 296–297
translocation processes, 295–296

R

Radius of gyration
chain configurations, 94, 96
concentration effects, 42–43, 111
high salt limit, 101–102
$N = 100$, 94–95
polymer conformations, 15

shape factor, 94, 97
uncharged polymer, 182–183
Restricted primitive model, 53
Rouse model
    bead friction coefficient, 184
    connectivity force, 183
    diffusion coefficient, 185
    Einsteinian dynamics, 185
    equation of motion, 183
    frictional force, 184
    hydrodynamic interactions, 186
    longest relaxation time, 185
    relaxation time, 185
    Rouse equation, 184
    Rouse matrix, 184

S

Salty solutions
    capacitance of interface, 65
    Debye–Hückel law, 63
    electric double layer, 65–66
    electric potential, 62–63
    electrolyte concentration, 64–65
    GC vs. DH results, 63
    Grahame equation, 65
    ion density distributions, 62
    surface charge density, 63–64
Semidilute solutions and hydrodynamic
        screening
    cooperative diffusion coefficient,
        192
    Debye–Hückel form, 190
    diffusion constant, 190–191
    entangled regime, 189
    flux and longest relaxation time, 191
    local monomer concentration, 191
    mean square displacement, 190
    overlap concentration, 189
    polymer dynamics, 193
    Rouse model, 192
    tracer diffusion constant, 191
    Zimm dynamics, 190
Single-molecule nanopore
    fundamental molecular
        understanding, 4
    α-hemolysin pore, 4–5
    ionic current bear information, 5–6

nanoscopic channels, DNA
        translocation, 6
    polymer chains, α HL pore, 4–5
    single-stranded DNA/RNA
        threading, 6
    solid-state nanopore, 4–5
    stochastic process, 6–7
Solid-state nanopores
    average translocation time, 305
    dsDNA translocation, 276–277
    hairpins vs. unfolded conformations,
        304
    large capture region, 304
    net electrophoretic velocity, 276–277
    single-file events vs. voltage bias,
        304–305
    standard deviation, 276, 306
    systematic analysis of data, 306
    translocation time, 276–278
    variance, 305
Spherical cavity
    Gaussian chain (*see* Gaussian chain)
    geometry, 115–116
    polyelectrolyte chain
        confinement free energy, 132–133
        electrostatic interaction
            energy, 132
        Poisson–Boltzmann–Edwards
            equations, 133
        self-consistent treatment, 132
        uncharged vs. charged chain,
            133–134
    translocation, two compartments,
        121–122
    uncharged polymer, excluded volume
        confinement free energy, 129
        Edwards equation, 128–129
        entropic barrier, 130–131
        intrachain role, 128–129
        local free energy minimum,
            131–132
        metastable state, 131–132
        segments, 128

T

Threading
    DNA, bacteriophage, 3
    mitochondrial transport, 321–322

nonequilibrium conformations
  memory effect, 314–315
  translocation vs. Zimm time,
    313–314
shallow channels, 306–307
single-stranded DNA/RNA, 6
Translocation kinetics
  computer simulations
    average translocation time, 281
    coil-stretch transition, velocity
      gradients, 283
    cubic cavity, 280
    data interpretation, 279
    diffusion coefficient, 280
    flexible polyelectrolyte chain, 281
    forward flux sampling method,
      283–284
    nucleation and growth, 281–282
    polymer escape, entropic
      traps, 280
    vivid details, molecular
      processes, 283
  drift and diffusion contribution, 287
  drift–diffusion equation, 290
  flux, 287–288
  Fokker–Planck equation, 287
  Fokker–Planck method, 285
  four elementary steps, 285
  free energy profile, 290
  generic free energy profile, 284
  α-hemolysin pore
    average translocation time,
      272–273, 302
    β-barrel pore, 303
    data analysis, 303–304
    dwell time, 272–273
    ionic current blockades, 3 event
      types, 272, 274
    N/V_m vs. dwell times, 273–274
    polymer orientation, 303
    pore current, 272
    shape, 303
    sodium translocation, event
      diagrams, 273–274
    trans compartment, 275–276
    vestibule, 303
  monomers translocation, 284–285
  net rate, 286
  nucleation time, 288–289

one absorbing and reflecting barriers,
    293–295
pore–polymer interactions
    membrane potential, 300–302
    narrow pore, 297–299
    spherical cavity, 296–297
    translocation processes, 295–296
shallow channels
    average translocation time,
      278–279
    critical values, 309
    deep region, 307
    DNA translocation, 310
    DNA transport, periodic array,
      278–279
    entropic barrier model, 306
    free energy, 307–308
    linear and hairpin mode, 307,
      309–310
    nanofluidic channel device,
      278–279
    nucleation and threading, 306–307
    periodic array, 306
    probability of nucleation, 309–310
    relative probability, 308–309
solid-state nanopores
    average translocation time, 305
    dsDNA translocation, 276–277
    hairpins vs. unfolded
      conformations, 304
    large capture region, 304
    net electrophoretic velocity,
      276–277
    single-file events vs. voltage bias,
      304–305
    standard deviation, 276, 306
    systematic analysis of data, 306
    translocation time, 276–278
    variance, 305
stochastic process, 290
two absorbing barriers
    average translocation time, 292
    conditional probability, 293
    mean FPT, 292
    probability distribution, 291–292
Translocation phenomenon
  capture
    conformational deformation, 10
    electroosmotic force, 10

pore mouth, 9, 11
    strong flow currents, 9–10
chain-end localization, 11
drift–diffusion, 9–10
nucleation and threading, 11
polymer, metastable state, 11
Translocon, 320–321

W

Weissenberg number, 205
White noise/Gaussian noise, 153
Wormlike chain model
    correlation function, 32, 34
    definition, 32
    Kuhn length, 33
    Kuhn steps, 32
    mean square end-to-end distance, 33
    orientational correlation function, 33
    persistence length, 32

quantum mechanical free particle, 34
space curve, 32–33
total energy, 34

Z

Zimm dynamics
    center-of-mass, 188
    Einsteinian relation, 188
    force, 186
    Gaussian statistics, 187
    hydrodynamic interaction, 186
    Kirkwood–Riseman–Zimm dynamics, 187
    Oseen tensor, 187–188
    preaveraging approximation, 187
    rod-like conformations, 188
    translational friction coefficient, 188
    velocity, 186–187
    Zimm time, 189

9 780367 382797